T0296076

LONDON MATHEMATICAL SOCIETY LECTURE NOTE SERIES

Managing Editor: Professor M. Reid, Mathematics Institute,
University of Warwick, Coventry CV4 7AL, United Kingdom

The titles below are available from booksellers, or from Cambridge University Press at
http://www.cambridge.org/mathematics

London Mathematical Society Lecture Note Series: 424

Surveys in Combinatorics 2015

Edited by

ARTUR CZUMAJ
University of Warwick

AGELOS GEORGAKOPOULOS
University of Warwick

DANIEL KRÁL'
University of Warwick

VADIM LOZIN
University of Warwick

OLEG PIKHURKO
University of Warwick

CAMBRIDGE
UNIVERSITY PRESS

CAMBRIDGE
UNIVERSITY PRESS

University Printing House, Cambridge CB2 8BS, United Kingdom

Cambridge University Press is part of the University of Cambridge.

It furthers the University's mission by disseminating knowledge in the pursuit of education, learning and research at the highest international levels of excellence.

www.cambridge.org
Information on this title: www.cambridge.org/9781107462502

© Cambridge University Press 2015

This publication is in copyright. Subject to statutory exception and to the provisions of relevant collective licensing agreements, no reproduction of any part may take place without the written permission of Cambridge University Press.

First published 2015

A catalogue record for this publication is available from the British Library

ISBN 978-1-107-46250-2 Paperback

Cambridge University Press has no responsibility for the persistence or accuracy of URLs for external or third-party internet websites referred to in this publication, and does not guarantee that any content on such websites is, or will remain, accurate or appropriate.

Contents

Preface

The 25th British Combinatorial Conference is to take place at the University of Warwick in July 2015. The British Combinatorial Committee has invited nine distinguished combinatorialists to give survey lectures in areas of their expertise. This volume contains the survey articles on which these lectures are to be based.

We would like to thank the authors for preparing the excellent and interesting surveys, and the anonymous referees for careful reading, in some cases under tight time constraints. Also, we are grateful to the team at the Cambridge University Press, in particular Sam Harrison, for their help, advice, and professionalism. Finally, our task was much simpler thanks to the experience of the editors of earlier *Surveys* and the guidance of the British Combinatorial Committee.

Artur Czumaj
Agelos Georgakopolous
Dan Král'
Vadim Lozin
Oleg Pikhurko

University of Warwick

February 2015

vii

Ramsey classes: examples and constructions

Manuel Bodirsky[1]

Abstract

This article is concerned with classes of relational structures that are closed under taking substructures and isomorphism, that have the joint embedding property, and that furthermore have the *Ramsey property*, a strong combinatorial property which resembles the statement of Ramsey's classic theorem. Such classes of structures have been called *Ramsey classes*. Nešetřil and Rödl showed that they have the *amalgamation property*, and therefore each such class has a homogeneous Fraïssé limit. Ramsey classes have recently attracted attention due to a surprising link with the notion of extreme amenability from topological dynamics. Other applications of Ramsey classes include reduct classification of homogeneous structures.

We give a survey of the various fundamental Ramsey classes and their (often tricky) combinatorial proofs, and about various methods to derive new Ramsey classes from known Ramsey classes. Finally, we state open problems related to a potential classification of Ramsey classes.

Contents

[1]The author has received funding from the European Research Council under the European Community's Seventh Framework Programme (FP7/2007-2013 Grant Agreement no. 257039).

1

1 Introduction

Let $\mathcal{C}$ be a class of finite relational structures. Then $\mathcal{C}$ has the *Ramsey property* if it satisfies a property that resembles the statement of Ramsey's theorem: for all $\mathfrak{A}, \mathfrak{B} \in \mathcal{C}$ there exists $\mathfrak{C} \in \mathcal{C}$ such that for every colouring of the embeddings of $\mathfrak{A}$ into $\mathfrak{C}$ with finitely many colours there exists a 'monochromatic copy' of $\mathfrak{B}$ in $\mathfrak{C}$, that is, an embedding e of $\mathfrak{B}$ into $\mathfrak{C}$ such that all embeddings of $\mathfrak{A}$ into the image of e have the same colour. An example of a class of structures with the Ramsey property is the class of all finite linearly ordered sets; this is Ramsey's theorem [45]. Another example of a class with the Ramsey property is the class of all ordered finite graphs, that is, structures $(V; E, \preceq)$ where V is a finite set, E the undirected edge relation, and $\preceq$ a linear order on V; this result has been discovered by Nešetřil and Rödl [41], and, independently, Abramson and Harrington [1].

In this article we will be concerned exclusively with classes $\mathcal{C}$ that are closed under taking substructures and isomorphism, and that have the *joint embedding property*: whenever $\mathfrak{A}, \mathfrak{B} \in C$, then there exists a $\mathfrak{C} \in \mathcal{C}$ such that both $\mathfrak{A}$ and $\mathfrak{B}$ embed into $\mathfrak{C}$. These are precisely the classes $\mathcal{C}$ for which there exists a countably infinite structure Γ such that a structure belongs to $\mathcal{C}$ if and only if it embeds into Γ. In this case, following Fraïssé's terminology, we say that $\mathcal{C}$ is the *age* of Γ. Our structures always have an at most countable signature. A class $\mathcal{C}$ will be called a *Ramsey class* [38] if it has the Ramsey property, is closed under isomorphisms and substructures, and has the joint embedding property. It is an open research problem, raised in [38], whether Ramsey classes can be *classified* in some sense that needs to be specified.

It has been shown by Nešetřil [38] that Ramsey classes have the *amalgamation property*, a central property in model theory. A class of structures $\mathcal{C}$ has the amalgamation property if for all $\mathfrak{A}, \mathfrak{B}_1, \mathfrak{B}_2 \in \mathcal{C}$ with embeddings e_i of $\mathfrak{A}$ into $\mathfrak{B}_i$, for $i \in \{1, 2\}$, there exist $\mathfrak{C} \in \mathcal{C}$ and embeddings f_i of $\mathfrak{B}_i$ into $\mathfrak{C}$ such that $f_1(e_1(a)) = f_2(e_2(a))$ for all elements a of $\mathfrak{A}$. A class of finite structures $\mathcal{C}$ is an *amalgamation class* if it is closed under induced substructures, isomorphism, and has the amalgamation property. By Fraïssé's theorem (which will be recalled in Section 2.5) for every amalgamation class $\mathcal{C}$ there exists a countably infinite structure Γ of age $\mathcal{C}$ which is *homogeneous*, that is, any isomorphism between finite substructures of Γ can be extended to an automorphism of Γ. The structure Γ is in fact unique up to isomorphism, and called the *Fraïssé limit of* $\mathcal{C}$. In our example above where $\mathcal{C}$ is the class of all finite linearly ordered sets $(V; <)$, the Fraïssé limit is isomorphic to $(\mathbb{Q}; <)$, that is, the linear order of the rationals.

The age of a homogeneous structure with a finite relational signature is in general not Ramsey. However, quite surprisingly, homogeneous structures with finite relational signature typically have a homogeneous *expansion* by finitely many relations such that the age of the resulting structure is Ramsey. The question whether we can replace in the previous sentence the word 'typically' by the word 'always' appeared in discussions of the author with Michael Pinsker and Todor Tsankov in 2010, and has been asked, first implicitly in a conference publication [11], then explicitly in the journal version. The question motivates much of the material present in this article, so we prominently state it here as follows.

Conjecture 1.1 (Ramsey expansion conjecture) *Let Γ be a homogeneous structure with finite relational signature. Then Γ has a homogeneous expansion by finitely many relations whose age has the Ramsey property.*

This conjecture has explicitly been confirmed for all countable homogeneous directed graphs in [31] (those graphs have been classified by Cherlin [17]), and other homogeneous structures of interest [26]. The Ramsey expansion conjecture has several variants that are formally unrelated, but related in spirit; we will come back to this in the final section of the article. There we also discuss that the conjecture can be translated into questions in topological dynamics which are of independent interest.

This text has its focus on the combinatorial aspects of the theory, rather than the links with topological dynamics. What we do find convenient, though, is the usage of concepts from model theory to present the results: instead of manipulating classes of structures $\mathcal{C}$ that are closed under substructures, isomorphism, and that have the joint embedding and the amalgamation property, it is often more convenient to directly manipulate homogeneous structures of age $\mathcal{C}$.

Outline of the article. In Section 2.2 we give a self-contained introduction to the basics of Ramsey classes, including the proofs of some well-known and easy observations about them. In Section 3 we show how to derive new Ramsey classes from known ones; this section contains various facts or proofs that have not explicitly appeared in the literature yet.

- In Section 3.3 we have basic results about the Ramsey properties of interpreted structures that have not been formulated previously in this form, but that are not difficult to show via variations of the so-called *product Ramsey theorem*.

- In Section 3.4 we present a new non-topological proof, due to Miodrag Sokic, of a known fact from [11] about expanding Ramsey classes with constants.

- In Section 3.5 and 3.6 we present generalisations of results from [6] about the Ramsey properties of model-companions and model-complete cores of ω-categorical structures.

Some fundamental Ramsey classes cannot be constructed by the general construction principles from Section 3. The most powerful tool that we have to prove Ramsey theorems from scratch is the *partite method*, developed in the 70s and 80s, most notably by Nešetřil and Rödl, which we present in Section 4. With this method we will show that the following classes are Ramsey: the class of all ordered graphs, the class of all ordered triangle-free graphs, or more generally the class of all ordered structures given by a set of homomorphically forbidden irreducible substructures.

There are also Ramsey classes with finite relational signature where it is not clear how to show the Ramsey property with the partite method, to the best of my knowledge. We will see such an example, based on Ramsey theorems for tree-like structures, in Section 5.

When we want to make progress on Conjecture 1.1, we need a better understanding of the type of expansion needed to turn a homogeneous structure in a finite language into a Ramsey structure. Very often, this can be done by adding a linear ordering to the signature (a partial explanation for this is given in Section 2.8). But not any linear ordering might do the job; a crucial property for finding the right ordering is the so-called *ordering property*, which is a classical notion in structural Ramsey theory. We will present in Section 6 a powerful condition that implies that a Ramsey class has the ordering property with respect to some given ordering.

Finally, in Section 7, we discuss the mentioned link between Ramsey theory and topological dynamics, then present an application of Ramsey theory for classifying reducts of homogeneous structures, and conclude with some open problems related to Conjecture 1.1.

2 Ramsey classes: definition, examples, background

The definition of Ramsey classes is inspired by the statement of the classic theorem of Ramsey, which we therefore recall in the next subsection, before defining the Ramsey property in Section 2.2 and Ramsey classes in Section 2.3.

There are two important necessary conditions for a class to be Ramsey: *rigidity* (Section 2.4) and *amalgamation* (Section 2.5). We will see examples that show that these two conditions are not sufficient (Section 2.6). The Ramsey property of a Ramsey class $\mathcal{C}$ can be seen as a property of the automorphism group of the Fraïssé limit of $\mathcal{C}$; this perspective is discussed in Sections 2.7 and 2.8.

2.1 Ramsey's theorem

The set of positive integers is denoted by $\mathbb{N}$, and the set $\{1, \ldots, n\}$ is denoted by $[n]$. For $M, S \subseteq \mathbb{N}$ we write $\binom{M}{S}$ for the set of all order-preserving maps from S into M. When f is a map, and $\mathcal{S}$ is a set of maps whose range equals the domain of f, then $f \circ \mathcal{S}$ denotes the set $\{f \circ e \mid e \in \mathcal{S}\}$. A proof of Ramsey's theorem can be found in almost any textbook on combinatorics.

Theorem 2.1 (Ramsey's theorem [45]) *For all $r, m, k \in \mathbb{N}$ there is a positive integer g such that for every $\chi \colon \binom{[g]}{[k]} \to [r]$ there exists an $f \in \binom{[g]}{[m]}$ such that $|\chi(f \circ \binom{[m]}{[k]})| \leq 1$.*

2.2 The Ramsey property

In this section we define the Ramsey property for classes of structures. All structures in this article have an at most countable domain, and have an at most countable signature. Typically, the signature will be relational and even finite; but many results generalise to signatures that are infinite and also contain function symbols. In Section 3.4 it will be useful to consider signatures that also contain constant symbols (i.e., function symbols of arity zero).

Let τ be a relational signature, let $\mathfrak{B}$ be a τ-structure. For $R \in \tau$, we write $R^{\mathfrak{B}}$ for the corresponding relation of $\mathfrak{B}$. Typically, the domain of $\mathfrak{A}, \mathfrak{B}, \mathfrak{C}$ will be denoted by A, B, C, respectively. Let A be a subset of the domain B of $\mathfrak{B}$. Then the *substructure* of $\mathfrak{B}$ *induced by* A is the τ-structure $\mathfrak{A}$ with domain A such that for every relation symbol $R \in \tau$ of arity k we have $R^{\mathfrak{A}} = R^{\mathfrak{B}} \cap A^k$.

If τ is not a purely relational signature, but also contains constant symbols, then every substructure $\mathfrak{A}$ of $\mathfrak{B}$ must contain for every constant symbol c in τ the element $c^{\mathfrak{B}}$, and $c^{\mathfrak{A}} = c^{\mathfrak{B}}$. An *embedding* of $\mathfrak{B}$ into $\mathfrak{A}$ is a mapping f from B to A which is an isomorphism between $\mathfrak{B}$ and the substructure induced by the image of f in $\mathfrak{B}$. This substructure will also be called a *copy* of $\mathfrak{A}$ in $\mathfrak{B}$. We write $\binom{\mathfrak{B}}{\mathfrak{A}}$ for the set of all embeddings of $\mathfrak{A}$ into $\mathfrak{B}$.

Definition 2.2 (The partition arrow) When $\mathfrak{A}, \mathfrak{B}, \mathfrak{C}$ are τ-structures, and $r \in \mathbb{N}$, then we write $\mathfrak{C} \to (\mathfrak{B})^{\mathfrak{A}}_r$ if for all $\chi \colon \binom{\mathfrak{C}}{\mathfrak{A}} \to [r]$ there exists an $f \in \binom{\mathfrak{C}}{\mathfrak{B}}$ such that $|\chi(f \circ \binom{\mathfrak{B}}{\mathfrak{A}})| \leq 1$.

We would like to mention that in some papers, the partition arrow is defined for the situation where $\binom{\mathfrak{B}}{\mathfrak{A}}$ does not denote the set of embeddings of $\mathfrak{A}$ into $\mathfrak{B}$, but the set of copies of $\mathfrak{A}$ in $\mathfrak{B}$. These two definitions are closely related; the article [36] is specifically about this difference. Also [27] and [55] treat the relationship between the two definitions.

In analogy to the statement of Ramsey's theorem, we can now define the Ramsey property for a class of relational structures.

Definition 2.3 (The Ramsey property) A class $\mathcal{C}$ of finite structures has the *Ramsey property* if for all $\mathfrak{A}, \mathfrak{B} \in \mathcal{C}$ and $k \in \mathbb{N}$ there exists a $\mathfrak{C} \in \mathcal{C}$ such that $\mathfrak{C} \to (\mathfrak{B})^{\mathfrak{A}}_k$.

Example 2.4 The class of all finite linear orders, denoted by $\mathcal{LO}$, has the Ramsey property. This is a reformulation of Theorem 2.1.

The following well-known fact shows that we can always work with 2-colourings instead of general colourings when we want to prove that a certain class has the Ramsey property.

Lemma 2.5 *Let $\mathcal{C}$ be a class of structures, and $\mathfrak{A} \in \mathcal{C}$. Then for every $\mathfrak{B} \in \mathcal{C}$ and $r \in \mathbb{N}$ there exists a $\mathfrak{C} \in \mathcal{C}$ such that $\mathfrak{C} \to (\mathfrak{B})^{\mathfrak{A}}_r$ if and only if for every $\mathfrak{B} \in \mathcal{C}$ there exists a $\mathfrak{C} \in \mathcal{C}$ such that $\mathfrak{C} \to (\mathfrak{B})^{\mathfrak{A}}_2$.*

Proof Suppose that for every $\mathfrak{B} \in \mathcal{C}$ there exists a $\mathfrak{C} \in \mathcal{C}$ such that $\mathfrak{C} \to (\mathfrak{B})_2^{\mathfrak{A}}$. We inductively define a sequence $\mathfrak{C}_1, \ldots, \mathfrak{C}_{r-1}$ of structures in $\mathcal{C}$ as follows. Let $\mathfrak{C}_1$ be such that $\mathfrak{C}_1 \to (\mathfrak{B})_2^{\mathfrak{A}}$. For $i \in \{2, \ldots, r-1\}$, let $\mathfrak{C}_i$ be such that $\mathfrak{C}_i \to (\mathfrak{C}_{i-1})_2^{\mathfrak{A}}$. We leave it to the reader to verify that $\mathfrak{C}_{r-1} \to (\mathfrak{B})_r^{\mathfrak{A}}$. $\qquad\qquad\square$

2.3 The joint embedding property and Ramsey classes

We say that a class of structures $\mathcal{C}$ is *closed under substructures* if for every $\mathfrak{B} \in \mathcal{C}$, all substructures of $\mathfrak{B}$ are also in $\mathcal{C}$. The class $\mathcal{C}$ is *closed under isomorphism* if for every $\mathfrak{B} \in \mathcal{C}$, all structures that are isomorphic to $\mathfrak{B}$ are also in $\mathcal{C}$. In this article, we will focus on classes of finite structures that are closed under induced substructures and isomorphism, and that have the joint embedding property. Recall from the introduction that $\mathcal{C}$ has the joint embedding property if for every $\mathfrak{A}, \mathfrak{B} \in C$, there exists a $\mathfrak{C} \in \mathcal{C}$ such that both $\mathfrak{A}$ and $\mathfrak{B}$ embed into $\mathfrak{C}$. Such classes of structures naturally arise as follows; see e.g. [25].

Proposition 2.6 *A class of finite structures $\mathcal{C}$ is closed under substructures, isomorphism, and has the joint embedding property if and only if there exists a countable structure Γ whose age equals $\mathcal{C}$.*

Proposition 2.6 is the main motivation why we exclusively work with classes of structures that are closed under substructures; however, as demonstrated in a recent paper by Zucker [55], several Ramsey results and techniques can meaningfully be extended to isomorphism-closed classes that only satisfy the joint embedding property and amalgamation, but that are not necessarily closed under substructures.

Definition 2.7 (Ramsey class) Let τ be an at most countable relational signature. A class of finite τ-structures is called a *Ramsey class* if it is closed under substructures, isomorphism, and has the joint embedding and the Ramsey property.

Examples of Ramsey classes will be presented below, in Example 2.12, or more generally, in Example 2.13. The following can be shown by a simple compactness argument.

Proposition 2.8 *Let Γ be a structure of age $\mathcal{C}$. Then $\mathcal{C}$ is a Ramsey class if and only if for all $\mathfrak{A}, \mathfrak{B} \in \mathcal{C}$ and $r \in \mathbb{N}$ we have that $\Gamma \to (\mathfrak{B})_r^{\mathfrak{A}}$.*

Proof Let $\mathfrak{A}, \mathfrak{B} \in \mathcal{C}$, and $r \in \mathbb{N}$ an integer. When k is the cardinality of $\binom{\mathfrak{B}}{\mathfrak{A}}$, then for any structure $\mathfrak{C}$ the fact that $\mathfrak{C} \to (\mathfrak{B})_r^{\mathfrak{A}}$ can equivalently be expressed in terms of r-colourability of a certain k-uniform hypergraph, defined as follows. Let $G = (V; E)$ be the structure whose vertex set V is $\binom{\mathfrak{C}}{\mathfrak{A}}$, and where $(e_1, \ldots, e_k) \in E$ if there exists an $f \in \binom{\mathfrak{C}}{\mathfrak{B}}$ such that $f \circ \binom{\mathfrak{B}}{\mathfrak{A}} = \{e_1, \ldots, e_k\}$. Let $H = ([r]; E)$ be the structure where E contains all tuples except for the tuples $(1, \ldots, 1), \ldots, (r, \ldots, r)$. Then $\mathfrak{C} \not\to (\mathfrak{B})_r^{\mathfrak{A}}$ if and only if G does not homomorphically map to H. An easy and well-known compactness argument (see e.g. Lemma 3.1.5 in [5]) shows that this is the case if and only if some finite substructure of G does not homomorphically map to H. Thus, $\Gamma \to (\mathfrak{B})_r^{\mathfrak{A}}$ if and only if $\mathfrak{C} \to (\mathfrak{B})_r^{\mathfrak{A}}$ for all finite substructures $\mathfrak{C}$ of Γ. $\square$

2.4 Ramsey degrees and rigidity

Let $\mathcal{C}$ be a class of structures with the Ramsey property. In this section we will see that each structure in $\mathcal{C}$ must be *rigid*, that is, it has no automorphism other than the identity.

Definition 2.9 (Ramsey degrees) Let $\mathcal{C}$ be a class of structures and let $\mathfrak{A} \in \mathcal{C}$. We say that $\mathfrak{A}$ has *Ramsey degree k (in $\mathcal{C}$)* if $k \in \mathbb{N}$ is least such that for any $\mathfrak{B} \in \mathcal{C}$ and for any $r \in \mathbb{N}$ there exists a $\mathfrak{C} \in \mathcal{C}$ such that for any r-colouring χ of $\binom{\mathfrak{C}}{\mathfrak{A}}$ there is an $f \in \binom{\mathfrak{C}}{\mathfrak{B}}$ such that $|\chi(f \circ \binom{\mathfrak{B}}{\mathfrak{A}})| \leq k$.

Hence, by definition, $\mathcal{C}$ has the Ramsey property if every $\mathfrak{A} \in \mathcal{C}$ has Ramsey degree one.

Lemma 2.10 *Let $\mathcal{C}$ be a class of finite structures. Then for every $\mathfrak{A} \in \mathcal{C}$, the Ramsey degree of $\mathfrak{A}$ in $\mathcal{C}$ is at least $|\operatorname{Aut}(\mathfrak{A})|$.*

Proof We have to show that for some $\mathfrak{B} \in \mathcal{C}$ and $r \in \mathbb{N}$, every $\mathfrak{C} \in \mathcal{C}$ can be r-coloured such that for all $f \in \binom{\mathfrak{C}}{\mathfrak{B}}$ we have $|\chi(f \circ \binom{\mathfrak{B}}{\mathfrak{A}})| \geq |\operatorname{Aut}(\mathfrak{A})|$. We choose $\mathfrak{B} := \mathfrak{A}$ and $r := |\operatorname{Aut}(\mathfrak{A})|$.

Let $\mathfrak{C} \in \mathcal{C}$ be arbitrary. Define an equivalence relation $\sim$ on $\binom{\mathfrak{C}}{\mathfrak{A}}$ by setting $f \sim g$ if there exists an $h \in \operatorname{Aut}(\mathfrak{A})$ such that $f = g \circ h$. Let $f_1, \ldots, f_t$ be a list of representatives for the equivalence classes of $\sim$. Define $\chi \colon \binom{\mathfrak{C}}{\mathfrak{A}} \to \operatorname{Aut}(\mathfrak{A})$ as follows. For $f \in \binom{\mathfrak{C}}{\mathfrak{A}}$, let i be the unique i such that $f_i \sim f$. Define $\chi(f) = h$ if $f = f_i \circ h$. Now let $e \in \binom{\mathfrak{C}}{\mathfrak{A}}$ be arbitrary. Then $|\chi(e \circ \binom{\mathfrak{A}}{\mathfrak{A}})| = |\operatorname{Aut}(\mathfrak{A})|$. $\square$

Corollary 2.11 *Let $\mathcal{C}$ be a class with the Ramsey property. Then all $\mathfrak{A}$ in $\mathcal{C}$ are rigid.*

It follows that in particular the class of all finite graphs does *not* have the Ramsey property. Frequently, a class without the Ramsey property can be made Ramsey by expanding its members appropriately with a linear ordering (the expanded structures are clearly rigid).

Example 2.12 Abramson and Harrington [1] and independently Nešetřil and Rödl [39] showed that for any relational signature τ, the class $\mathcal{C}$ of all finite *linearly ordered* τ-structures has the Ramsey property. That is, the members of $\mathcal{C}$ are finite structures $\mathfrak{A} = (A; \preceq, R_1, R_2, \dots)$ for some fixed signature $\tau = \{\preceq, R_1, R_2, \dots\}$ where $\preceq$ denotes a linear order of A.

A shorter and simpler proof of this substantial result, based on the *partite method*, can be found in [40] and [37] and will be presented in Section 4.

For a class of finite τ-structures $\mathcal{N}$, we write $\mathrm{Forb}(\mathcal{N})$ for the class of all finite τ-structures that does not admit a homomorphism from any structure in $\mathcal{N}$.

Example 2.13 The classes from Example 2.12 have been further generalised by Nešetřil and Rödl [39] as follows. Suppose that $\mathcal{N}$ is a (not necessarily finite) class of structures $\mathfrak{F}$ with finite relational signature τ such that for all elements u, v of $\mathfrak{F}$ there is a tuple in a relation $R^{\mathfrak{F}}$ for $R \in \tau$ that contains both u and v. Such structures have been called *irreducible* in the Ramsey theory literature. Then the class of all expansions of the structures in $\mathcal{C} := \mathrm{Forb}(\mathcal{N})$ by a linear order has the Ramsey property. Again, there is a proof based on the partite method, which will be presented in Section 4. This is indeed a generalization since we obtain the classes from Example 2.12 by taking $\mathcal{N} = \emptyset$.

2.5 The amalgamation property

The Ramsey classes we have seen so far will look familiar to model theorists. As mentioned in the introduction, the fact that all of the above Ramsey classes could be described as the age of a homogeneous structure is not a coincidence.

Theorem 2.14 ([38]) *Let τ be a relational signature, and let $\mathcal{C}$ be a class of finite τ-structures that is closed under isomorphism, and has the joint embedding property. If $\mathcal{C}$ has the Ramsey property, then it also has the amalgamation property.*

Proof Let $\mathfrak{A}, \mathfrak{B}_1, \mathfrak{B}_2$ be members of $\mathcal{C}$ such that there are embeddings $e_i \in \binom{\mathfrak{B}_i}{\mathfrak{A}}$ for $i = 1$ and $i = 2$. Since $\mathcal{C}$ has the joint embedding property, there exists a structure $\mathfrak{C} \in \mathcal{C}$ with embeddings f_1, f_2 of $\mathfrak{B}_1$ and $\mathfrak{B}_2$ into

$\mathfrak{C}$. If $f_1 \circ e_1 = f_2 \circ e_2$, then $\mathfrak{C}$ shows that $\mathfrak{B}_1$ and $\mathfrak{B}_2$ amalgamate over $\mathfrak{A}$, so assume otherwise.

Let $\mathfrak{D} \in \mathcal{C}$ be such that $\mathfrak{D} \to (\mathfrak{C})_2^{\mathfrak{A}}$. Define a colouring $\chi \colon \binom{\mathfrak{D}}{\mathfrak{A}} \to [2]$ as follows. For $g \in \binom{\mathfrak{D}}{\mathfrak{A}}$, let $\chi(g) = 1$ if there is a $t \in \binom{\mathfrak{D}}{\mathfrak{C}}$ such that $g = t \circ f_1 \circ e_1$, and $\chi(g) = 0$ otherwise. Since $\mathfrak{D} \to (\mathfrak{C})_2^{\mathfrak{A}}$, there exists a $t_0 \in \binom{\mathfrak{D}}{\mathfrak{C}}$ such that $|\chi(t_0 \circ \binom{\mathfrak{C}}{\mathfrak{A}})| = 1$. Note that $\chi(t_0 \circ f_1 \circ e_1) = 1$ by the definition of χ. It follows that $\chi(t_0 \circ h) = 1$ for all $h \in \binom{\mathfrak{C}}{\mathfrak{A}}$. In particular $\chi(t_0 \circ f_2 \circ e_2) = 1$, because $f_2 \circ e_2 \in \binom{\mathfrak{C}}{\mathfrak{A}}$. Thus, by the definition of χ, there exists a $t_1 \in \binom{\mathfrak{D}}{\mathfrak{C}}$ such that $t_1 \circ f_1 \circ e_1 = t_0 \circ f_2 \circ e_2$ (here we use that the structure $\mathfrak{A}$ must be rigid, by Corollary 2.11). This shows that $\mathfrak{D}$ together with the embeddings $t_1 \circ f_1 \colon \mathfrak{B}_1 \to \mathfrak{D}$ and $t_0 \circ f_2 \colon \mathfrak{B}_2 \to \mathfrak{D}$ is an amalgam of $\mathfrak{B}_1$ and $\mathfrak{B}_2$ over $\mathfrak{A}$. $\square$

Definition 2.15 (Amalgamation class) An isomorphism-closed class of finite structures with an at most countable relational signature that has the amalgamation property (defined in the introduction), and that is closed under taking induced substructures, is called an *amalgamation class*.

Theorem 2.16 (Fraïssé [20, 21]; see [25]) *Let τ be a countable relational signature and let $\mathcal{C}$ be an amalgamation class of τ-structures. Then there is a homogeneous and at most countable τ-structure $\mathfrak{C}$ whose age equals $\mathcal{C}$. The structure $\mathfrak{C}$ is unique up to isomorphism, and called the* Fraïssé limit *of $\mathcal{C}$.*

Example 2.17 The Fraïssé limit of the class of all finite linear orders is isomorphic to $(\mathbb{Q}; <)$, the order of the rationals. The Fraïssé limit of the class of all graphs is the so-called random graph (or Rado graph); see e.g. [15].

We also have the following converse of Theorem 2.16.

Theorem 2.18 (Fraïssé; see [25]) *Let Γ be a homogeneous relational structure. Then the age of Γ is an amalgamation class.*

As we have seen, there is a close connection between amalgamation classes and homogeneous structures, and we therefore make the following definition.

Definition 2.19 (Ramsey structure) A homogeneous structure Γ is called *Ramsey* if the age of Γ has the Ramsey property.

2.6 Counterexamples

We have so far seen two important necessary conditions for a class $\mathcal{C}$ to be a Ramsey class: rigidity of the members of $\mathcal{C}$ (Corollary 2.11) and amalgamation (Theorem 2.14). As we will see in the examples in this section, these conditions are not sufficient for being Ramsey.

Example 2.20 Let $\mathcal{C}$ be the class of all finite $\{E, <\}$-structures where E denotes an equivalence relation and $<$ denotes a linear order. It is easy to verify that $\mathcal{C}$ has the amalgamation property. Moreover, all automorphisms of structures in $\mathcal{C}$ have to preserve $<$ and hence must be the identity. But $\mathcal{C}$ does not have the Ramsey property: let $\mathfrak{A}$ be the structure with domain $\{u, v\}$ such that $<^{\mathfrak{A}} = \{(u, v)\}$, and such that u and v are not E-equivalent. Let $\mathfrak{B}$ be the structure with domain $\{a, b, c, d\}$ such that $b <^{\mathfrak{B}} c <^{\mathfrak{B}} a <^{\mathfrak{B}} d$ and such that $\{a, b\}$ and $\{c, d\}$ are the equivalence classes of $E^{\mathfrak{B}}$. There are four copies of $\mathfrak{A}$ in $\mathfrak{B}$.

Suppose for contradiction that there is $\mathfrak{C} \in \mathcal{C}$ such that $\mathfrak{C} \to (\mathfrak{B})^{\mathfrak{A}}_2$. Let $\prec$ be a *convex* linear ordering of the elements of C, that is, a linear ordering such that $E(x, z)$ and $x < y < z$ implies that $E(x, y)$ and $E(y, z)$. Let $g \in \binom{\mathfrak{C}}{\mathfrak{A}}$. Define $\chi(g) = 1$ if $g(u) \prec g(v)$, and $\chi(g) = 2$ otherwise. Note that there are only two convex linear orderings of $\mathfrak{B}$, and that $|\chi(f \circ \binom{\mathfrak{B}}{\mathfrak{A}})| = 2$ for all $f \in \binom{\mathfrak{C}}{\mathfrak{B}}$.

However, the class of all equivalence relations with a *convex* linear order is Ramsey; see [27]. Moreover, as we will see in Example 3.25 in Section 3.7, the Fraïssé limit of the class $\mathcal{C}$ from Example 2.20 can be expanded by a convex linear order $\prec$ so that the resulting structure is homogeneous and Ramsey.

Example 2.21 The class of finite trees is not closed under taking substructures. If we close it under substructures, we obtain the class of all finite forests, a class which does not have the amalgamation property. The solution for a proper model-theoretic treatment of trees and forests is to use the concept of C-relations.

Formally, a ternary relation C is said to be a *C-relation*[2] on a set L if for all $a, b, c, d \in L$ the following conditions hold:

C1 $C(a; b, c) \to C(a; c, b)$;

C2 $C(a; b, c) \to \neg C(b; a, c)$;

C3 $C(a; b, c) \to C(a; d, c) \vee C(d; b, c)$;

---

[2] Terminology of Adeleke and Neumann [2].

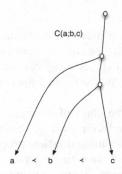

Figure 1: Illustration of a convexly ordered C-relation.

C4 $a = b \prec C(a; b\ b)$.

A C-relation on a set L is called *binary branching* if for all pairwise distinct $a\ b\ c\ \ L$ we have $C(a; b\ c)$ or $C(b; a\ c)$ or $C(c; a\ b)$.

The intuition here is that the elements of L denote the leaves of a rooted binary tree, and $C(a; b\ c)$ holds if in the tree, the shortest path from b to c does not intersect the shortest path from a to the root; see Figure 1. For nite L, this property is actually equivalent to the axiomatic de nition above [2].

The class of structures $(L; C)$ where L is a nite set and C is a binary branching C-relation on L is of course not a Ramsey class, since $(L; C)$ has nontrivial automorphisms, unless $L = 1$. The same argument does not work for the class $\mathcal{C}$ of all structures $(L; C\ <)$ where L is nite set, C is a binary branching C-relation on L, and $<$ is a linear ordering of L. In fact, $\mathcal{C}$ is an amalgamation class (a well-known fact; for a proof, see [5]), but not a Ramsey class. To see how the Ramsey property fails, consider the structure $\mathfrak{B}\quad \mathcal{C}$ with domain $a\ b\ c\ d$ where $a < c < b < d$ such that $C(a; c\ d)\ C(b; c\ d)\ C(d; a\ b)\ C(c; a\ b)$, and the structure $\mathfrak{A}\quad \mathcal{C}$ with domain $u\ v$ where $u < v$ let $\mathfrak{C}\quad \mathcal{C}$ be arbitrary. Let be a *convex* ordering of $\mathfrak{C}$, that is, a linear ordering such that for all $u\ v\ w\quad L$, if $C(u; v\ w)$ and $v\quad w$, then either $u\quad v\quad w$ or $v\quad w\quad u$. De ne $\chi: \begin{smallmatrix}\mathfrak{C}\\\mathfrak{A}\end{smallmatrix} \prec [2]$ as follows. For $g\quad \begin{smallmatrix}\mathfrak{C}\\\mathfrak{A}\end{smallmatrix}$ de ne $\chi(g) = 1$ if $g(u)\quad g(v)$, and $\chi(g) = 2$ otherwise. Note that for every convex ordering of B there exists an $e_1\quad \begin{smallmatrix}\mathfrak{B}\\\mathfrak{A}\end{smallmatrix}$ such that $e_1(u)\quad e_1(v)$, and an $e_2\quad \begin{smallmatrix}\mathfrak{B}\\\mathfrak{A}\end{smallmatrix}$ such that $e_2(v)\quad e_2(u)$. Hence, for every $f\quad \begin{smallmatrix}\mathfrak{C}\\\mathfrak{B}\end{smallmatrix}$ we have $\chi(f\quad \begin{smallmatrix}\mathfrak{B}\\\mathfrak{A}\end{smallmatrix}) = 2$.

Again, the class of all convexly ordered binary branching C-relations over a finite set is an amalgamation class (Theorem 5.1). Moreover, by the results from Section 3.7, the Fraïssé limit of the class $\mathcal{C}$ from Example 2.21 can be expanded by a convex linear order so that the resulting structure is homogeneous and Ramsey; see Example 5.4.

2.7 Automorphism groups

Let $f\colon D \to D$ be a function and $t \in D^m$ a tuple. Then $f(t)$ denotes the tuple $(f(t_1), \ldots, f(t_m))$. We say that a relation $R \subseteq D^m$ is *preserved* by a function $f\colon D \to D$ if $f(t) \in R$ for all $t \in R$. An automorphism of a structure Γ with domain D is a permutation α such that both α and α^{-1} preserve all relations (and if the signature contains constant symbols, α must fix the constants).

The equivalent formulation of the Ramsey property in Proposition 2.22 will be useful later, for instance to prove that for every homogeneous Ramsey structure Γ there exists a linear order on the domain of Γ that is preserved by all automorphisms of Γ.

Proposition 2.22 *Let Γ be a homogeneous structure. Then the following are equivalent.*

1. *Γ is Ramsey.*

2. *For every finite substructure $\mathfrak{B}$ of Γ and $r \in \mathbb{N}$ there exists a finite substructure $\mathfrak{C}$ of Γ such that for all substructures $\mathfrak{A}_1, \ldots, \mathfrak{A}_\ell$ of $\mathfrak{B}$ and all $\chi_i\colon \binom{\mathfrak{C}}{\mathfrak{A}_i} \to [r]$ there exists an $e \in \binom{\mathfrak{C}}{\mathfrak{B}}$ such that $|\chi_i(e \circ \binom{\mathfrak{B}}{\mathfrak{A}_i})| = 1$ for all $i \in [\ell]$.*

Proof (1) $\Rightarrow$ (2). We only show the forward implication, the backward implication being trivial. Let $\mathfrak{B}$ be a finite substructure of Γ and $r \in \mathbb{N}$. Let $\mathfrak{A}_1, \ldots, \mathfrak{A}_\ell$ be an enumeration of the substructures of $\mathfrak{B}$. We are going to construct a sequence of structures $\mathfrak{C}_1, \ldots, \mathfrak{C}_\ell$. Since Γ is Ramsey, there exists a substructure $\mathfrak{C}_1$ of Γ such that $\mathfrak{C}_1 \to (\mathfrak{B})_r^{\mathfrak{A}_1}$. Inductively, for $i \in \{2, \ldots, \ell\}$ there exists a substructure $\mathfrak{C}_i$ of Γ such that $\mathfrak{C}_i \to (\mathfrak{C}_{i-1})_r^{\mathfrak{A}_i}$. Define $\mathfrak{C} := \mathfrak{C}_\ell$.

For all $i \in [\ell]$, let $\chi_i\colon \binom{\mathfrak{C}}{\mathfrak{A}_i} \to [r]$ be arbitrary. Since $\mathfrak{C}_\ell \to (\mathfrak{C}_{\ell-1})_r^{\mathfrak{A}_\ell}$, there exists an $e_\ell \in \binom{\mathfrak{C}_\ell}{\mathfrak{C}_{\ell-1}}$ with $|\chi(e_\ell \circ \binom{\mathfrak{C}_{\ell-1}}{\mathfrak{A}_\ell})| \leq 1$. Inductively, suppose we have already defined $e_i \in \binom{\mathfrak{C}_\ell}{\mathfrak{C}_{i-1}}$ for an $i \in \{2, \ldots, \ell\}$ such that for all $j \in \{i, \ldots, \ell\}$ we have $|\chi(e_i \circ \binom{\mathfrak{C}_{i-1}}{\mathfrak{A}_j})| \leq 1$. Then there exists an $e_{i-1} \in \binom{\mathfrak{C}_{i-1}}{\mathfrak{C}_{i-2}}$ such that $|\chi(e_{i-1} \circ \binom{\mathfrak{C}_{i-2}}{\mathfrak{A}_{i-1}})| \leq 1$. Hence, for all $j \in \{i-1, \ldots, \ell\}$

we have $|\chi(e_{i-1} \circ (\begin{smallmatrix} \mathfrak{C}_{i-2} \\ \mathfrak{A}_j \end{smallmatrix}))| \leq 1$. Then the map $e_1 \in (\begin{smallmatrix} \mathfrak{C}_k \\ \mathfrak{B} \end{smallmatrix})$ has the desired properties from the statement of the proposition.   □

Proposition 2.23 *Let Γ be a homogeneous Ramsey structure with domain D. Then there exists a linear order on D that is preserved by all automorphisms of Γ.*

Proof Let $d_1, d_2, \ldots$ be an enumeration of D, and let $<$ be the linear order on D given by this enumeration, that is, $d_i < d_j$ if and only if $i < j$. Let $\mathfrak{T}$ be a tree whose vertices on level n are linear orders $\prec$ of $D_n := \{d_1, \ldots, d_n\}$ with the property that for all $a, b \in D_n$ and $\alpha \in \operatorname{Aut}(\Gamma)$ such that $\alpha(a), \alpha(b) \in D_n$, we have that $a \prec b$ if and only if $\alpha(a) \prec \alpha(b)$. Note that when a linear order satisfies this condition, then also restrictions of the linear order to subsets satisfy this condition. Adjacency in $\mathfrak{T}$ is defined by restriction. Clearly, $\mathfrak{T}$ is finitely branching. We will show that $\mathfrak{T}$ has vertices on each level. By König's lemma, there is an infinite path in $\mathfrak{T}$, which defines a linear ordering on D that is preserved by $\operatorname{Aut}(\Gamma)$.

To show that there is a linear order $\prec$ on D_n that satisfies the condition, let $\mathfrak{B}$ be the structure induced by D_n in Γ, and let $\mathfrak{A}_1, \ldots, \mathfrak{A}_\ell$ list the substructures of Γ that are induced by the two-element subsets of D_n. By Proposition 2.22, there exists a finite substructure $\mathfrak{C}$ of Γ such that for all $\chi_i \colon (\begin{smallmatrix} \mathfrak{C} \\ \mathfrak{A}_i \end{smallmatrix}) \to [r]$ there exists an $e \in (\begin{smallmatrix} \mathfrak{C} \\ \mathfrak{B} \end{smallmatrix})$ such that $|\chi_i(e \circ (\begin{smallmatrix} \mathfrak{B} \\ \mathfrak{A}_i \end{smallmatrix}))| = 1$ for all $i \in [l]$. Let $\chi_i \colon (\begin{smallmatrix} \mathfrak{C} \\ \mathfrak{A}_i \end{smallmatrix}) \to [2]$ be defined as follows. For $e \in (\begin{smallmatrix} \mathfrak{C} \\ \mathfrak{A}_i \end{smallmatrix})$, we define $\chi_i(e) = 1$ if e preserves $<$, and $\chi_i(e) = 2$ otherwise. By the property of $\mathfrak{C}$, there is an $e \in (\begin{smallmatrix} \mathfrak{C} \\ \mathfrak{B} \end{smallmatrix})$ be such that $|\chi_i(e \circ (\begin{smallmatrix} \mathfrak{B} \\ \mathfrak{A}_i \end{smallmatrix}))| = 1$ for all $i \in [l]$.

Let $\prec$ be the linear order on D_n given by $a \prec b$ if $e(a) < e(b)$. Suppose now that $a, b \in D_n$ and $\alpha \in \operatorname{Aut}(\Gamma)$ such that $\alpha(a), \alpha(b) \in D_n$. Let $i \in [l]$ be such that $\{a, b\}$ induce $\mathfrak{A}_i$ in Γ. Let f_1 be the identity on $\{a, b\}$, and let f_2 be the restriction of α to $\{a, b\}$; then $f_1, f_2 \in (\begin{smallmatrix} \mathfrak{B} \\ \mathfrak{A}_i \end{smallmatrix})$, and $\chi_i(e \circ f_1) = \chi_i(e \circ f_2)$. By the definition of χ_i, we have that $e(a) < e(b)$ if and only if $e(\alpha(a)) < e(\alpha(b))$. By the definition of $\prec$ we obtain that $a \prec b$ if and only if $\alpha(a) \prec \alpha(b)$.   □

2.8 Countably categorical structures

In this subsection we present a generalization of the class of all homogeneous structures with a finite relational signature that still satisfies a certain finiteness condition, namely the class of all countable ω-*categorical* structures.

Definition 2.24 A countable structure is said to be ω-*categorical* if all countable structures that satisfy the same first-order sentences as Γ are isomorphic to Γ.

Theorem 2.26 below explains why ω-categoricity can be seen as a finiteness condition. It will be easy to see from Theorem 2.26 that all structures that are homogeneous in a finite relational signature are ω-categorical. But we first show an example where ω-categoricity can be seen directly.

Example 2.25 All countably infinite vector spaces $\mathfrak{V}$ over a fixed finite field $\mathbb{F}$ are isomorphic. Since the isomorphism type of $\mathbb{F}$, the axioms of vector spaces, and having infinite dimension can be expressed by first-order sentences it follows that $\mathfrak{V}$ is ω-categorical. These structures are homogeneous; however, their signature is not relational. The relational structure with the same domain that contains all relations that are first-order definable over $\mathfrak{V}$ is homogeneous, too (this follows from Theorem 2.26 below). It is easy to see that all relational structures obtained from those examples by dropping all but finitely many relations, but have the same automorphism group as $\mathfrak{V}$, are *not* homogeneous. For example, the structure that just contains the ternary relation defined by $x = y + z$ has the same automorphism group as $\mathfrak{V}$, but is *not* homogeneous. The Ramsey properties of those examples are beyond the scope of this survey, but are discussed in [27].

The following theorem of Engeler, Ryll-Nardzewski, and Svenonius shows that whether a structure is ω-categorical can be seen from the automorphism group $\mathrm{Aut}(\Gamma)$ of Γ (as a permutation group).

Theorem 2.26 (see e.g. [25]) *Let Γ be a countably infinite structure with a countably infinite signature. Then the following are equivalent.*

1. *Γ is ω-categorical;*

2. *$\mathrm{Aut}(\Gamma)$ is* oligomorphic, *that is, for all $n \geq 1$, the componentwise action of $\mathrm{Aut}(\Gamma)$ on n-tuples from Γ has finitely many orbits;*

3. *all orbits of n-tuples in Γ are first-order definable in Γ;*

4. *all relations preserved by $\mathrm{Aut}(\Gamma)$ are first-order definable in Γ.*

Theorem 2.26 implies that when Γ is ω-categorical, then the expansion Γ' of Γ by all first-order definable relations is homogeneous. We therefore make the following definition.

Definition 2.27 An ω-categorical structure Γ is called *Ramsey* if the expansion of Γ by all relations with a first-order definition in Γ is Ramsey (as a homogeneous structure).

This definition is compatible with Definition 2.19, since expansions by first-order definable relations do not change the automorphism group, and since the Ramsey property only depends on the automorphism group, as reflected in the next proposition. For subsets S and M of the domain of an ω-categorical structure Γ, we write $\binom{M}{S}$ for the set of all maps from S to M that can be extended to an automorphism of Γ. The following is immediate from the definitions, Theorem 2.26, and Proposition 2.8.

Proposition 2.28 *Let Γ be an ω-categorical structure with domain D. Then the following are equivalent.*

1. *Γ is Ramsey;*

2. *For all $r \in \mathbb{N}$ and finite $M \subset D$ and $S \subset M$ there exists a finite $L \subseteq D$ such that for every map χ from $\binom{L}{S}$ to $[r]$ there exists $f \in \binom{L}{M}$ such that $|\chi(f \circ \binom{M}{S}))| = 1$.*

3. *For all $r \in \mathbb{N}$ and finite $M \subset D$ and $S \subset M$ and every map χ from $\binom{D}{S}$ to $[r]$ there exists $f \in \binom{D}{M}$ such that $|\chi(f \circ \binom{M}{S}))| = 1$.*

The following is a direct consequence of Proposition 2.23.

Corollary 2.29 *Let Γ be an ω-categorical Ramsey structure. Then there is a linear order with a first-order definition in Γ.*

Proof Let Γ^* be the homogeneous expansion of Γ by all first-order definable relations. By Proposition 2.23, there exists a linear ordering of the domain of Γ^* which is preserved by all automorphisms of Γ. By ω-categoricity of Γ and Γ^*, Theorem 2.26, this linear order is first-order definable in Γ^*. Since all first-order definable relations of Γ^* are first-order definable in Γ, they are present in the signature of Γ^*, and the statement follows. $\square$

3   New Ramsey classes from old

The class of ω-categorical Ramsey structures is remarkably robust with respect to basic model-theoretic constructions. We will consider the following model-theoretic constructions to obtain new structures from given structures $\Gamma, \Gamma_1, \Gamma_2$:

- disjoint unions and products of Γ_1 and Γ_2;

- structures with a first-order interpretation in Γ;

- expansions of Γ by finitely many constants;

- the model companion of Γ;

- the model-complete core of Γ;

- superpositions of Γ_1 and Γ_2.

If the structures $\Gamma, \Gamma_1, \Gamma_2$ we started from are ω-categorical (or homogeneous in a finite relational signature), the structure we thus obtain will be again ω-categorical (or homogeneous in a finite relational signature). In this section we will see that if the original structures have good Ramsey properties, then the new structures also do.

3.1 Disjoint unions

One of the simplest operations on structures is the formation of disjoint unions: when $\mathfrak{A}_1$ and $\mathfrak{A}_2$ are structures with the same relational signature τ and disjoint domains, then the disjoint union of $\mathfrak{A}_1$ and $\mathfrak{A}_2$ is the structure $\mathfrak{B}$ with domain $B := A_1 \cup A_2$ where for each $R \in \tau$ we set $R^{\mathfrak{B}} := R^{\mathfrak{A}_1} \cup R^{\mathfrak{A}_2}$. The disjoint union of two ω-categorical structures is always ω-categorical. The disjoint union of two homogeneous structures Γ_1 and Γ_2 might not be homogeneous; but it clearly becomes homogeneous when we add an additional new unary predicate P to the disjoint union which precisely contains the vertices from Γ_1. We denote the resulting structure by $\Gamma_1 \uplus_P \Gamma_2$. The transfer of the Ramsey property is a triviality in this case.

Lemma 3.1 *Let Γ_1 and Γ_2 be ω-categorical Ramsey structures. Then $\Gamma := \Gamma_1 \uplus_P \Gamma_2$ is an ω-categorical Ramsey structure, too. If Γ_1 and Γ_2 are homogeneous with finite relational signature, then so is Γ.*

While this lemma looks innocent, it still has interesting applications in combination with the other constructions that we present; see Example 3.26.

3.2 Products

When G_1 and G_2 are permutation groups acting on the sets D_1 and D_2, respectively, then the direct product $G_1 \times G_2$ of G_1 and G_2 naturally acts on $D_1 \times D_2$: the element (g_1, g_2) of $G_1 \times G_2$ maps (x_1, x_2) to $(g_1(x_1), g_2(x_2))$. When G_1 and G_2 are the automorphism groups of relational structures Γ_1 and Γ_2, then the following definition yields a structure whose automorphism group is precisely $G_1 \times G_2$. (The direct product $\Gamma_1 \times \Gamma_2$ does not have this property.)

Definition 3.2 (Full product) Let $\Gamma_1, \ldots, \Gamma_d$ be structures with domains $D_1, \ldots, D_d$ and pairwise disjoint signatures $\tau_1, \ldots, \tau_d$. Then the *full product structure* $\Gamma_1 \boxtimes \cdots \boxtimes \Gamma_d$ is the structure with domain $D_1 \times \cdots \times D_d$ that contains for every $i \leq d$ and m-ary $R \in (\tau_i \cup \{=\})$ the relation defined by $\{((x_1^1, \ldots, x_1^d), \ldots, (x_m^1, \ldots, x_m^d)) : (x_1^i, \ldots, x_m^i) \in R^{\Gamma_i}\}$.

The following proposition is known as the *product Ramsey theorem* to combinatorists.

Proposition 3.3 *Let* $\Gamma_1, \ldots, \Gamma_d$ *be* ω-*categorical Ramsey structures with pairwise disjoint signatures. Then* $\Gamma := \Gamma_1 \boxtimes \cdots \boxtimes \Gamma_d$ *is* ω-*categorical and Ramsey. If* $\Gamma_1, \ldots, \Gamma_d$ *are homogeneous with finite relational signature, then so is* Γ.

Proof It follows from Theorem 2.26 that if $\Gamma_1, \ldots, \Gamma_d$ are ω-categorical, then $\Gamma_1 \boxtimes \cdots \boxtimes \Gamma_d$ is ω-categorical.

For the homogeneity of $\Gamma_1 \boxtimes \Gamma_2$, let $u_1 := (u_1^1, \ldots, u_1^d)$, $\ldots$, $u_m := (u_m^1, \ldots, u_m^d)$ and $v_1 := (v_1^1, \ldots, v_1^d)$, $\ldots$, $v_m := (v_m^1, \ldots, v_m^d)$ be elements of Γ such that the map a that sends $(u_1, \ldots, u_m)$ to $(v_1, \ldots, v_m)$ is an isomorphism between substructures of Γ. For $i \leq d$, define a_i as the map that sends u_j^i to v_j^i for all $j \leq m$; this is well-defined since a preserves the relation $\{(x^1, \ldots, x^d, y^1, \ldots, y^d) : x^i = y^i\}$. By homogeneity of Γ_i, there exists an extension α_i of a_i to an automorphism of Γ_i. Then the map α given by $\alpha(x_1, \ldots, x_d) := (\alpha_1(x_1), \ldots, \alpha_d(x_d))$ is an automorphism of Γ and extends α.

To prove that Γ is Ramsey, we show the statement for $d = 2$; the general case then follows by induction on d. Let $\mathfrak{A}, \mathfrak{B}$ be substructures of $\Gamma = \Gamma_1 \boxtimes \Gamma_2$ and $r \in \mathbb{N}$ be arbitrary. We will show that $\Gamma \to (\mathfrak{B})_r^{\mathfrak{A}}$, so let $\chi \colon \binom{\mathfrak{B}}{\mathfrak{A}} \to [r]$ be arbitrary. If $\binom{\mathfrak{B}}{\mathfrak{A}}$ is empty, then the statement is trivial, so in the following we assume that $\mathfrak{A}$ embeds into $\mathfrak{B}$. For $i \in \{1, 2\}$, let $\mathfrak{A}_i$ be the structure induced in Γ_i by $\{a_i : (a_1, a_2) \in A\}$, and define $\mathfrak{B}_i$ analogously with B instead of A. Since Γ_2 is Ramsey there exists a finite substructure $\mathfrak{C}_2$ of Γ_2 such that $\mathfrak{C}_2 \to (\mathfrak{B}_2)_r^{\mathfrak{A}_2}$. Define $s := |\binom{\mathfrak{C}_2}{\mathfrak{A}_2}|$. Since Γ_1 is Ramsey there exists a finite substructure $\mathfrak{C}_1$ of Γ_1 such that $\mathfrak{C}_1 \to (\mathfrak{B}_1)_{r^s}^{\mathfrak{A}_1}$. We identify the elements of $[r^s]$ with functions from $\binom{\mathfrak{C}_2}{\mathfrak{A}_2}$ to $[r]$. Define $\chi_1 \colon \binom{\mathfrak{C}_1}{\mathfrak{A}_1} \to [r^s]$ as follows. Let $e_1 \in \binom{\mathfrak{C}_1}{\mathfrak{A}_1}$, let $e_2 \in \binom{\mathfrak{C}_2}{\mathfrak{A}_2}$, and let $e \in \binom{\mathfrak{C}}{\mathfrak{A}}$ be the embedding such that $e(a_1, a_2) = (e_1(a_1), e_2(a_2))$. Let $\xi \colon \binom{\mathfrak{C}_2}{\mathfrak{A}_2} \to [r]$ be the function that maps $e_2 \in \binom{\mathfrak{C}_2}{\mathfrak{A}_2}$ to $\chi(e)$. Define $\chi_1(e_1) = \xi$. Then there exists an $f_1 \in \binom{\mathfrak{B}_1}{\mathfrak{A}_1}$ such that $\chi_1(f_1 \circ \binom{\mathfrak{A}_1}{\mathfrak{A}_1})) = \{\chi_2\}$ for some $\chi_2 \in \binom{\mathfrak{C}_2}{\mathfrak{A}_2} \to [r]$. As $\mathfrak{C}_2 \to (\mathfrak{B}_2)_r^{\mathfrak{A}_2}$, there exists an $f_2 \in \binom{\mathfrak{C}_2}{\mathfrak{B}_2}$ such that $|\chi_2(f_2 \circ \binom{\mathfrak{B}_2}{\mathfrak{A}_2}))| = 1$. Let $f \in \binom{\mathfrak{C}_1 \boxtimes \mathfrak{C}_2}{\mathfrak{B}}$ be given by $b \mapsto (f_1(b), f_2(b))$.

We claim that $|\chi(f \circ \binom{\mathfrak{B}}{\mathfrak{A}})| = 1$. Arbitrarily choose $e, e' \in \binom{\mathfrak{B}}{\mathfrak{A}}$. Then there are $e_i, e'_i \colon \binom{\mathfrak{B}_i}{\mathfrak{A}_i}$ for $i \in \{1,2\}$ such that $e(A) \subseteq (e_1(A_1), e_2(A_2))$ and $e'(A) \subseteq (e'_1(A_1), e'_2(A_2))$. Then $\chi_1(f_1 \circ e_1) = \chi_1(f_1 \circ e'_1) = \chi_2$, and $\chi_2(f_2 \circ e_1) = \chi_2(f_2 \circ e'_2)$. Then $\chi(e) = \chi_2(f_2 \circ e_2) = \chi_2(f_2 \circ e'_2) = \chi(e')$, which is what we had to show.   □

The special case of Proposition 3.3 where $\Gamma_1 = \cdots = \Gamma_d = (\mathbb{Q}; <)$ can be found in [22] (page 97). The general case can also be shown inductively, see e.g. [11]. One may also derive it using the results in Kechris–Pestov–Todorcevic [27], since the direct product of extremely amenable groups is extremely amenable (also see [6]).

3.3 Interpretations

The concept of *first-order interpretations* is a powerful tool to construct new structures. A simple example of an interpretation is the line graph of a graph G, which has a first-order interpretation over G. By passing to the age of the constructed structure, they are also a great tool to define new *classes* of structures.

Definition 3.4 A relational σ-structure $\mathfrak{B}$ has a *(first-order) interpretation* I in a τ-structure $\mathfrak{A}$ if there exists a natural number d, called the *dimension* of I, and

- a τ-formula $\delta_I(x_1, \ldots, x_d)$ – called the *domain formula*,

- for each atomic σ-formula $\phi(y_1, \ldots, y_k)$ a τ-formula

$$\phi_I(y_{1,1}, \ldots, y_{1,d}, y_{2,1}, \ldots, y_{2,d}, \ldots, y_{k,1}, \ldots, y_{k,d})$$

  – the *defining formulas*;

- a surjective map h from $\{\bar{a} : \mathfrak{A} \models \delta_I(\bar{a})\}$ to B – called the *coordinate map*,

such that for all atomic σ-formulas ϕ and all elements $a_{1,1}, \ldots, a_{k,d}$ with $\mathfrak{A} \models \delta_I(a_{i,1}, \ldots, a_{i,d})$ for all $i \leq k$

$$\mathfrak{B} \models \phi(h(a_{1,1}, \ldots, a_{1,d}), \ldots, h(a_{k,1}, \ldots, a_{k,d}))$$
$$\Leftrightarrow \quad \mathfrak{A} \models \phi_I(a_{1,1}, \ldots, a_{k,d}) \, .$$

We give illustrating examples.

Example 3.5 When $(V; E)$ is an undirected graph, then the *line graph* of $(V; E)$ is the undirected graph $(E; F)$ where $F := \{\{u, v\} : |u \cap v| = 1\}$. Undirected graphs can be seen as structures where the signature contains a single binary relation denoting a symmetric irreflexive relation. Then the line graph of $(V; E)$ has the following 2-dimensional interpretation I over $(V; E)$: the domain formula $\delta_I(x_1, x_2)$ is $E(x_1, x_2)$, the defining formula for the atomic formula $y_1 = y_2$ is

$$(y_{1,1} = y_{2,1} \wedge y_{1,2} = y_{2,2}) \vee (y_{1,1} = y_{2,2} \wedge y_{1,2} = y_{2,1}),$$

and the defining formula for the atomic formula $F(y_1, y_2)$ is

$$((y_{1,1} \neq y_{2,1} \wedge y_{1,1} \neq y_{2,2}) \vee (y_{1,2} \neq y_{2,2} \wedge y_{1,2} \neq y_{2,2}))$$
$$\wedge (y_{1,1} = y_{2,1} \vee y_{1,1} = y_{2,2} \vee y_{1,2} = y_{2,1} \vee y_{1,2} = y_{2,2}).$$

The coordinate map is the identity.

Example 3.6 A poset $(P; \leq)$ has *poset dimension at most k* if there are k linear extensions $\leq_1, \ldots, \leq_k$ of $\leq$ such that $x \leq y$ if and only if $x \leq_i y$ for all $i \in [k]$. The class of all finite posets of poset dimension at most k is the age of $(\mathbb{Q}; \leq)^k$, which clearly has a k-dimensional interpretation in $(\mathbb{Q}; <)$.

Lemma 3.7 (Theorem 7.3.8 in [24]) *Let $\mathfrak{A}$ be an ω-categorical structure. Then every structure $\mathfrak{B}$ that is first-order interpretable in $\mathfrak{A}$ is countably infinite ω-categorical or finite.*

Note that in particular all reducts (defined in the introduction) of an ω-categorical structure Γ have an interpretation in Γ and are thus again ω-categorical. On the other hand, being homogeneous with finite relational signature is not inherited by the interpreted structures. An example of a structure which is not interdefinable with a homogeneous structure in a finite relational signature, but which has a first-order interpretation over $(\mathbb{N}; =)$, has been found by Cherlin and Lachlan [16].

Proposition 3.8 *Suppose that Γ is ω-categorical Ramsey. Then every structure with a first-order interpretation in Γ has an ω-categorical Ramsey expansion Δ. Furthermore, if Γ is homogeneous with a finite relational signature, then we can choose Δ to be homogeneous in a finite relational signature, too.*

Corollary 3.9 *Conjecture 1.1 is true for countable stable[3] homogeneous structures with finite relational signature.*

---

[3]For the definition of stability we refer to any text book in model theory.

Proof Lachlan [30] proved that every stable homogeneous structure with a finite relational signature has a first-order interpretation over $(\mathbb{Q}; <)$. The statement follows from the fact that $(\mathbb{Q}; <)$ is Ramsey, and Proposition 3.8. □

3.4 Adding constants

Let Γ be homogeneous. It is clear that the expansion $(\Gamma, d_1, \ldots, d_n)$ by finitely many constants $d_1, \ldots, d_n$ is again homogeneous. Similarly, if Γ is ω-categorical, then $(\Gamma, d_1, \ldots, d_n)$ is ω-categorical, as a consequence of Theorem 2.26. We will show here that if Γ is Ramsey, then $(\Gamma, d_1, \ldots, d_n)$ remains Ramsey. The original proof [11] went via a more general fact from topological dynamics (open subgroups of extremely amenable groups are extremely amenable). We give an elementary proof here, due to Miodrag Sokic.

Theorem 3.10 *Let Γ be homogeneous and Ramsey. Let $d_1, \ldots, d_n$ be elements of Γ. Then $(\Gamma, d_1, \ldots, d_n)$ is also Ramsey.*

Proof Let τ be the signature, and D the domain of Γ. We write d for $(d_1, \ldots, d_n)$. Let $\mathfrak{A}^*, \mathfrak{B}^*$ be two finite substructures of Γ^*, let $r \in \mathbb{N}$, and let $\chi^* : \binom{\Gamma^*}{\mathfrak{A}^*} \to [r]$ be arbitrary. We have to show that there exists an $f \in \binom{\Gamma^*}{\mathfrak{B}^*}$ such that $|\chi(f \circ \binom{\mathfrak{B}^*}{\mathfrak{A}^*})| = 1$. We write $\mathfrak{A}$ and $\mathfrak{B}$ for the τ-reducts of $\mathfrak{A}^*$ and $\mathfrak{B}^*$, respectively.

Define $\chi : \binom{\Gamma}{\mathfrak{A}} \to [r]$ as follows. First, we fix for each tuple $a \in D^n$ that lies in the same orbit as d in $\mathrm{Aut}(\Gamma)$ an automorphism α_a of Γ such that $\alpha_a(a) = d$. Let $e \in \binom{\Gamma}{\mathfrak{A}}$, and $a := e(d)$. By the homogeneity of Γ, the tuples a and d lie in the same orbit of $\mathrm{Aut}(\Gamma)$. Note that $\alpha_a \circ e$ fixes d and is an embedding of $\mathfrak{A}^*$ into Γ^*. Define $\chi(e) := \chi^*(\alpha_a \circ e)$.

Since Γ is Ramsey, there is an $f \in \binom{\Gamma}{\mathfrak{B}}$ such that $\chi(f \circ \binom{\mathfrak{B}}{\mathfrak{A}}) = \{c\}$ for some $c \in [r]$. Let b be $f(d)$. By the homogeneity of Γ, the tuples b and d lie in the same orbit of $\mathrm{Aut}(\Gamma)$. Observe that $f' := \alpha_b \circ f$ fixes d and is an embedding of $\mathfrak{B}^*$ into Γ^*.

We claim that $|\chi^*(f' \circ \binom{\mathfrak{B}^*}{\mathfrak{A}^*})| = 1$. To prove this, let $g \in \binom{\mathfrak{B}^*}{\mathfrak{A}^*}$ be arbitrary. Since g is in particular from $\binom{\mathfrak{B}}{\mathfrak{A}}$ we have $\chi(f \circ g) = c$. By the definition of χ we have that $\chi(f \circ g) = \chi^*(\alpha_b \circ f \circ g) = \chi^*(f' \circ g)$. Hence, $\chi^*(f' \circ g) = c$, which proves the claim. □

In this article, we work mostly with relational signatures. It is therefore important to note that the relational structure $(\Gamma, \{d_1\}, \ldots, \{d_n\})$ is in general *not* homogeneous even if Γ is. Consider for example the Fraïssé limit $\Gamma = (\mathbb{V}; E)$ of the class of all finite graphs, and an arbitrary $d_1 \in \mathbb{V}$.

Let $p \in \mathbb{V} \setminus \{d_1\}$ be such that $E(p, d_1)$ and $q \in \mathbb{V} \setminus \{d_1\}$ be such that $\neg E(d_1, q)$. Then the mapping that sends p to q is an isomorphism between (one-element) substructures of $(\Gamma, \{d_1\})$ which cannot be extended to an automorphism of $(\Gamma, \{d_1\})$. (The difference to (Γ, d_1) is that all substructures of (Γ, d_1) must contain d_1.) Note, however, that $(\Gamma, d_1, \ldots, d_n)$ and $(\Gamma, \{d_1\}, \ldots, \{d_n\})$ have the same automorphism group.

The solution to stating the result about expansions of homogeneous structures with constants in the relational setting is linked to the following definition.

Definition 3.11 Let Γ be a relational structure with signature τ, and $d_1, \ldots, d_n$ elements of Γ. Then $\Gamma_{d_1, \ldots, d_n}$ denotes the expansion of Γ which contains for every $R \in (\tau \cup \{=\})$ of arity $k \geq 2$, every $i \in [k]$ and $j \in [n]$, the $(k-1)$-ary relation $\{(x_1, \ldots, x_{i-1}, x_{i+1}, \ldots, x_k) : (x_1, \ldots, x_k) \in R \text{ and } x_i = d_j\}$.

Note that if the signature of Γ is finite, then the signature of $\Gamma_{d_1, \ldots, d_n}$ is also finite, and the maximal arity is unaltered. Also note that $\Gamma_{d_1, \ldots, d_n}$ has in particular the unary relations $\{d_1\}, \ldots, \{d_n\}$.

Lemma 3.12 *Let Γ be a homogeneous relational structure, and $d_1, \ldots, d_n$ elements of Γ. Then $\Gamma_{d_1, \ldots, d_n}$ is homogeneous.*

Proof Let a be an isomorphism between two finite substructures A_1, A_2 of $\Gamma_{d_1, \ldots, d_n}$. Since $\Gamma_{d_1, \ldots, d_n}$ contains for all $i \leq n$ the relation $\{d_i\}$ which is preserved by a, it follows that if A_1 or A_2 contains c_i, then both A_1 and A_2 must contain d_i, and $a(d_i) = d_i$. If d_i is contained in neither A_1 nor A_2, then a can be extended to a partial isomorphism a' of $\Gamma_{d_1, \ldots, d_n}$ with domain $A_1 \cup \{d_i\}$ by setting $a(d_i) = d_i$: this follows directly from the definition of the signature of $\Gamma_{d_1, \ldots, d_n}$. By the homogeneity of Γ, the map a' can be extended to an automorphism of Γ. This automorphism fixes $d_1, \ldots, d_n$ pointwise, and hence is an automorphism of $\Gamma_{d_1, \ldots, d_n}$. $\square$

Corollary 3.13 *Let Γ be homogeneous, ω-categorical, and Ramsey, and let $d_1, \ldots, d_n$ be elements of Γ. Then $\Gamma_{d_1, \ldots, d_n}$ is also Ramsey.*

Proof The statement follows from Theorem 3.10 from the observation that $\Gamma_{d_1, \ldots, d_n}$ and $(\Gamma, d_1, \ldots, d_n)$ have the same automorphism group, and that whether an ω-categorical structure has the Ramsey property only depends on its automorphism group (Proposition 2.28). $\square$

3.5 Passing to the model companion

A structure Γ is called *model-complete* if all embeddings between models of the first-order theory of Γ preserve all first-order formulas. It is well-known that this is equivalent to every first-order formula being equivalent to an existential formula over Γ (see e.g. [25]). It is also known (see Theorem 3.6.7 in [5]) that an ω-categorical structure Γ is model-complete if and only if for every finite tuple t of elements of Γ and for every self-embedding e of Γ into Γ there exists an automorphism α of Γ such that $e(t) = \alpha(t)$.

A *model companion* of Γ is a model-complete structure Δ with the same age as Γ. If Γ has a model companion, then the model companion is unique up to isomorphism [25]. Every ω-categorical structure has a model companion, and the model companion is again ω-categorical [46].

Example 3.14 We write $\mathbb{Q}_0^+$ for $\{q \in \mathbb{Q} : q \geq 0\}$. The structure $\Gamma := (\mathbb{Q}_0^+; <)$ is ω-categorical, but not model-complete: for instance the map $x \mapsto x + 1$ is an embedding of Γ into Γ which does not preserve the unary relation $\{0\}$ with the first-order definition $\forall y(y \geq x)$ over Γ. The model companion of Γ is $(\mathbb{Q}; <)$.

In this subsection we prove the following.

Theorem 3.15 *Let Γ be ω-categorical and Ramsey, and let Δ be the model companion of Γ. Then Δ is also Ramsey.*

Proof Let e be an embedding of Γ into Δ, and let i be an embedding of Δ into Γ; such embeddings exist by ω-categoricity of Δ and Γ, see Section 3.6.2 in [5]. We will work with the equivalent characterisation of the Ramsey property given in item 2 of Proposition 2.28.

Let S and M be finite subsets of the domain D of Δ and $r \in \mathbb{N}$, and let $\chi \colon \binom{D}{S} \to [r]$ be arbitrary. Let D' be the domain of Γ. We define a map $\chi' \colon \binom{D'}{i(S)} \to [r]$ as follows. For $q' \in \binom{D'}{i(S)}$, note that $e \circ q' \circ i \in \binom{D}{S}$. We define $\chi'(q') := \chi(e \circ q' \circ i)$.

Since Γ is Ramsey, there exists an $f' \in \binom{D'}{i(M)}$ and $c \in [r]$ such that for all $g' \in \binom{i(M)}{i(S)}$ we have $\chi'(f' \circ g') = c$. Let $\alpha' \in \mathrm{Aut}(\Gamma)$ be an extension of f'. Note that $e \circ \alpha' \circ i$ is an embedding of Δ into Δ, and since Δ is model-complete there exists an $\alpha \in \mathrm{Aut}(\Delta)$ that extends the restriction f of $e \circ \alpha' \circ i$ to M.

Let $g \in \binom{M}{S}$ be arbitrary. We claim that $\chi(f \circ g) = c$. Since $e \circ i$ is an embedding of Δ into Δ and Δ is model-complete, there exists an automorphism β of Δ such that $\beta(e(i(x))) = x$ for all $x \in S$. Note that

$g' := i \circ g \circ \beta \circ e \in \binom{i(M)}{i(S)}$, and hence $\chi'(f' \circ g') = c$. Also note that by the definition of χ' we have

$$\chi'(f' \circ g') = \chi(e \circ f' \circ g' \circ i) = \chi(e \circ f' \circ i \circ g \circ \beta \circ e \circ i) = \chi(f \circ g).$$

Hence, $\chi(f \circ g) = c$, and $|\chi(f \circ \binom{M}{S})| \leq 1$, and thus Δ is Ramsey. $\square$

3.6 Passing to the model-complete core

Cores play an important role in finite combinatorics. The concept of model-complete cores can be seen as an existential-positive analog of model-companions, where embeddings are replaced by homomorphisms and self-embeddings are replaced by endomorphisms. We state here results that are analogous to the results for model companions that we have seen in the previous section.

Definition 3.16 Let $\mathfrak{A}$ and $\mathfrak{B}$ be two structures with domain A and B, respectively, and the same relational signature τ. Then a *homomorphism* from $\mathfrak{A}$ to $\mathfrak{B}$ is a function $f \colon A \to B$ such that for all $(a_1, \ldots, a_n) \in R^{\mathfrak{A}}$ we have $(f(a_1), \ldots, f(a_n)) \in R^{\mathfrak{B}}$. An *endomorphism* of a structure Γ is a homomorphism from Γ to Γ. A structure Γ is called a *core* if every endomorphism of Γ is an embedding.

An ω-categorical structure Γ is a model-complete core if and only if for every finite tuple t of elements of Γ and for every endomorphism e of Γ there exists an automorphism α of Γ such that $e(t) = \alpha(t)$ (Theorem 3.6.11 in [5]). The following has been shown in [4] (also see [7]). Two structures Γ and Δ are *homomorphically equivalent* if there is a homomorphism from Γ to Δ and a homomorphism from Δ to Γ.

Theorem 3.17 *Every ω-categorical structure is homomorphically equivalent to a model-complete core Δ, which is unique up to isomorphism, and again countably infinite ω-categorical or finite. The expansion of Δ by all existential positive definable relations is homogeneous.*

The structure Δ in Theorem 3.17 will be called *the model-complete core of* Γ.

Theorem 3.18 *Let Γ be ω-categorical and Ramsey, and let Δ be the model-complete core of Γ. Then Δ is also Ramsey.*

Proof The proof is similar to the proof of Theorem 3.15. $\square$

3.7 Superimposing signatures

An amalgamation class $\mathcal{C}$ is called a *strong amalgamation class* if, informally, we can amalgamate structures $\mathfrak{B}_1, \mathfrak{B}_2 \in \mathcal{C}$ over $\mathfrak{A} \in \mathcal{C}$ in such a way that no points of $\mathfrak{B}_1$ and $\mathfrak{B}_2$ other than the elements of $\mathfrak{A}_1$ will be identified in the amalgam. Formally, we require that for all $\mathfrak{A}, \mathfrak{B}_1, \mathfrak{B}_2 \in \mathcal{C}$ and embeddings $e_i : \mathfrak{A} \to \mathfrak{B}_i$, $i \in \{1,2\}$, there exists a structure $\mathfrak{C} \in \mathcal{C}$ and embeddings $f_i : \mathfrak{B}_i \to \mathfrak{C}$ such that $e_1(f_1(x)) = e_2(f_2(x))$ for all $x \in A$, and additionally $f_1(B_1) \cap f_2(B_2) = f_1(e_1(A)) = f_2(e_2(A))$. When an amalgamation class $\mathcal{C}$ even has strong amalgamation, then this can be seen from the automorphism group of the Fraïssé limit of $\mathcal{C}$.

Definition 3.19 ([14]) We say that a permutation group has *no algebraicity* if for every finite tuple $(a_1, \ldots, a_n)$ of the domain the set of all permutations of the group that fix each of $a_1, \ldots, a_n$ fixes no other elements of the domain.

For automorphism groups of ω-categorical structures Γ, having no algebraicity coincides with the model-theoretic notion of Γ having no algebraicity (see, e.g., [25]).

Lemma 3.20 (see (2.15) in [14]) *Let $\mathcal{C}$ be an amalgamation class of relational structures and Γ its Fraïssé limit. Then $\mathcal{C}$ has strong amalgamation if and only if Γ has no algebraicity.*

For strong amalgamation classes there is a powerful construction to obtain new strong amalgamation classes from known ones.

Definition 3.21 Let $\mathcal{C}_1$ and $\mathcal{C}_2$ be classes of finite structures with disjoint relational signatures τ_1 and τ_2, respectively. Then the *free superposition of $\mathcal{C}_1$ and $\mathcal{C}_2$*, denoted by $\mathcal{C}_1 * \mathcal{C}_2$, is the class of $(\tau_1 \cup \tau_2)$-structures $\mathfrak{A}$ such that the τ_i-reduct of $\mathfrak{A}$ is in $\mathcal{C}_i$, for $i \in \{1,2\}$.

The following lemma has a straightforward proof by combining amalgamation in $\mathcal{C}_1$ with amalgamation in $\mathcal{C}_2$.

Lemma 3.22 *If $\mathcal{C}_1$ and $\mathcal{C}_2$ are strong amalgamation classes, then $\mathcal{C}_1 * \mathcal{C}_2$ is also a strong amalgamation class.*

When Γ_1 and Γ_2 are homogeneous structures with no algebraicity, then $\Gamma_1 * \Gamma_2$ denotes the (up to isomorphism unique) Fraïssé limit of the free superposition of the age of Γ_1 and the age of Γ_2.

Example 3.23 For $i \in \{1, 2\}$, let $\tau_i = \{<_i\}$, let $\mathcal{C}_i$ be the class of all finite τ_i-structures where $<_i$ denotes a linear order, and let Γ_i be the Fraïssé limit of $\mathcal{C}_i$. Then $\Gamma_1 * \Gamma_2$ is known as the *random permutation* (see e.g. [12, 33, 47]).

We have the following result about free superpositions.

Theorem 3.24 ([6]) *Let Γ_1 and Γ_2 be homogeneous ω-categorical structures with no algebraicity such that both Γ_1 and Γ_2 are Ramsey. Then $\Gamma_1 * \Gamma_2$ is Ramsey.*

We mention that the proof of Theorem 3.24 from [6] uses Theorem 3.18 about model-complete cores. An alternative proof can be found in [49].

Example 3.25 Recall from Example 2.20 that the amalgamation class of all finite structures $(V; E, <)$ where E denotes an equivalence relation and $<$ denotes a linear order, is *not* Ramsey. In the light of Conjecture 1.1 for the Fraïssé limit Γ of this class, we therefore look for a homogeneous Ramsey expansion of Γ. Let $\mathcal{C}$ be the class of all finite structures $(V; E, \prec)$ where E is an equivalence relation and $\prec$ is a linear order that is convex with respect to E. We have mentioned before that $\mathcal{C}$ is Ramsey, and by Theorem 3.24 the class $\mathcal{C} * \mathcal{LO}$ is Ramsey. Then the Fraïssé limit of $\mathcal{C} * \mathcal{LO}$ is isomorphic to a homogeneous Ramsey expansion of Γ.

Example 3.26 The directed graph $S(2)$ is one of the homogeneous directed graphs that figures in the classification of all homogeneous directed graphs of Cherlin [17]. In fact, it is a homogeneous tournament and therefore already appeared in the classification of homogeneous tournaments of Lachlan [29]. It has many equivalent definitions, one of them being the following: the vertices of $S(2)$ are a countable dense set of points on the unit circle without antipodal points. We add an edge from x to y if and only if the line from x to y has the origin on the left; that is, x, y, and $(0, 0)$ lie in clockwise order in the plane.

We will show that $S(2)$ has a Ramsey expansion which is homogeneous and has a finite relational signature. This can be derived from general principles and Ramsey's theorem as follows. In the following, $(\mathbb{Q}; <_1)$ and $(\mathbb{Q}; <_2)$ both denote the order of the rationals, but have disjoint signature. Let Γ be the disjoint union $(\mathbb{Q}; <_1) \uplus_P (\mathbb{Q}; <_1)$, which has the Ramsey property by Theorem 2.1 and Lemma 3.1 (Example 2.4). Then the free superposition Δ of $(\mathbb{Q}; <_2)$ with Γ is Ramsey by Theorem 3.24, and homogeneous with finite relational signature. The structure $S(2)$ is a reduct of Δ: for elements $x, y \in S(2)$, we define $x \prec y$ if

$$\big(x <_2 y \wedge (P(x) \Leftrightarrow P(y))\big) \vee \big(y <_2 x \wedge (P(x) \not\Leftrightarrow P(y))\big) .$$

Let Γ_1 and Γ_2 be two ω-categorical Ramsey structures. Note that since there is a linear order with a first-order definition in Γ_1, and a first-order definition of a linear order in Γ_2, the structure $\Gamma_1 * \Gamma_2$ must carry two independent linear orders.

To prove the Ramsey property for structures that do not have a second independent linear order, we have the following variant.

Theorem 3.27 ([6]) *Let $\mathcal{C}_1$ and $\mathcal{C}_2$ be classes of structures such that $\mathcal{C}_1$, $\mathcal{C}_2$ and $\mathcal{LO}$ have pairwise disjoint signatures. Also suppose that $\mathcal{C}_1$ and $\mathcal{LO} * \mathcal{C}_2$ are Ramsey classes with strong amalgamation and ω-categorical Fraïssé limits. Then $\mathcal{C}_1 * \mathcal{C}_2$ is also a Ramsey class.*

4   The partite method

There are some homogeneous Ramsey structures where no proof of the Ramsey property from general principles is known. One of the most powerful methods to prove the Ramsey property is such situations is the *partite method*. The first result that we see in this section is that for any finite relational signature τ, the class of all finite ordered τ-structures is a Ramsey class. This is due to Nešetřil and Rödl [39] and independently to Abramson and Harrington [1]; in these original papers, the statement is made for hypergraphs only, but it holds for relational structures in general. We then apply the partite method to classes that are characterised by forbidding finite structures as induced substructures.

4.1   The class of all ordered structures

We will prove the following theorem, due to Nešetřil and Rödl, and, independently, Abramson and Harrington.

Theorem 4.1 ([1, 39]) *For every relational signature τ the class of all $\tau \cup \{\preceq\}$-structures, where $\preceq$ denotes a linear order, is a Ramsey class.*

The construction to prove the Ramsey property is due to Nešetřil and Rödl [40], with only minor modifications in the presentation. It relies on the concept of *n-partite structures*. We formalize this slightly differently than Nešetřil and Rödl in [40].

Definition 4.2 Let $n \in \mathbb{N}$, and τ a relational signature. An *n-partite structure* is a finite $(\tau \cup \{\preceq\})$-structure $(\mathfrak{A}, \preceq)$ where $\preceq$ is a weak linear order (that is, a linear quasi-order) such that the equivalence relation $\approx$ on A defined by $x \approx y \Leftrightarrow (x \preceq y \land y \preceq x)$ has n equivalence classes. An

n-partite structure is called a *transversal* if each equivalence class of $\approx$ has size one.

Note that the elements of a finite n-partite structure $(\mathfrak{A}, \preceq)$ are partitioned into *levels* $A_1, \ldots, A_n$ which are uniquely given by the property that for $u \in A_i$ and $v \in A_j$ we have $u \preceq v$ if and only if $i \leq j$.

Lemma 4.3 (Partite Lemma) *Let $\mathfrak{A}$ be an n-partite transversal, $\mathfrak{B}$ an arbitrary n-partite structure, and $r \in \mathbb{N}$. Then there exists an n-partite structure $\mathfrak{C}$ such that $\mathfrak{C} \to (\mathfrak{B})_r^{\mathfrak{A}}$.*

The idea of the proof of Lemma 4.3 is to use the theorem of Hales–Jewett (see [22]), which we quickly recall here to fix some terminology.

Definition 4.4 Let $m, d \in \mathbb{N}$. A *combinatorial line* is a set $L \subseteq [m]^d$ of the form
$$\{(\alpha_1^1, \ldots, \alpha_d^1), \ldots, (\alpha_1^m, \ldots, \alpha_d^m)\}$$
such that there exists a non-empty set $P_L \subseteq [d]$ satisfying

- $\alpha_p^k = \alpha_p^l$ for all $k, l \in [m]$ and $p \in [d] \setminus P_L$, and

- $\alpha_p^k = k$ for all $k \in [m]$ and $p \in P_L$.

Note that for every $k \in [m]$ there exists exactly one $\alpha = (\alpha_1, \ldots, \alpha_d) \in L$ with $\alpha_p = k$ for all $p \in P_L$; we write $L(k)$ for this α.

Theorem 4.5 (Hales–Jewett; see [22]) *For any $m, r \in \mathbb{N}$ there exists $d \in \mathbb{N}$ such that for every function $\xi \colon [m]^d \to [r]$ there exists a combinatorial line L such that ξ is constant on L.*

We write $HJ(m, r)$ for the smallest $d \in \mathbb{N}$ that satisfies the condition in Theorem 4.5. See Figure 4.1 for an illustration that shows that $HJ(2, 2) = 2$: if we colour the vertices of $[2]^2$ with two colours, we always find a monochromatically coloured combinatorial line.

Proof of Lemma 4.3 We assume that every vertex of $\mathfrak{B}$ is contained in a copy of $\mathfrak{A}$ in $\mathfrak{B}$. This is without loss of generality: if $\mathfrak{B}^*$ is the substructure of $\mathfrak{B}$ induced by the elements of the copies of $\mathfrak{A}$ in $\mathfrak{B}$, and $\mathfrak{C}^*$ is such that $\mathfrak{C}^* \to (\mathfrak{B}^*)_r^{\mathfrak{A}^*}$, then we can construct $\mathfrak{C}$ such that $\mathfrak{C} \to (\mathfrak{B})_r^{\mathfrak{A}}$ from $\mathfrak{C}^*$ by amalgamating at every copy of $\mathfrak{B}^*$ in $\mathfrak{C}^*$ a copy of $\mathfrak{B}$. So assume in the following that $\mathfrak{B} = \mathfrak{B}^*$.

Let $g_1, \ldots, g_m$ be an enumeration of $\binom{\mathfrak{B}}{\mathfrak{A}}$. Let d be $HJ(m, r)$ (according to Theorem 4.5). The idea of the construction in the proof of Lemma 4.3

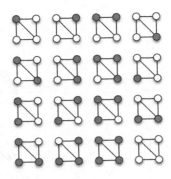

Figure 2: Illustration for $HJ(2\ 2) = 2$.

is to construct $\mathfrak{C}$ in such a way that for every element of $[m]^d$ there exists a copy of $\mathfrak{A}$ in $\mathfrak{C}$ such that monochromatically coloured lines in $[m]^d$ correspond to monochromatic copies of $\mathfrak{B}$ in $\mathfrak{C}$. The direct product $\mathfrak{B}^d$ has many copies of $\mathfrak{A}$, but in general does not have enough copies of $\mathfrak{B}$. The following ingenious construction, named after the initials of its inventors, is a modi cation of the direct product that overcomes the mentioned problem by creating su ciently many copies of $\mathfrak{B}$.

De nition 4.6 (The NR-power) Let $\mathfrak{A}$, $\mathfrak{B}$ be n-partite structures with signature ϕ. Then the d-th *NR-power of* $\mathfrak{B}$ *over* $\mathfrak{A}$ is the n-partite structure $\mathfrak{C}$ de ned as follows. Write B_i for the i-th level of B, for $i \quad [n]$. The domain of $\mathfrak{C}$ is $C_1 \subset \quad \subset C_n$ where $C_i := (B_i)^d$. For $R \quad \phi$ of arity h, and $u^1 \quad u^h \quad C$, we de ne $(u^1 \quad u^h) \quad R^{\mathfrak{C}}$ i there is a non-empty set $P \approx [d]$ and $(w^1 \quad w^h) \quad R^{\mathfrak{B}}$ such that $u_q^s = w^s$ for $q \quad P$ and $s \quad [h]$.

For an illustration of the NR-power, see Figure 3.

Let $\mathfrak{C}$ be the d-th NR-power of $\mathfrak{B}$ over $\mathfrak{A}$. To prove the partite lemma, it su ces to show that $\mathfrak{C} \prec (\mathfrak{B})_r^{\mathfrak{A}}$. Let $\chi\colon \frac{\mathfrak{C}}{\mathfrak{A}} \prec [r]$ be arbitrary. We are going to de ne a function $\colon [m]^d \prec [r]$.

Claim 4.7 *For* $\tau = (\tau_1 \quad \tau_d) \quad [m]^d$, *the map* $g\ \colon A \prec C$ *given by* $a \leadsto (g_{\ 1}(a) \quad g_{\ d}(a))$ *is an embedding of* $\mathfrak{A}$ *into* $\mathfrak{C}$.

Proof of Claim 4.7 Suppose that $(a_1 \quad a_h) \quad R^{\mathfrak{A}}$. Then

$$(g_{\ p}(a_1) \quad g_{\ p}(a_h)) \quad R^{\mathfrak{B}}$$

A   B   C

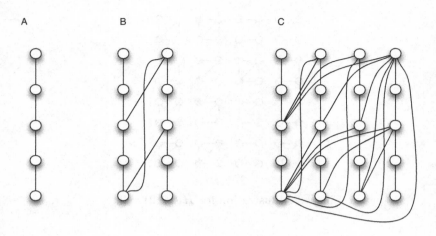

Figure 3: Illustration of two five-partite graphs $\mathfrak{A}$, $\mathfrak{B}$, where $\mathfrak{A}$ is transversal. On the right, we see the second NR-power of $\mathfrak{B}$ over $\mathfrak{A}$.

for all $p \in [d]$ since g_p preserves R. By the definition of $R^{\mathfrak{C}}$ we have that $(g(a_1), \ldots, g(a_h)) \in R^{\mathfrak{C}}$ (arbitrarily choose $i \in [d]$ and verify Definition 4.6 for $P = \{i\}$). Conversely, suppose that $(g(a_1), \ldots, g(a_h)) \in R^{\mathfrak{C}}$. Then there exists a non-empty set $P \approx [d]$ and $(w_1, \ldots, w_h) \in R^{\mathfrak{B}}$ such that for all $q \in P$ and $s \in [h]$ we have $(g(a_s))_q = g_q(a_s) = w_s$. Since g_q is an embedding of $\mathfrak{A}$ into $\mathfrak{B}$, we obtain in particular that $(a_1, \ldots, a_h) \in R^{\mathfrak{A}}$, proving the claim.

Define $\iota(\tau) := \chi(g)$. By the theorem of Hales-Jewett (Theorem 4.5), there exists a combinatorial line $L \approx [m]^d$ and $c \in [r]$ such that $\iota(\tau) = c$ for all $\tau \in L$. We describe how L gives rise to an embedding g_L of $\mathfrak{B}$ into $\mathfrak{C}$. For $u \in B$, we write $\sigma(u)$ for the unique element of A that lies on the same level as u. Observe that $\sigma(g_k(a)) = a$ for all $k \in [m]$ and for all $a \in A$ since $\mathfrak{A}$ is transversal. Recall our assumption that every $u \in B$ appears in a copy of $\mathfrak{A}$ in $\mathfrak{B}$, and hence there exists a $k \in [m]$ such that $u \in g_k(A)$.

Claim 4.8 *The map $g_L \colon B \prec C$ given by $g_L(u) := g_{L(k)}(\sigma(u))$, for some $k \in [m]$ such that $u \in g_k(A)$, is well-defined, and an embedding of $\mathfrak{B}$ into $\mathfrak{C}$.*

Proof of Claim 4.8 In order to show that the value of g_L does not depend on the choice of k, we have to show that if there are $k \neq l \in [m]$

such that $u \in B$ appears in both $g_k(A)$ and in $g_l(A)$, then $g_{L(k)}(\pi(u)) = g_{L(l)}(\pi(u))$, that is, $g_{L(k)_p}(\pi(u)) = g_{L(l)_p}(\pi(u))$ for all $p \in [d]$. This is clear when $p \in [d] \setminus P_L$ since we then have $L(k)_p = L(l)_p$. So consider the case $p \in P_L$. Then

$$g_{L(k)_p}(\pi(u)) = g_k(\pi(u)) = u = g_l(\pi(u)) = g_{L(l)_p}(\pi(u))$$

where the equation $g_k(\pi(u)) = u = g_l(\pi(u))$ holds since A is transversal and $u \in g_k(A) \cap g_l(A)$.

To show that g_L is an embedding, let $R \in \tau$ be of arity h, and let $u_1, \ldots, u_h \in B$ be arbitrary. Let $s \in [h]$ and k be such that $u_s \in g_k(A)$. Let $p \in P_L$ be arbitrary. Then

$$(g_L(u_s))_p = (g_{L(k)}(\pi(u_s)))_p = g_{L(k)_p}(\pi(u_s)) = g_k(\pi(u_s)) = u_s . \quad (4.1)$$

Hence, if $(u_1, \ldots, u_s) = ((g_L(u_1))_p, \ldots, (g_L(u_h))_p) \in R^{\mathfrak{B}}$, then by the definition of $R^{\mathfrak{C}}$ for $P := P_L$ and $w^s := u_s$ for all $s \in [h]$ we have that $(g_L(u_1), \ldots, g_L(u_h)) \in R^{\mathfrak{C}}$, and g_L preserves R.

Conversely, suppose that $(g_L(u_1), \ldots, g_L(u_h)) \in R^{\mathfrak{C}}$. Then there is a non-empty set $P \subseteq [d]$ and $(w^1, \ldots, w^h) \in R^{\mathfrak{B}}$ such that for $q \in P$ and $s \in [h]$ we have $g_L(u_s)_q = w^s$, and for $q \in [d] \setminus P$, all of $g_L(u_1)_q, \ldots, g_L(u_h)_q$ lie in the same copy of $\mathfrak{A}$ in $\mathfrak{B}$. For $p \in P$ we have $w^s = (g_L(u_s))_p = u_s$, and thus $(u_1, \ldots, u_h) \in R^{\mathfrak{B}}$. Applied to the case where R is the equality relation (for proving Ramsey results, we can assume without loss of generality that the signature contains a symbol for equality), this also shows injectivity of g_L. Hence, g_L is an embedding, which concludes the proof of the claim. □

Since $|L| = m$ and since the embeddings g_α, g_β are distinct whenever α, β are distinct elements of L, we conclude that all of the m copies of $\mathfrak{A}$ in the structure induced by $h(B)$ in $\mathfrak{C}$ have the same colour under χ, which completes the proof. □

To finally prove Theorem 4.1, we combine the partite lemma (Lemma 4.3) with the so-called *partite construction*; again, we follow [41].

Proof of Theorem 4.1 Let $\mathfrak{A}, \mathfrak{B}$ be $\tau \cup \{\preceq\}$-structures where $\preceq$ denotes a linear order, and $r \in \mathbb{N}$ be arbitrary. Set $a := |A|$ and $b := |B|$. We view $\mathfrak{A}$ as an a-partite transversal and $\mathfrak{B}$ as a b-partite transversal. Let $p \in \mathbb{N}$ be such that $([p], <) \to ([b], <)_r^{([a], <)}$ which exists since $\mathcal{LO}$ is a Ramsey class (Example 2.4). Let $q := \binom{p}{q}$, and $\binom{([p], <)}{([a], <)} = \{g_1, \ldots, g_q\}$. Construct p-partite $\tau \cup \{\preceq\}$-structures $\mathfrak{P}_0, \mathfrak{P}_1, \ldots, \mathfrak{P}_q$ inductively as follows. Let $\mathfrak{P}_0$ be such that for any b parts $P_{0,i_1}, \ldots, P_{0,i_b}$ of $\mathfrak{P}_0$ there is an embedding of

$\mathfrak{B}$ into the substructure of $\mathfrak{P}_0$ induced by those parts. It is clear that such a $(\tau \cup \{\preceq\})$-structure $\mathfrak{P}_0$ exists; one may for instance take an appropriate quasi-ordering $\preceq$ on a disjoint union of the τ-reduct of $\mathfrak{B}$.

Now suppose that we have already constructed the p-partite structure $\mathfrak{P}_{k-1}$, with parts $P_{k-1,1}, \ldots, P_{k-1,p}$; to construct $\mathfrak{P}_k$, let $\mathfrak{D}_{k-1}$ be the a-partite system induced in $\mathfrak{P}_{k-1}$ by $\bigcup_{i \in [a]} P_{k-1,g_k(i)}$. By the partite lemma (Lemma 4.3) there exists an a-partite structure $\mathfrak{E}_k$ such that $\mathfrak{E}_k \to (\mathfrak{D}_{k-1})_r^{\mathfrak{A}}$. We construct the p-partite structure $\mathfrak{P}_k$ by amalgamating $\mathfrak{E}_k$ with $\mathfrak{P}_{k-1}$ over $\mathfrak{D}_{k-1}$, for each occurrence of $\mathfrak{D}_{k-1}$ in $\mathfrak{E}_k$.

Finally, let $\mathfrak{C}$ be the structure obtained from $\mathfrak{P}_q$ by replacing the linear quasi-order $\preceq$ by a (total) linear extension. We claim that $\mathfrak{C} \to (\mathfrak{B})_r^{\mathfrak{A}}$. Let $\chi \colon \binom{\mathfrak{C}}{\mathfrak{A}} \to [r]$ be arbitrary. For $k \in \{0, \ldots, q\}$ and $l \in \{k, \ldots, q\}$, we will construct embeddings $h_{l,k} \in \binom{\mathfrak{P}_l}{\mathfrak{P}_k}$ such that for all $m \in \{k, \ldots, l\}$

- $h_{l,m} \circ h_{m,k} = h_{l,k}$, and

- $|\chi(h_{q,m} \circ \binom{\mathfrak{D}_m}{\mathfrak{A}}))| \leq 1$.

Our construction is by induction on k, starting with $k = q$. For $k = l = q$ we can choose $h_{q,q}$ to be the identity. Now suppose that $h_{l',k'}$ has already been defined for all k' such that $k \leq k' \leq l' \leq q$. We want to define $h_{k,k-1}$. Since $\mathfrak{E}_k \to (\mathfrak{D}_{k-1})_c^{\mathfrak{A}}$, there exists an $e_{k-1} \in \binom{\mathfrak{E}_k}{\mathfrak{D}_{k-1}}$ such that $|\chi(h_{q,k} \circ e_{k-1} \circ \binom{\mathfrak{D}_{k-1}}{\mathfrak{A}}))| \leq 1$. By construction of $\mathfrak{P}_k$, the embedding e_{k-1} can be extended to an embedding $h_{k,k-1} \in \binom{\mathfrak{P}_k}{\mathfrak{P}_{k-1}}$. For $m \in \{k, \ldots, l\}$, we define $h_{m,k-1} := h_{m,k} \circ h_{k,k-1}$, completing the inductive construction.

For all $m \in [q]$ there exists a $c_m \in [r]$ such that for all $f \in \binom{\mathfrak{D}_m}{\mathfrak{A}}$ we have $\chi(h_{q,m} \circ f) = c_m$. Define $\xi(g_m) := c_m$. Since $([p], <) \to ([b], <)_r^{([a],<)}$, there exists an $h \in \binom{([p],<)}{([a],<)}$ and $c \in [r]$ such that for all $h' \in \binom{([b],<)}{([a],<)}$ we have $\xi(h \circ h') = c$. By construction of $\mathfrak{P}_0$, there exists a $g \in \binom{\mathfrak{P}_0}{\mathfrak{B}}$ such that $g(B) \subseteq \bigcup_{i \in [a]} P_{0,h(i)}$. To show the claim it suffices to prove that $\chi(g_{k,0} \circ g \circ \binom{\mathfrak{B}}{\mathfrak{A}})) \leq 1$. Let $g' \in \binom{\mathfrak{B}}{\mathfrak{A}}$ be arbitrary. Note that $g_{k,0} \circ g \circ g' \in \binom{\mathfrak{D}_k}{\mathfrak{A}}$ for some $k \in [q]$. Hence, $\chi(g_{q,0} \circ g \circ g') = \chi(g_{q,k} \circ g_{k,0} \circ g \circ g') = c$, finishing the proof of the claim. $\qquad \square$

4.2 Irreducible homomorphically forbidden structures

For every $n \geq 2$, the class of all ordered K_n-free graphs is Ramsey. In fact, something more general is true; in order to state the result in full generality, we need the following concept.

A structure $\mathfrak{F}$ is called *irreducible* (in the terminology of [40]) if for any pair of distinct elements $x, y \in F$ there exists an $R \in \tau$ and $z_1, \ldots, z_h \in F$

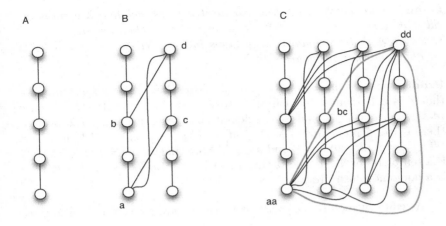

Figure 4: The partite lemma (Lemma 4.3) can create triangles from triangle-free 5-partite A and B.

such that $(z_1 \ldots z_h) \in R^{\mathcal{F}}$ and $x \, y \, z_1 \ldots z_h$. It is straightforward to verify that for a set $\mathcal{F}$ of irreducible structures, the class Forb($\mathcal{F}$) has (strong) amalgamation, is closed under substructures, isomorphism, and has the joint embedding property, and therefore is an amalgamation class.

Theorem 4.9 (Nesetril Rodl) *Let $\mathcal{F}$ be a set of nite irreducible ϕ-structures. Then $\mathcal{C} := \text{Forb}(\mathcal{F}) \, \mathcal{LO}$ is a Ramsey class.*

This theorem can be shown by a variant of the partite method as presented in the previous section. However, it is important to note that the proof from the previous section cannot be applied without an important modi cation. More concretely, already for the class of triangle-free graphs, NR-powers of $\mathfrak{B}$ over $\mathfrak{A}$ might contain triangles even if the n-partite structures $\mathfrak{A}$ and $\mathfrak{B}$ are triangle-free; see Figure 4. To overcome this problem, we need the following de nition. Let $\mathfrak{A} \, \mathfrak{B}$ be two n-partite $\phi \subset \preceq$ -structures, and suppose that $\mathfrak{A}$ is transversal. Recall that for $u \quad \mathfrak{B}$, we write $\sigma(u)$ for the unique element of $\mathfrak{A}$ that lies on the same level as u.

De nition 4.10 We say that $\mathfrak{A}$ is a *template for* $\mathfrak{B}$ if for all $R \quad \phi$, $(b_1 \ldots b_h) \in R^{\mathfrak{B}}$ implies that $(\sigma(b_1) \ldots \sigma(b_h)) \in R^{\mathfrak{A}}$.

We state an important property of the NR-powers of $\mathfrak{B}$ over $\mathfrak{A}$ when $\mathfrak{A}$ is a template for $\mathfrak{B}$.

Lemma 4.11 *Let $\mathfrak{A}$ and $\mathfrak{B}$ be n-partite structures such that $\mathfrak{A}$ is transversal and $\mathfrak{A}$ is a template for $\mathfrak{B}$, and let $r \in \mathbb{N}$. Then every irreducible structure $\mathfrak{F}$ that homomorphically maps into an NR-power of $\mathfrak{B}$ over $\mathfrak{A}$ also homomorphically maps into $\mathfrak{A}$.*

Proof Let $\mathfrak{C}$ be the d-th NR-power of $\mathfrak{B}$ over $\mathfrak{A}$ for some $d \in \mathbb{N}$. Suppose that e is a homomorphism from $\mathfrak{F}$ to $\mathfrak{C}$. Let $(z_1, \ldots, z_h) \in R^{\mathfrak{F}}$. Since $(e(z_1), \ldots, e(z_h)) \in \mathfrak{C}$ and by the definition of the NR-power of $\mathfrak{B}$ over $\mathfrak{A}$, there exists a non-empty set $P \subseteq [d]$ and $(w_1, \ldots, w_h) \in R^{\mathfrak{B}}$ such that $(e(z_s))_q = w_s$ for all $q \in P$ and $s \in [h]$. Note that $\pi(w_s) = \pi(z_s)$. Since $\mathfrak{A}$ is a template for $\mathfrak{B}$, it follows that $(\pi(w_1), \ldots, \pi(w_h)) \in R^{\mathfrak{A}}$. Hence, $\pi \circ e$ is a homomorphism from $\mathfrak{F}$ to $\mathfrak{A}$. $\quad\square$

We can now modify the partite construction from Section 4.1 as follows.

Proof of Theorem 4.9 Let $\mathfrak{A}, \mathfrak{B} \in \mathcal{C}$ and $r \in \mathbb{N}$ be arbitrary. By Theorem 4.1, there exists a $\tau \cup \{\preceq\}$-structure $\mathfrak{C}$ where $\preceq$ denotes a linear order (but which need not be from $\mathcal{C}$) such that $\mathfrak{C} \to (\mathfrak{B})_r^{\mathfrak{A}}$. Let $q := |\binom{\mathfrak{C}}{\mathfrak{A}}|$, and $\binom{\mathfrak{C}}{\mathfrak{A}} = \{g_1, \ldots, g_q\}$. Let $p := |\binom{\mathfrak{C}}{\mathfrak{B}}|$, and $\binom{\mathfrak{C}}{\mathfrak{B}} = \{f_1, \ldots, f_p\}$. Let $\mathfrak{C}_i$ be the substructure of $\mathfrak{C}$ induced by $f_i(B)$. We inductively construct a sequence of $|C|$-partite τ-structures $\mathfrak{P}_0, \mathfrak{P}_1, \ldots, \mathfrak{P}_q$. Let $\mathfrak{P}_0$ be the $(\tau \cup \{\preceq\})$-structure obtained as follows: define the relation $\preceq$ on the disjoint union of all the $\mathfrak{C}_i$ by setting $x \preceq y$ if x is a copy of a vertex x' in $\mathfrak{C}$, y is a copy of a vertex y' in $\mathfrak{C}$, and $x' \preceq y'$ in $\mathfrak{C}$. Note that $\mathfrak{C}$ is a template for $\mathfrak{P}_0$.

The construction of $\mathfrak{P}_k$ for $k > 0$ is as in the partite construction in the proof of Theorem 4.1: suppose that we have already constructed $\mathfrak{P}_{k-1}$; to construct $\mathfrak{P}_k$, let $\mathfrak{D}_{k-1}$ be the $|A|$-partite system induced in $\mathfrak{P}_{k-1}$ by $\bigcup_{i \in [a]} P_{k, g_k(i)}$. By the partite lemma (Lemma 4.3) there exists an $|A|$-partite structure $\mathfrak{E}_k$ such that $\mathfrak{E}_k \to (\mathfrak{D}_{k-1})_r^{\mathfrak{A}}$. Note that $\mathfrak{A}$ is a template for $\mathfrak{D}_{k-1}$, and hence, by Lemma 4.11, none of the structures from $\mathfrak{F}$ embeds into $\mathfrak{E}_k$. We construct the p-partite structure $\mathfrak{P}_k$ by amalgamating $\mathfrak{E}_k$ with $\mathfrak{P}_{k-1}$ over $\mathfrak{D}_{k-1}$, for each occurrence of $\mathfrak{D}_{k-1}$ in $\mathfrak{E}_k$. The proof that $\mathfrak{P}_q \to (\mathfrak{B})_r^{\mathfrak{A}}$ is as in the proof of Theorem 4.1. $\quad\square$

For a recent application of the partite method to prove the Ramsey property for classes of structures given by homomorphically forbidden trees, see [19].

5 An inductive proof

In this section we present a Ramsey class with finite relational signature for which (to the best of my knowledge) no proof with the partite method is known. Recall the definition of C-relations, and of convex linear orders of C-relations from Example 2.21.

Theorem 5.1 (see [8, 32, 35]) *The class of all finite binary branching convexly ordered C-relations is a Ramsey class.*

This is a consequence of a more powerful theorem due to Milliken [35], and follows also from results of Leeb [32]. A weaker version of this theorem has been shown by Deuber [18] (my academic grand-father). A direct proof for the statement in the above form can be found in [8].

Throughout this section, $\mathcal{C}$ denotes the class of all finite binary branching convexly ordered C-relations. Recall that the members of $\mathcal{C}$ are in one-to-one correspondence to rooted binary trees, and in the proof it will be convenient to use this perspective.

If $\mathfrak{T}$ is a tree with more than one vertex, then the root of $\mathfrak{T}$ has exactly two children; we denote the subtree of $\mathfrak{T}$ rooted at the left child (with respect to the convex linear ordering) by $\mathfrak{T}_{\swarrow}$, and the subtree of $\mathfrak{T}$ rooted at the right child by $\mathfrak{T}_{\searrow}$ (and we speak of the *left subtree of $\mathfrak{T}$* and the *right subtree of $\mathfrak{T}$*, respectively). Finally, suppose that $e_1 \in \binom{\mathfrak{T}_{\swarrow}}{\mathfrak{A}_{\swarrow}}$ and $e_2 \in \binom{\mathfrak{T}_{\searrow}}{\mathfrak{A}_{\searrow}}$, then $\langle e_1, e_2 \rangle$ is the embedding e of $\mathfrak{A}$ into $\mathfrak{T}$ defined by $e(a) := e_1(a)$ if $a \in \mathfrak{A}_{\swarrow}$ and $e(a) := e_2(a)$ if $a \in \mathfrak{A}_{\searrow}$. We write $\bullet$ for the up to isomorphism unique structure from $\mathcal{C}$ with one element.

For $\mathfrak{T}_1, \mathfrak{T}_2 \in \mathcal{C}$, define $\mathfrak{T}_1[\mathfrak{T}_2]$ to be the element of $\mathcal{C}$ with elements $T_1 \times T_2$ where $C((u_1, v_1); (u_2, v_2), (u_3, v_3))$ holds if $C(u_1; u_2, u_3)$ or $u_1 = u_2 = u_3$ and $C(v_1; v_2, v_3)$, and similarly $(u_1, v_1) \prec (u_2, v_2)$ if $(u_1 \prec u_2)$ or $u_1 = u_2$ and $v_1 \prec v_2$. Informally, $\mathfrak{T}_1[\mathfrak{T}_2]$ the the tree obtained from $\mathfrak{T}_1$ by replacing every leaf of $\mathfrak{T}_1$ by a copy of $\mathfrak{T}_2$. We also write $\mathfrak{T}^{(1)}$ for $\mathfrak{T}_1$, and inductively define $\mathfrak{T}^{(i)} := \mathfrak{T}^{(i-1)}[\mathfrak{T}]$ for $i \geq 2$. The following is easy to prove.

Lemma 5.2 *For all $\mathfrak{B} \in \mathcal{C}$ and $r \in \mathbb{N}$ we have $\mathfrak{B}^{(r)} \to (\mathfrak{B})_r^{\bullet}$.*

Proof of Theorem 5.1 Let $\mathfrak{A}, \mathfrak{B} \in \mathcal{C}$, and $r \in \mathbb{N}$; we have to show that there is a $\mathfrak{C} \in \mathcal{C}$ such that $\mathfrak{C} \to (\mathfrak{B})_r^{\mathfrak{A}}$. We prove the statement by induction on the size of $\mathfrak{A}$. For $\mathfrak{A} = \bullet$, the statement follows from Lemma 5.2.

Claim 5.3 *For every $\mathfrak{D} \in \mathcal{C}$ there exists an $\mathfrak{F} \in \mathcal{C}$ such that for any $\chi \colon \binom{\mathfrak{F}}{\mathfrak{A}} \to [r]$ there are $f_1 \in \binom{\mathfrak{F}_{\swarrow}}{\mathfrak{D}}$, $f_2 \in \binom{\mathfrak{F}_{\searrow}}{\mathfrak{D}}$, and $c \in [r]$ such that for all $e_1 \in \binom{\mathfrak{D}}{\mathfrak{A}_{\swarrow}}$ and $e_2 \in \binom{\mathfrak{D}}{\mathfrak{A}_{\searrow}}$ we have $\chi(\langle f_1 \circ e_1, f_2 \circ e_2 \rangle) = c$.*

Proof By the inductive assumption, there are structures $\mathfrak{F}_1, \mathfrak{F}_2 \in \mathcal{C}$ such that $\mathfrak{F}_2 \to (\mathfrak{D})_r^{\mathfrak{A}_{\searrow}}$ and $\mathfrak{F}_1 \to (\mathfrak{D})_s^{\mathfrak{A}_{\swarrow}}$ where $s := \left| [r]^{\binom{\mathfrak{F}_2}{\mathfrak{A}_{\searrow}}} \right|$. Let $\mathfrak{F}$ be such that $\mathfrak{F}_{\swarrow} = \mathfrak{F}_1$ and $\mathfrak{F}_{\searrow} = \mathfrak{F}_2$. For a given $\chi\colon \binom{\mathfrak{F}}{\mathfrak{A}} \to [r]$, define $\psi\colon \binom{\mathfrak{F}_{\swarrow}}{\mathfrak{A}_{\swarrow}} \to [r]^{\binom{\mathfrak{F}_{\searrow}}{\mathfrak{A}_{\searrow}}}$ as follows.

$$\psi(e_1) := \left(e_2 \mapsto \chi(\langle e_1, e_2 \rangle)\right)$$

By the choice of $\mathfrak{F}_{\swarrow} = \mathfrak{F}_1$ there exists an $f_1 \in \binom{\mathfrak{F}_{\swarrow}}{\mathfrak{D}}$ and $\phi\colon \binom{\mathfrak{F}_{\searrow}}{\mathfrak{A}_{\searrow}} \to [r]$ such that $\psi(f_1 \circ \binom{\mathfrak{D}}{\mathfrak{A}_{\swarrow}}) = \{\phi\}$. By the choice of $\mathfrak{F}_{\searrow} = \mathfrak{F}_2$ there exists an $f_2 \in \binom{\mathfrak{F}_2}{\mathfrak{D}}$ and a $c \in [r]$ such that $\phi(f_2 \circ \binom{\mathfrak{D}}{\mathfrak{A}_{\searrow}}) = \{c\}$. Let $g_1 \in \binom{\mathfrak{D}}{\mathfrak{A}_{\swarrow}}$ and $g_2 \in \binom{\mathfrak{D}}{\mathfrak{A}_{\searrow}}$. Note that $\psi(f_1 \circ g_1) = \phi$ and $\phi(f_2 \circ g_2) = c$. By definition, $\phi(f_2 \circ g_2) = \chi(\langle f_1 \circ g_1, f_2 \circ g_2 \rangle)$, and hence $\chi(\langle f_1 \circ g_1, f_2 \circ g_2 \rangle) = c$ as desired. $\square$

Let h be the height of B (that is, the maximal distance from the root of $\mathfrak{B}$ to one of its leaves), and let n be h^r. Define $\mathfrak{C}_1, \mathfrak{C}_2, \ldots$ inductively as follows. Set $\mathfrak{C}_1 := \bullet$, and for $i \geq 2$ let $\mathfrak{C}_i$ be the structure $\mathfrak{F}$ that has been constructed for $\mathfrak{D} := \mathfrak{C}_{i-1}$ in Claim 5.3. Set $\mathfrak{C} := \mathfrak{C}_n$.

We claim that $\mathfrak{C} \to (\mathfrak{B})^{\mathfrak{A}}_r$. So let $\chi\colon \binom{\mathfrak{C}}{\mathfrak{A}} \to [r]$ be given. For all words w over the alphabet $[2]$ of length $i \in \{0, \ldots, n-1\}$ we define $g_w \in \binom{\mathfrak{C}}{\mathfrak{C}_{n-i}}$, $f_{w1} \in \binom{(\mathfrak{C}_i)_{\swarrow}}{\mathfrak{C}_{i-1}}$, $f_{w2} \in \binom{(\mathfrak{C}_i)_{\searrow}}{\mathfrak{C}_{i-1}}$, and $c_w \in [r]$ as follows. For $i = 0$ and $w = \epsilon$, the empty word of length 0, Claim 5.3 asserts the existence of $f_1 \in \binom{(\mathfrak{C}_n)_{\swarrow}}{\mathfrak{C}_{n-1}}$, $f_2 \in \binom{(\mathfrak{C}_n)_{\searrow}}{\mathfrak{C}_{n-1}}$, and $c \in [r]$ such that for all $e_1 \in \binom{\mathfrak{C}_{n-1}}{\mathfrak{A}_{\swarrow}}$ and $e_2 \in \binom{\mathfrak{C}_{n-1}}{\mathfrak{A}_{\searrow}}$ we have $\chi(\langle f_1 \circ e_1, f_2 \circ e_2 \rangle) = c_\epsilon$. Set $g_1 := f_1$ and $g_2 := f_2$.

Now suppose that f_w and g_w are already defined for a word w of length $i \in \{1, \ldots, n-1\}$. Let $\psi\colon \binom{\mathfrak{C}_{n-i}}{\mathfrak{A}} \to [r]$ be the map defined by $\psi(e) := \chi(g_w \circ e)$ for all $e \in \binom{\mathfrak{C}_{n-i}}{\mathfrak{A}}$. Then Claim 5.3 asserts the existence of $f_{w1} \in \binom{(\mathfrak{C}_{n-i})_{\swarrow}}{\mathfrak{C}_{n-i-1}}$, $f_{w2} \in \binom{(\mathfrak{C}_{n-i})_{\searrow}}{\mathfrak{C}_{n-i-1}}$, and $c_w \in [r]$ such that

$$\psi(\langle f_{w1} \circ e_1, f_{w2} \circ e_2 \rangle) = c_w \tag{5.1}$$

for all $e_1 \in \binom{\mathfrak{C}_{n-i-1}}{\mathfrak{A}_{\swarrow}}$ and $e_2 \in \binom{\mathfrak{C}_{n-i-1}}{\mathfrak{A}_{\searrow}}$. Set $g_{w1} := g_w \circ f_{w1}$ and $g_{w1} := g_w \circ f_{w2}$.

We claim that there exists an injection $\beta\colon B \to [2]^{h^r}$ and a $c \in [r]$ such that

- the map m given by $x \mapsto g_{\beta(x)}(\bullet)$ is from $\binom{\mathfrak{C}}{\mathfrak{B}}$, and

- for all $b_1, b_2 \in B$ we have that $c_w = c$ when w is the longest common prefix of $\beta(b_1)$ and $\beta(b_2)$.

We show this claim by induction on r. The statement is true if $r = 1$ since we can certainly find an injection $\beta\colon B \to [2]^h$ such that $x \mapsto g_{\beta(x)}$ is from $\binom{\mathfrak{C}}{\mathfrak{B}}$, since h is the height of $\mathfrak{B}$.

Otherwise we distinguish two possibilities. We write S_w for the set of words of length at most $|w| + h^{r-1}$ that start with w. Then either

1. for every word w of length at most $n' := n - h^{r-1} = (h-1)h^{r-1}$ there exists a word $u_w \in S_w$ such that $c_{u_w} = r$. In this case, we construct the desired map β recursively as follows. If $\mathfrak{B} = \bullet$ then define $\beta(b_1) = \epsilon 1 \cdots 1 \in [2]^n$ (where ϵ denotes the empty word). Otherwise, $h \geq 1$, and there exists a word $v := u_\epsilon$ with $c_v = r$. We repeat this procedure for $\mathfrak{B}_{\swarrow}$, with $v1$ instead of ϵ, and for $\mathfrak{B}_{\searrow}$, with $v2$ instead of ϵ, until β is defined on all elements of $\mathfrak{B}$.

2. there exists a word w of length at most n' such that $\{c_u \mid u \in S_w\}$ does not contain the colour r. In this case we find an injection $\beta\colon B \to S_w$ with the desired properties by the inductive hypothesis (since we only consider $r - 1$ colours instead of r, and S_w still has size h^{r-1}).

We claim that $\chi(m \circ e) = c$ for all $e \in \binom{\mathfrak{B}}{\mathfrak{A}}$. Since $|A| \geq 2$, there are $a_1 \in \mathfrak{A}_{\swarrow}$ and $a_2 \in \mathfrak{A}_{\searrow}$. Let w be the longest common prefix of $\beta(e(a_1))$ and $\beta(e(a_2))$. We then have $c_w = c$. Write e as $\langle e_1, e_2 \rangle$ where e_1 is the restriction of e to $\mathfrak{A}_{\swarrow}$ and e_2 is the restriction of e to $\mathfrak{A}_{\searrow}$. Let $k_1 \in \binom{\mathfrak{C}_{n-i-1}}{\mathfrak{A}_{\swarrow}}$ be

$$x \mapsto g_{w1}^{-1} g_{\beta(e_1(x))}(\bullet) \,.$$

Similarly, let $k_2 \in \binom{\mathfrak{C}_{n-i-1}}{\mathfrak{A}_{\searrow}}$ be $x \mapsto g_{w2}^{-1} g_{\beta(e_2(x))}(\bullet)$. We then have

$$\chi(m \circ e) = \chi(\langle g_w \circ f_{w1} \circ k_1, g_w \circ f_{w2} \circ k_2 \rangle) = c_w = c$$

due to Equation (5.1).   □

In Example 2.21, we have seen that the class of finite ordered binary branching C-relations does not have the Ramsey property. In the context of Conjecture 1.1, we want to show how to expand the class to make it Ramsey.

Example 5.4 The class $\mathcal{C}$ of all finite structures $(L; C, <, \prec)$, where $<$ is an arbitrary linear order, and $\prec$ is convex with respect to C, is a Ramsey class. This is an immediate consequence of Theorem 3.24: the class $\mathcal{C}$ can be described as the superposition of the Ramsey class $\mathcal{LO}$ with the class of all convexly ordered C-relations, which is Ramsey by Theorem 5.1.

6   The ordering property

There are strong links between the Ramsey property and the *ordering property* (as defined in [27, 38]).

Definition 6.1 (Ordering property) Let $\mathcal{C}'$ be a class of finite structures over the signature $\tau \cup \{\prec\}$ where $\prec$ denotes a linear order, and let $\mathcal{C}$ be the class of all τ-reducts of structures from $\mathcal{C}'$. Then $\mathcal{C}'$ has the *ordering property with respect to* $\prec$ if for every $\mathfrak{X} \in \mathcal{C}$ there exists a $\mathfrak{Y} \in \mathcal{C}$ such that for all expansions $\mathfrak{X}' \in \mathcal{C}'$ of $\mathfrak{X}$ and $\mathfrak{Y}' \in \mathcal{C}'$ of $\mathfrak{Y}$ there is an embedding of $\mathfrak{X}'$ into $\mathfrak{Y}'$.

Many examples of classes with the ordering property can be obtained from Theorem 6.4 below, so we rather start with an example of a Ramsey class *without* the ordering property.

Example 6.2 Let $\mathcal{C}$ be the class of all finite sets that are linearly ordered by two linear orders $<_1$ and $<_2$ (see Example 3.23). Then $\mathcal{C}$ does *not* have the ordering property with respect to $<_1$. Indeed, let $\mathfrak{A} \in \mathcal{C}$ be the structure $(\{0, 1, 2\}; \{(0, 1), (1, 2), (0, 2)\}, \{(1, 0), (0, 2), (1, 2)\})$, and let $\mathfrak{B}$ be an arbitrary $\{<_1\}$-reduct of a structure from $\mathcal{C}$, that is, an arbitrary finite linearly ordered set. Then the expansion of $\mathfrak{B}$ where $<_2$ denotes the same relation as $<_1$ is in $\mathcal{C}$, but certainly contains no copy of $\mathfrak{A}$.

Proposition 6.3 *Let Γ be a homogeneous relational τ-structure with domain D, and suppose that Γ has an ω-categorical homogeneous expansion Γ' with signature $\tau \cup \{\prec\}$ where $\prec$ denotes a linear order. Then the following are equivalent.*

- *the age of Γ' has the ordering property with respect to $\prec$;*

- *for every finite $X \subseteq D$ there exists a finite $Y \subseteq D$ such that for every $\beta \in \mathrm{Aut}(\Gamma)$ there exists an $\alpha \in \mathrm{Aut}(\Gamma')$ such that $\alpha(X) \subseteq \beta(Y)$.*

Proof First suppose that the age $\mathcal{C}'$ of Γ' has the ordering property. Let $X \subset D$ be finite, and let $\mathfrak{X}$ be the structure induced by X in Γ. Then there exists a $\mathfrak{Y}$ in $\mathrm{Age}(\Gamma)$ such that for all expansions $\mathfrak{X}' \in \mathcal{C}'$ of $\mathfrak{X}$ and $\mathfrak{Y}' \in \mathcal{C}'$ of $\mathfrak{Y}$ there exists an embedding of $\mathfrak{X}'$ into $\mathfrak{Y}'$. Suppose without loss of generality that $\mathfrak{Y}$ is a substructure of Γ with domain Y. Let $\beta \in \mathrm{Aut}(\Gamma)$ be arbitrary. Let $\mathfrak{X}'$ be the structure induced by $\beta(X)$ in Γ', and $\mathfrak{Y}'$ the structure induced by $\beta(Y)$ in Γ'. Since $\beta \in \mathrm{Aut}(\Gamma)$, $\mathfrak{X}'$ is isomorphic to an expansion of $\mathfrak{X}$, and $\mathfrak{Y}'$ is isomorphic to an expansion of $\mathfrak{Y}$. By assumption, $\mathfrak{X}'$ embeds into $\mathfrak{Y}'$. By homogeneity of Γ', this embedding can be extended to an automorphism α of Γ', and α has the desired property.

For the converse, let $\mathfrak{X}$ be an arbitrary structure in Age(Γ). Let $Z \subseteq D$ be inclusion-wise minimal with the property that for every embedding e of $\mathfrak{X}$ into Γ there exists an automorphism α of Γ' such that $\alpha(e(X)) \subseteq Z$. Since Γ' is ω-categorical, there is a finite number m of orbits of $|X|$-tuples, and therefore Z has cardinality at most $m|X|$. Let $Y \subseteq D$ be such that for every $\beta \in \operatorname{Aut}(\Gamma)$ there exists an $\alpha \in \operatorname{Aut}(\Gamma')$ such that $\alpha(Z) \subseteq \beta(Y)$. Let $\mathfrak{Y}$ be the structure induced by Y in Γ. Now let $\mathfrak{X}' := (\mathfrak{X}, \prec) \in \mathcal{C}'$ and $\mathfrak{Y}' := (\mathfrak{Y}, \prec) \in \mathcal{C}'$ be order expansions of $\mathfrak{X}$ and $\mathfrak{Y}$. Let g be an embedding of $\mathfrak{Y}$ into Γ'. By the definition of Z, there is an embedding ρ of $\mathfrak{X}'$ into the substructure induced by Z in Γ'. By homogeneity of Γ, there is a $\beta \in \operatorname{Aut}(\Gamma)$ that maps Y to $g(Y)$. By the choice of Y there exists an $\alpha \in \operatorname{Aut}(\Gamma')$ such that $\alpha(Z) \subseteq \beta(Y)$. Now, $\beta^{-1} \circ \alpha \circ \rho$ is an embedding of $\mathfrak{X}'$ into $\mathfrak{Y}'$, which concludes the proof of the ordering property for $\mathcal{C}'$ with respect to $\prec$. $\square$

Our next theorem gives a sufficient condition for ω-categorical structures to have the ordering property with respect to a given ordering; this condition covers most structures of interest and generalises many previous isolated results [27,38,44,49].

An *orbital* of a permutation group G on a set D is an orbit O of the componentwise action of G on D^2. An *O-cycle* is a sequence of pairs $(u_1, u_2), (u_2, u_3), \ldots, (u_n, u_1)$ from O, for some n. We say that O is *cyclic* if it contains an O-cycle, and *acyclic* otherwise.

Theorem 6.4 *Let Γ be a homogeneous τ-structure with domain D, and $\prec$ an order on D such that $\Gamma' := (\Gamma, \prec)$ is ω-categorical homogeneous Ramsey. Suppose furthermore that every acyclic orbital of $\operatorname{Aut}(\Gamma)$ is also an orbital of $\operatorname{Aut}(\Gamma')$. Then Age($\Gamma'$) has the ordering property with respect to $\prec$.*

Proof Let $X \subset D$ be finite. By Proposition 6.3 we have to show that there exists a finite $Y \subset D$ such that for all $\beta \in \operatorname{Aut}(\Gamma)$ there exists an $\alpha \in \operatorname{Aut}(\Gamma')$ such that $\alpha(X) \subseteq \beta(Y)$. Since Γ is ω-categorical, there is a finite number m of orbits of $|X|$-tuples, and hence there exists a finite $Z \subset D$ with the following properties:

- for every $\gamma \in \operatorname{Aut}(\Gamma)$ there is a $\delta \in \operatorname{Aut}(\Gamma')$ such that $\delta(X) \subseteq \gamma(Z)$;

- for every cyclic orbital O of $\operatorname{Aut}(\Gamma)$, Z contains an O-cycle.

Since Γ' is Ramsey, there exists by Proposition 2.22 a finite set $L \subset D$ such that for all 2-element subsets $S_1, \ldots, S_\ell$ of Z and every $\chi_i \colon \binom{L}{S_i} \to [2]$ there exists a $\theta \in \operatorname{Aut}(\Gamma')$ such that $|\chi_i(\theta \circ \binom{Z}{S_i})| = 1$ for all $i \in [\ell]$.

Let $\beta \in \mathrm{Aut}(\Gamma)$ be arbitrary. Define the $\chi_i \colon \binom{L}{S_i} \to [2]$ as follows. For $g \in \binom{L}{S_i}$, put $\chi(g) := 0$ if $\beta|_{g(S_i)}$ preserves $\prec$, and $\chi(g) := 1$ otherwise. Let $\theta \in \mathrm{Aut}(\Gamma')$ be the automorphism that exists for these colourings $\chi_1, \ldots, \chi_\ell$ according to the choice of L.

We claim that $Y := \theta(Z)$ has the desired properties, that is, we show that there is an $\alpha \in \mathrm{Aut}(\Gamma')$ mapping X into $\beta(Y)$. By the definition of Z, there exists a $\delta_1 \in \mathrm{Aut}(\Gamma')$ that maps X into Z. By the definition of Z, there also exists a $\delta_2 \in \mathrm{Aut}(\Gamma')$ that maps $\beta(\theta(\delta_1(X)))$ into Z.

We claim that the restriction g of $\beta \circ \theta \circ \delta_2 \circ \beta \circ \theta \circ \delta_1$ to X can be extended to an automorphism α of Γ'. Since $\beta, \theta, \delta_1, \delta_2 \in \mathrm{Aut}(\Gamma)$, and by homogeneity of Γ', it suffices to show that g preserves $\prec$. So let $x_1, x_2 \in X$ be distinct, and let T be $\theta \circ \delta_1(\{x_1, x_2\})$.

- If $\chi(\theta \circ \delta_1) = 0$, then $\beta|_T$ preserves $\prec$. It follows that the restriction of $\theta \circ \delta_2 \circ \beta$ to T can be extended to an automorphism η of Γ'. By the property of θ, this means that $\chi(\theta \circ \delta_2 \circ \beta \circ \theta \circ \delta_1) = \chi(\theta \circ \delta_1) = 0$. By the definition of χ, it follows that $\beta|_{\theta \delta_2 \circ \beta(T)}$ preserves $\prec$, and so does the restriction of g to $\{x_1, x_2\}$.

- Otherwise, if $\chi(\theta \circ \delta_1) = 1$, then $\beta|_T$ reverses $\prec$. Note that in this case the orbital O of (x_1, x_2) in $\mathrm{Aut}(\Gamma)$ cannot be acyclic, since it consists of the orbit of (x_1, x_2) and the orbit of $(\beta(\theta(\delta_1(x_1))), \beta(\theta(\delta_1(x_2)))$ in $\mathrm{Aut}(\Gamma')$, which are distinct, contrary to our assumption for acyclic orbitals. Therefore, Z contains an O-cycle, and so does $\theta(Z)$ since θ preserves O. Let $(u_1, u_2), (u_2, u_3), \ldots, (u_n, u_1)$ be this O-cycle in $\theta(Z)$. Suppose for contradiction that $\chi(\theta \circ \delta_2 \circ \beta \circ \theta \circ \delta_1) = 0$. Then $u_i \prec u_{i+1}$ (coefficients are modulo n) if and only if $\beta(u_i) \prec \beta(u_{i+1})$. Hence, there is a directed cycle in $\prec$, a contradiction since $\prec$ is a linear order. We conclude that $\chi(\theta \circ \delta_2 \circ \beta \circ \theta \circ \delta_1) = 1$, and thus $\beta|_{\theta(\delta_2(\beta(T)))}$ also reverses $\prec$. Reversing $\prec$ twice means preserving $\prec$, and so we conclude that the restriction of $g = \beta \circ \theta \circ \delta_2 \circ \beta \circ \theta \circ \delta_1$ to $\{x_1, x_2\}$ preserves $\prec$.

So g indeed preserves $\prec$ on all of X, which proves the claim about the existence of $\alpha \in \mathrm{Aut}(\Gamma')$. Note that by the properties of g we also have that $g(X) \subseteq \beta(Y)$, and this concludes the proof. $\qquad\square$

Corollary 6.5 *The following classes have the ordering property with respect to $\prec$:*

- *the class of all finite $\prec$-ordered graphs;*

- *the class of all C-relations over finite sets which are convexly ordered by $\prec$;*

- *the class of all finite $\prec$-ordered directed graphs;*

- *the class of all finite partially ordered sets with a linear order $\prec$ that extends the partial order.*

Proof The Fraïssé limit Γ of the first three classes do not have acyclic orbitals. The Ramsey property for those classes has been established earlier in this text, so the statement follows from Theorem 6.4.

Now, let $\Gamma' = (\Gamma, \prec)$ be the Fraïssé limit of the class from the last item. A proof of the Ramsey property for this class can be found in [47, 48]. There is one acyclic orbital in Γ, namely the strict order relation of the poset. But since $\prec$ is a linear extension of the poset relation, this orbital is also an orbital of Γ'. The statement therefore follows again from Theorem 6.4. $\qquad\square$

The following example shows that the sufficient condition for the ordering property that we gave in Theorem 6.4 is not necessary.

Example 6.6 Let $(D; \prec)$ be any countable dense linear order without endpoints. By Theorem 3.24, the structure $\Gamma := (\mathbb{Q}; \mathrm{Betw}, <) * (D; \prec)$ is Ramsey. Note that the reduct of Γ with signature $\{\mathrm{Betw}, <\}$ is isomorphic to $(\mathbb{Q}; \mathrm{Betw}, <)$ (this can be shown by a simple back-and-forth argument), so we assume that Γ has domain $\mathbb{Q}$. Observe that $<$ is certainly an acyclic orbital in $(\mathbb{Q}; \mathrm{Betw}, <)$, but it splits into the orbital $\{(x, y) : x \prec y \wedge x < y\}$ and the orbital $\{(x, y) : y \prec x \wedge x < y\}$ of $\mathrm{Aut}(\Gamma)$. Hence, the condition from Theorem 6.4 does not apply. But nonetheless, the age $\mathcal{C}$ of $(\mathbb{Q}; \mathrm{Betw}, \prec)$ has the ordering property with respect to $<$. To see this, note that for every finite substructure $\mathfrak{A}$ of $(\mathbb{Q}; \mathrm{Betw}, \prec)$ the only two expansions of $\mathfrak{A}$ by a linear order such that the expansion is isomorphic to a structure from $(\mathbb{Q}; \mathrm{Betw}, \prec, <)$ are $(\mathfrak{A}, <)$ and $(\mathfrak{A}, >)$. With this observation it is straightforward to adapt the proof given in Theorem 6.4 to show the ordering property of $\mathcal{C}$.

7 Concluding remarks and open problems

7.1 An application

Ramsey classes are an important tool in classifications of *reducts* of structures. When Γ is a structure, a *reduct* of Γ is a relational structure Δ with the same domain as Γ such that all the relations of Δ have a *first-order definition* over Γ, that is, for every relation R of Δ there exists a first-order formula ϕ over the signature of Γ (without parameters) such that a tuple t is in R if and only $\phi(t)$ holds in Γ. We say that two reducts are *interdefinable* if they are reducts of each other. Quite surprisingly, countable structures Γ that are homogeneous in a finite relational language tend to have finitely many reducts, up to interdefinability (see e.g. [3, 10, 13, 42, 43, 52, 53]), and Thomas [52] conjectured that this is always the case. If the age of Γ is Ramsey, or a homogeneous expansion of the structure is Ramsey, then this helps in classifying reducts; we refer to the survey article [9] for the technical details. Note that this application provides another motivation to study Conjecture 1.1: if the conjecture is true, then this means that Ramsey classes can be used to attack Thomas' conjecture in general.

7.2   Link with topological dynamics

Whether an amalgamation class $\mathcal{C}$ has the Ramsey property only depends on the (topological) automorphism group $\mathrm{Aut}(\Gamma)$ of the Fraïssé limit Γ of $\mathcal{C}$: by a theorem of Kechris, Pestov, and Todorcevic [27], the class $\mathcal{C}$ is Ramsey if and only if $\mathrm{Aut}(\Gamma)$ is *extremely amenable*, that is, if every continuous action of $\mathrm{Aut}(\Gamma)$ on a compact Hausdorff space has a fixed point. This result has attracted considerable attention [23, 36, 50, 51, 54]. We would like to mention that recently, Melleray, Van Thé, and Tsankov [34] showed that a variant of Conjecture 1.1 (namely Question 7.1 in Section 7) is equivalent to the so-called *universal minimal flow* of $\mathrm{Aut}(\Gamma)$ being metrizable and having a G_δ orbit. Even more recently, Zucker proved that $\mathrm{Aut}(\Gamma)$ does have a G_δ orbit [55] provided that it is metrisable. Hence, the task that remains to prove Conjecture 1.1 with the topological approach is to prove that the universal minimal flow of $\mathrm{Aut}(\Gamma)$ is metrisable (also see Theorem 8.14 in [55]).

These developments in topological dynamics are promising, but so far every single combinatorial result about Ramsey classes that can be proved using topological dynamics also has a direct combinatorial proof. The converse is not true: we have seen in this introductory article several combinatorial proofs where no topological proof is known.

7.3   Variants of the Ramsey expansion conjecture

We have seen that many classes of homogeneous structures can be expanded so that the class of expanded structures becomes Ramsey. Note that if we are allowed to use any expansion, we can trivially turn every class into a Ramsey class, simply by adding unary predicates such that for every element of every structure in the class there is a unary predicate that just contains this element of the structure. This is why it is important to require in Conjecture 1.1 that the expansion has a finite signature.

A weaker finiteness condition than being homogeneous in a finite relational language is the requirement that the expansion is ω-categorical (recall Theorem 2.26). Therefore, a natural variant of Conjecture 1.1 is the following.

Question 7.1 *Is it true that every ω-categorical structure has an ω-categorical expansion which is Ramsey? Equivalently, is it true that every closed oligomorphic subgroup of S_ω has an extremely amenable closed oligomorphic subgroup?*

Formally, Conjecture 1.1 and Question 7.1 are unrelated, since both the hypothesis and the conclusion are stronger in Conjecture 1.1. A common weakening is the question whether every structure which is homogeneous in a finite relational language has an ω-categorical Ramsey expansion. Time will show for which set of hypotheses we can obtain which positive results.

Also in Question 7.1, the assumption that the expansion be ω-categorical is important, since otherwise the answer is trivially positive since the trivial group that just consists of the identity element is extremely amenable. It is also true (Theorem 4.5 in [28]) that every closed oligomorphic subgroup of S_ω has a non-trivial extremely amenable subgroup (which is not ω-categorical, though).

7.4   More Ramsey classes from old?

We do not know the answer to the following question about model companions and model-complete cores in the context of Ramsey classes.

Question 7.2 *Suppose that Γ is a relational structure with a homogeneous Ramsey expansion with finite relational signature, and let Δ be the model companion of Γ. Is it true that Δ has a homogeneous Ramsey expansion with finite relational signature?*

A positive answer would be a strengthening of Theorem 3.15. We can ask the same question for the model-complete core Δ of Γ. Note that a positive answer to Conjecture 1.1 implies a positive answer to Question 7.2.

We also have a variant for ω-categorical expansions instead of homogeneous expansions in a finite relational language.

Question 7.3 *Suppose that Γ is a relational structure with an ω-categorical Ramsey expansion, and let Δ be the model companion of Γ. Is it true that Δ has an ω-categorical Ramsey expansion?*

Acknowledgements

I want to thank Miodrag Sokic for the proof about adding constants, Diana Piguet for the permission to include parts from our unpublished joint paper, and Lionel Van Thé and Miodrag Sokic for discussions about the partite method. I also want to thank Antoine Mottet and András Pongrácz for discussions around the topic of this survey. Many thanks to Trung Van Pham, András Pongrácz, Miodrag Sokic, and to the anonymous referee for many very helpful comments on earlier versions of the text.

References

[1] F. G. Abramson and L. Harrington, *Models without indiscernibles*, Journal of Symbolic Logic **43** (1978), no. 3, 572–600.

[2] S. A. Adeleke and P. M. Neumann, *Relations related to betweenness: their structure and automorphisms*, Memoirs of the AMS, vol. 623, American Mathematical Society, 1998.

[3] J. H. Bennett, *The reducts of some infinite homogeneous graphs and tournaments*, Ph.D. Thesis, 1997.

[4] M. Bodirsky, *Cores of countably categorical structures*, Logical Methods in Computer Science **3** (2007), no. 1, 1–16.

[5] M. Bodirsky, *Complexity classification in infinite-domain constraint satisfaction*, 2012.

[6] M. Bodirsky, *New Ramsey classes from old*, Electronic Journal of Combinatorics **21** (2014), no. 2. Preprint arXiv:1204.3258.

[7] M. Bodirsky, M. Hils, and B. Martin, *On the scope of the universal-algebraic approach to constraint satisfaction*, Logical Methods in Computer Science (LMCS) **8** (2012), no. 3:13. An extended abstract that announced some of the results appeared in the proceedings of Logic in Computer Science (LICS'10).

[8] M. Bodirsky and D. Piguet, *Finite trees are Ramsey with respect to topological embeddings*, 2010.

[9] M. Bodirsky and M. Pinsker, *Reducts of Ramsey structures*, AMS Contemporary Mathematics, vol. 558 (Model Theoretic Methods in Finite Combinatorics) (2011), 489–519.

[10] M. Bodirsky, M. Pinsker, and A. Pongrácz, *The 42 reducts of the random ordered graph*, 2013. Preprint arXiv:1309.2165.

[11] M. Bodirsky, M. Pinsker, and T. Tsankov, *Decidability of definability*, Journal of Symbolic Logic **78** (2013), no. 4, 1036–1054. A conference version appeared in the Proceedings of LICS 2011, pages 321–328.

[12] J. Böttcher and J. Foniok, *Ramsey properties of permutations*, Electronic Journal of Combinatorics **20** (2013), no. 1.

[13] P. J. Cameron, *Transitivity of permutation groups on unordered sets*, Mathematische Zeitschrift **148** (1976), 127–139.

[14] P. J. Cameron, *Oligomorphic permutation groups*, Cambridge University Press, Cambridge, 1990.

[15] P. J. Cameron, *The random graph revisited*, Proceedings of the european congress of mathematics, 2001, pp. 267–274.

[16] G. Cherlin and A. H. Lachlan, *Stable finitely homogeneous structures*, TAMS **296** (1986), 815–850.

[17] G. Cherlin, *The classification of countable homogeneous directed graphs and countable homogeneous n-tournaments*, AMS Memoir **131** (1998January), no. 621.

[18] W. Deuber, *A generalization of Ramsey's theorem for regular trees*, Journal of Combinatorial Theory, Series B **18** (1975), 18–23.

[19] J. Foniok, *On ramsey properties of classes with forbidden trees*, Logical Methods in Computer Science **10** (2014), no. 3.

[20] R. Fraïssé, *Sur l'extension aux relations de quelques propriétés des ordres*, Annales Scientifiques de l'École Normale Supérieure **71** (1954), 363–388.

[21] R. Fraïssé, *Theory of relations*, Elsevier Science Ltd, North-Holland, 1986.

[22] R. L. Graham, B. L. Rothschild, and J. H. Spencer, *Ramsey theory*, Wiley-Interscience Series in Discrete Mathematics and Optimization, John Wiley & Sons, Inc., New York, 1990. Second edition.

[23] Y. Gutman and L. N. V. Thé, *Relative extreme amenability and interpolation*, 2011. Preprint arXiv:1105.6221.

[24] W. Hodges, *Model theory*, Cambridge University Press, 1993.

[25] W. Hodges, *A shorter model theory*, Cambridge University Press, Cambridge, 1997.

[26] J. Hubička and J. Nešetřil, *Bowtie-free graphs have a Ramsey lift*, 2014. arXiv:1402.2700.

[27] A. Kechris, V. Pestov, and S. Todorcevic, *Fraïssé limits, Ramsey theory, and topological dynamics of automorphism groups*, Geometric and Functional Analysis **15** (2005), no. 1, 106–189.

[28] A. S. Kechris, *Dynamics of non-archimedean Polish groups*, Proceedings of the european congress of mathematics, krakow, July 2, pp. 375–397.

[29] A. H. Lachlan, *Countable homogeneous tournaments*, TAMS **284** (1984), 431–461.

[30] A. H. Lachlan, *Structures coordinatized by indiscernible sets*, Annals of Pure and Applied Logic **34** (1987), 245–273.

[31] C. Laflamme, J. Jasinski, L. N. V. Thé, and R. Woodrow, *Ramsey precompact expansions of homogeneous directed graphs* (2013).

[32] K. Leeb, *Vorlesungen über Pascaltheorie*, Arbeitsberichte des Instituts für Mathematische Maschinen und Datenverarbeitung, vol. 6, Friedrich-Alexander-Universität Erlangen-Nürnberg, 1973.

[33] J. Linman and M. Pinsker, *Permutations on the random permutation*, 2014.

[34] J. Melleray, L. N. V. Thé, and T. Tsankov, *Polish groups with metrizable universal minimal flows*, 2014. Preprint arXiv:1404.6167.

[35] K. R. Milliken, *A Ramsey theorem for trees*, Journal of Combinatorial Theory, Series A **26** (1979), no. 3, 215 –237.

[36] M. Müller and A. Pongrácz, *Topological dynamics of unordered ramsey structures*, 2014. To appear in Fundamenta Mathematicae. ArXiv:1401.7766.

[37] J. Nešetřil, *Ramsey theory*, Handbook of Combinatorics (1995), 1331–1403.

[38] J. Nešetřil, *Ramsey classes and homogeneous structures*, Combinatorics, Probability & Computing **14** (2005), no. 1-2, 171–189.

[39] J. Nešetřil and V. Rödl, *Ramsey classes of set systems*, Journal of Combinatorial Theory, Series A **34** (1983), no. 2, 183–201.

[40] J. Nešetřil and V. Rödl, *The partite construction and Ramsey set systems*, Discrete Mathematics **75** (1989), no. 1-3, 327–334.

[41] J. Nešetřil and V. Rödl, *Mathematics of Ramsey theory*, Springer, Berlin, 1998.

[42] P. P. Pach, M. Pinsker, G. Pluhár, A. Pongrácz, and C. Szabó, *Reducts of the random partial order*, Advances in Mathematics **267** (2014), 94–120.

[43] A. Pongrácz, *Reducts of the Henson graphs with a constant* (2011). Preprint.

[44] H. J. Prömel, *Ramsey theory for discrete structures*, Springer-Verlag, 2013.

[45] F. P. Ramsey, *On a problem of formal logic*, Proceedings of the LMS (2) **30** (1930), no. 1, 264–286.

[46] D. Saracino, *Model companions for $\aleph_0$-categorical theories*, Proceedings of the AMS **39** (1973), 591–598.

[47] M. Sokić, *Ramsey property of posets and related structures*, Ph.D. Thesis, 2010.

[48] M. Sokić, *Ramsey property, ultrametric spaces, finite posets, and universal minimal flows*, Israel Journal of Mathematics **194** (2013), no. 2, 609–640.

[49] M. Sokić, *Directed graphs and Boron trees*, 2015. Preprint available from http://www.its.caltech.edu/~msokic/SAP3.pdf.

[50] L. N. V. Thé, *More on the Kechris-Pestov-Todorcevic correspondence: precompact expansions*, Fund. Math. **222** (2013), no. 1, 19–47. Preprint arXiv:1201.1270.

[51] L. N. V. Thé, *Universal flows of closed subgroups of S_∞ and relative extreme amenability*, Asymptotic Geometric Analysis, Fields Institute Communications **68** (2013), 229–245.

[52] S. Thomas, *Reducts of the random graph*, Journal of Symbolic Logic **56** (1991), no. 1, 176–181.

[53] S. Thomas, *Reducts of random hypergraphs*, Annals of Pure and Applied Logic **80** (1996), no. 2, 165–193.

[54] A. Zucker, *Amenability and unique ergodicity of automorphism groups of Fraissé structures*, 2013. Preprint, arXiv:1304.2839.

[55] A. Zucker, *Topological dynamics of closed subgroups of S_ω*, 2014. Preprint, arXiv:1404.5057.

Institut für Algebra, TU Dresden,
01069 Dresden, Germany
manuel.bodirsky@tu-dresden.de

Recent developments in graph Ramsey theory

David Conlon,[1] Jacob Fox[2] and Benny Sudakov[3]

Abstract

Given a graph H, the Ramsey number $r(H)$ is the smallest natural number N such that any two-colouring of the edges of K_N contains a monochromatic copy of H. The existence of these numbers has been known since 1930 but their quantitative behaviour is still not well understood. Even so, there has been a great deal of recent progress on the study of Ramsey numbers and their variants, spurred on by the many advances across extremal combinatorics. In this survey, we will describe some of this progress.

1 Introduction

In its broadest sense, the term Ramsey theory refers to any mathematical statement which says that a structure of a given kind is guaranteed to contain a large well-organised substructure. There are examples of such statements in many areas, including geometry, number theory, logic and analysis. For example, a key ingredient in the proof of the Bolzano–Weierstrass theorem in real analysis is a lemma showing that any infinite sequence must contain an infinite monotone subsequence.

A classic example from number theory, proved by van der Waerden [212] in 1927, says that if the natural numbers are coloured in any fixed number of colours then one of the colour classes contains arbitrarily long arithmetic progressions. This result has many generalisations. The most famous, due to Szemerédi [206], says that any subset of the natural numbers of positive upper density contains arbitrarily long arithmetic progressions. Though proved in 1975, the influence of this result is still being felt today. For example, it was a key ingredient in Green and Tao's proof [131] that the primes contain arbitrarily long arithmetic progressions.

Though there are many further examples from across mathematics, our focus in this survey will be on graph Ramsey theory. The classic theorem in this area, from which Ramsey theory as a whole derives its name, is Ramsey's theorem [174]. This theorem says that for any graph H there

[1] Research supported by a Royal Society University Research Fellowship.
[2] Research supported by a Packard Fellowship, by NSF Career Award DMS-1352121 and by an Alfred P. Sloan Fellowship.
[3] Research supported by SNSF grant 200021-149111.

exists a natural number N such that any two-colouring of the edges of K_N
contains a monochromatic copy of H. The smallest such N is known as
the *Ramsey number* of H and is denoted $r(H)$. When $H = K_t$, we simply
write $r(t)$.

Though Ramsey proved his theorem in 1930 and clearly holds prece-
dence in the matter, it was a subsequent paper by Erdős and Szekeres [102]
which brought the matter to a wider audience. Amongst other things,
Erdős and Szekeres were the first to give a reasonable estimate on Ramsey
numbers.[4] To describe their advance, we define the *off-diagonal Ramsey number* $r(H_1, H_2)$ as the smallest natural number N such that any
red/blue-colouring of the edges of K_N contains either a red copy of H_1 or
a blue copy of H_2. If we write $r(s,t)$ for $r(K_s, K_t)$, then what Erdős and
Szekeres proved is the bound

$$r(s,t) \le \binom{s+t-2}{s-1}.$$

For $s = t$, this yields $r(t) = O(\frac{4^t}{\sqrt{t}})$, while if s is fixed, it gives $r(s,t) \le t^{s-1}$.
Over the years, much effort has been expended on improving these bounds
or showing that they are close to tight, with only partial success. However,
these problems have been remarkably influential in combinatorics, playing
a key role in the development of random graphs and the probabilistic
method, as well as the theory of quasirandomness (see [11]). We will
highlight some of these connections in Section 2.1 when we discuss the
current state of the art on estimating $r(s,t)$.

If we move away from complete graphs, a number of interesting phenomena start to appear. For example, a famous result of Chvátal, Rödl,
Szemerédi and Trotter [44] says that if H is a graph with n vertices and
maximum degree Δ, then the Ramsey number $r(H)$ is bounded by $c(\Delta)n$
for some constant $c(\Delta)$ depending only on Δ. That is, the Ramsey number
of bounded-degree graphs grows linearly in the number of vertices. This
and related developments will be discussed in Section 2.3, while other aspects of Ramsey numbers for general H will be explored in Sections 2.4,
2.5 and 2.6.

In full generality, Ramsey's theorem applies not only to graphs but
also to k-uniform hypergraphs. Formally, a *k-uniform hypergraph* is a pair
$H = (V, E)$, where V is a collection of vertices and E is a collection of
subsets of V, each of order k. We write $K_N^{(k)}$ for the complete k-uniform

---

[4]Ramsey's original paper mentions the bound $r(t) \le t!$, but he does not pursue the
matter further. It is an amusing exercise to find a natural proof that gives exactly this
bound.

hypergraph on N vertices, that is, V has order N and E contains all subsets of V of order k.

The full statement of Ramsey's theorem, which also allows for more than two colours, now says that for any natural number $q \geq 2$ and any k-uniform hypergraphs $H_1, \ldots, H_q$ there exists a natural number N such that any q-colouring of the edges of $K_N^{(k)}$ contains a copy of H_i in the ith colour for some i. The smallest such N is known as the *Ramsey number* of $H_1, \ldots, H_q$ and is denoted $r_k(H_1, \ldots, H_q)$. If $H_i = K_{t_i}^{(k)}$ for each i, we write $r_k(t_1, \ldots, t_q)$. Moreover, if $H_1 = \cdots = H_q = H$, we simply write $r_k(H; q)$, which we refer to as the *q-colour Ramsey number* of H. If $H = K_t^{(k)}$, we write $r_k(t; q)$. If either k or q is equal to two, it is omitted.

Even for complete 3-uniform hypergraphs, the growth rate of the Ramsey number is not well understood. Indeed, it is only known that

$$2^{c't^2} \leq r_3(t) \leq 2^{2^{ct}}.$$

Determining the correct asymptotic for this function is of particular importance, since it is known that an accurate estimate for $r_3(t)$ would imply an accurate estimate on $r_k(t)$ for all $k \geq 4$. This and related topics will be discussed in depth in Section 2.2, though we will make reference to hypergraph analogues of graph Ramsey problems throughout the survey. As we will see, these questions often throw up new and interesting behaviour which is strikingly different from the graph case.

While our focus in Section 2 will be on the classical Ramsey function, we will move on to discussing a number of variants in Section 3. These variants include well-established topics such as induced Ramsey numbers and size Ramsey numbers, as well as a number of more recent themes such as ordered Ramsey numbers. We will not try to give a summary of these variants here, instead referring the reader to the individual sections, each of which is self-contained.

We should note that this paper is not intended to serve as an exhaustive survey of the subject. Instead, we have focused on those areas which are most closely related to our own interests. For the most part, this has meant that we have treated problems of an asymptotic nature rather than being concerned with the computation of exact Ramsey numbers.[5] Even with this caveat, it has still been necessary to gloss over a number of interesting topics. We apologise in advance for any particularly glaring omissions.

We will maintain a number of conventions throughout the paper. For the sake of clarity of presentation, we will sometimes omit floor and ceiling

[5]For this and more, we refer the reader to the excellent dynamic survey of Radziszowski [173].

signs when they are not crucial. Unless specified otherwise, we use log to denote the logarithm taken to the base two. We will use standard asymptotic notation with a subscript indicating that the implied constant may depend on that subscript. All other notation will be explained in the relevant sections.

2   The classical problem

2.1   Complete graphs

As already mentioned in the introduction, the classical bound on Ramsey numbers for complete graphs is the Erdős–Szekeres bound

$$r(s,t) \leq \binom{s+t-2}{s-1}.$$

In particular, for $s = t$, this gives $r(t) = O(\frac{4^t}{\sqrt{t}})$. The proof of the Erdős–Szekeres bound relies on the simple inequality

$$r(s,t) \leq r(s,t-1) + r(s-1,t).$$

To prove this inequality, consider a red/blue-colouring of the edges of $K_{r(s,t)-1}$ containing no red copy of K_s and no blue copy of K_t. The critical observation is that the red degree of every vertex, that is, the number of neighbours in red, is at most $r(s-1,t)-1$. Indeed, if the red neighbourhood of any vertex v contained $r(s-1,t)$ vertices, it would contain either a blue K_t, which would contradict our choice of colouring, or a red K_{s-1}, which together with v would form a red K_s, again a contradiction. Similarly, the blue degree of every vertex is at most $r(s,t-1)-1$. Since the union of any particular vertex with its red and blue neighbourhoods is the entire vertex set, we see that

$$r(s,t) - 1 \leq 1 + (r(s-1,t)-1) + (r(s,t-1)-1).$$

The required inequality follows.

The key observation here, that in any graph containing neither a red K_s nor a blue K_t the red degree of any vertex is less than $r(s-1,t)$ and the blue degree is less than $r(s,t-1)$, may be generalised. Indeed, an argument almost exactly analogous to that above shows that in any graph containing neither a red K_s nor a blue K_t, any red edge must be contained in fewer than $r(s-2,t)$ red triangles and any blue edge must be contained in fewer than $r(s,t-2)$ blue triangles. Indeed, if a red edge uv were contained in at least $r(s-2,t)$ red triangles, then the set W of vertices

w joined to both u and v in red would have order at least $r(s-2,t)$. If this set contained a blue K_t, we would have a contradiction, so the set must contain a red K_{s-2}. But the union of this clique with u and v forms a K_s, again a contradiction. Together with Goodman's formula [124] for the number of monochromatic triangles in a two-colouring of K_N, this observation may be used to show that

$$r(t,t) \le 4r(t,t-2) + 2.$$

Using the idea behind this inequality, Thomason [210] was able to improve the upper bound for diagonal Ramsey numbers to $r(t) = O(\frac{4^t}{t})$, improving an earlier result of Rödl [127], who was the first to show that $r(t) = o(\frac{4^t}{\sqrt{t}})$.

As the observant reader may already have noted, the argument of the previous paragraph is itself a special case of the following observation.

Observation 2.1 *In any graph containing neither a red K_s nor a blue K_t, any red copy of K_p must be contained in fewer than $r(s-p,t)$ red copies of K_{p+1} and any blue copy of K_p must be contained in fewer than $r(s,t-p)$ blue copies of K_{p+1}.*

By using this additional information, Conlon [45] was able to give the following superpolynomial improvement on the Erdős–Szekeres bound.

Theorem 2.2 *There exists a positive constant c such that*

$$r(t) \le t^{-c \log t/ \log \log t} 4^t.$$

In broad outline, the proof of Theorem 2.2 proceeds by using the $p = 1$ and $p = 2$ cases of Observation 2.1 to show that any red/blue-colouring of the edges of a complete graph with at least $t^{-c \log t/ \log \log t} 4^t$ vertices which contains no monochromatic K_t is quasirandom. Through a delicate counting argument, this is then shown to contradict Observation 2.1 for p roughly $\log t/ \log \log t$.

The first significant lower bound for the diagonal Ramsey number $r(t)$ was proved by Erdős [78] in 1947. This was one of the first applications of the probabilistic method and most introductions to this beautiful subject begin with his simple argument. Though we run the risk of being repetitious, we will also include this argument.

Colour the edges of the complete graph K_N randomly. That is, we colour each edge red with probability $1/2$ and blue with probability $1/2$. Since the probability that a given copy of K_t has all edges red is $2^{-\binom{t}{2}}$, the expected number of red copies of K_t in this graph is $2^{-\binom{t}{2}}\binom{N}{t}$. Similarly, the expected number of blue copies of K_t is $2^{-\binom{t}{2}}\binom{N}{t}$. Therefore, the expected number of monochromatic copies of K_t is

$$2^{1-\binom{t}{2}}\binom{N}{t} \leq 2^{1-t(t-1)/2}\left(\frac{eN}{t}\right)^t.$$

For $N = (1 - o(1))\frac{t}{\sqrt{2}e}\sqrt{2}^t$, we see that this expectation is less than one. Therefore, there must be some colouring of K_N for which there are no monochromatic copies of K_t. This bound,

$$r(t) \geq (1 - o(1))\frac{t}{\sqrt{2}e}\sqrt{2}^t,$$

has been astonishingly resilient to improvement. Since 1947, there has only been one noteworthy improvement.[6] This was achieved by Spencer [198], who used the Lovász local lemma to show that

$$r(t) \geq (1 - o(1))\frac{\sqrt{2}t}{e}\sqrt{2}^t.$$

That is, he improved Erdős' bound by a factor of two! Any further improvement to this bound, no matter how tiny, would be of significant interest.

Problem 2.3 *Does there exist a positive constant ϵ such that*

$$r(t) \geq (1 + \epsilon)\frac{\sqrt{2}t}{e}\sqrt{2}^t$$

for all sufficiently large t?

For off-diagonal Ramsey numbers, where s is fixed and t tends to infinity, the Erdős–Szekeres bound shows that $r(s,t) \leq t^{s-1}$. In 1980, this bound was improved by Ajtai, Komlós and Szemerédi [1], who proved that for any s there exists a constant c_s such that

$$r(s,t) \leq c_s \frac{t^{s-1}}{(\log t)^{s-2}}.$$

When $s = 3$, this follows from the statement that any triangle-free graph on N vertices with average degree d contains an independent set of order

---

[6]Though we do not know of an explicit reference, a simple application of the deletion method which improves Erdős' bound by a factor of $\sqrt{2}$ was surely known before Spencer's work.

$\Omega(\frac{N}{d}\log d)$. Indeed, in a triangle-free graph, the neighbourhood of every vertex must form an independent set and so $d < t$. But then the graph must contain an independent set of order $\Omega(\frac{N}{t}\log t)$ and, hence, for c sufficiently large and $N \geq ct^2/\log t$, the graph contains an independent set of order t.

For $s = 3$, this result was shown to be sharp up to the constant by Kim [141]. That is, he showed that there exists a positive constant c' such that

$$r(3,t) \geq c'\frac{t^2}{\log t}.$$

This improved on earlier work of Erdős [79], who used an intricate probabilistic argument to show that $r(3,t) \geq c'(t/\log t)^2$, a result which was subsequently reproved using the local lemma [199].

Kim's proof of this bound was a landmark application of the so-called semi-random method. Recently, an alternative proof was found by Bohman [20] using the triangle-free process. This is a stochastic graph process where one starts with the empty graph on N vertices and adds one edge at a time to create a graph. At each step, we randomly select an edge which is not in the graph and add it to the graph if and only if it does not complete a triangle. The process runs until every non-edge is contained in a triangle. By analysing the independence number of the resulting graph, Bohman was able to reprove Kim's bound. More recently, Bohman and Keevash [22] and, independently, Fiz Pontiveros, Griffiths and Morris [104] gave more precise estimates for the running time of the triangle-free process and as a consequence proved the following result.

Theorem 2.4
$$r(3,t) \geq \left(\frac{1}{4} - o(1)\right)\frac{t^2}{\log t}.$$

This is within an asymptotic factor of 4 of the best upper bound, due to Shearer [190], who showed that

$$r(3,t) \leq (1 + o(1))\frac{t^2}{\log t}.$$

This is already a very satisfactory state of affairs, though it would be of great interest to improve either bound further.

For general s, the best lower bound is due to Bohman and Keevash [21] and uses the analogous K_s-free process. Their analysis shows that for any s there exists a positive constant c'_s such that

$$r(s,t) \geq c'_s\frac{t^{\frac{s+1}{2}}}{(\log t)^{\frac{s+1}{2}-\frac{1}{s-2}}}.$$

Even for $s = 4$, there is a polynomial difference between the upper and lower bounds. Bringing these bounds closer together remains one of the most tantalising open problems in Ramsey theory.

Before concluding this section, we say a little about the multicolour generalisations of these problems. An easy extension of the Erdős–Szekeres argument gives an upper bound for the multicolour diagonal Ramsey number of the form $r(t; q) \leq q^{qt}$. On the other hand, an elementary product argument shows that, for any positive integers p and d, we have $r(t; pd) > (r(t; p) - 1)^d$. In particular, taking $p = 2$, we see that $r(t; q) > (r(t; 2) - 1)^{q/2} > 2^{qt/4}$ for q even and $t \geq 3$. To prove the bound, suppose that χ is a p-colouring of the edges of the complete graph on vertex set $[r(t; p) - 1] = \{1, 2, \dots, r(t; p) - 1\}$ with no monochromatic K_t and consider the lexicographic d^{th} power of χ. This is a pd-colouring of the edges of the complete graph with vertex set $[r(t; p) - 1]^d$ such that the colour of the edge between two distinct vertices $(u_1, \dots, u_d)$ and $(v_1, \dots, v_d)$ is $(i, \chi(u_i, v_i))$, where i is the first coordinate for which $u_i \neq v_i$. It is easy to check that this colouring contains no monochromatic K_t. Since the set has $(r(t; p) - 1)^d$ vertices, the result follows.

The key question in the multicolour case is to determine the dependence on the number of colours. Even for $t = 3$, we only know that

$$2^{c'q} \leq r(3; q) \leq cq!,$$

where $c \leq e$ and $c' \geq 1$ are constants whose values have each been improved a little over time. It is a major open problem to improve these bounds by a more significant factor.

In the off-diagonal case, less seems to be known, but we would like to highlight one result. While it is easy to see that

$$r(\underbrace{K_3, \dots, K_3}_{q-1}, K_t) = O(t^q),$$

it was an open question for many years to even show that the ratio $r(K_3, K_3, K_t)/r(K_3, K_t)$ tends to infinity with t. Alon and Rödl [9] solved this problem in a strong form by showing that the bound quoted above is tight up to logarithmic factors for all q. Their elegant construction involves overlaying a collection of random shifts of a sufficiently pseudo-random triangle-free graph.

2.2 Complete hypergraphs

Although there are already significant gaps between the lower and upper bounds for graph Ramsey numbers, our knowledge of hypergraph Ramsey numbers is even weaker. Recall that $r_k(s, t)$ is the minimum N such that every red/blue-colouring of the k-tuples of an N-element set contains a red $K_s^{(k)}$ or a blue $K_t^{(k)}$. While a naive extension of the Erdős–Szekeres argument gives extremely poor bounds for hypergraph Ramsey numbers when $k \geq 3$, a more careful induction, discovered by Erdős and Rado [99], allows one to bound Ramsey numbers for k-uniform hypergraphs using estimates for the Ramsey number of $(k-1)$-uniform hypergraphs. Quantitatively, their result says the following.

Theorem 2.5 $r_k(s,t) \leq 2^{\left(^{r_{k-1}\binom{s-1,t-1}{k-1}}\right)} + k - 2.$

Together with the standard exponential upper bound on $r(t)$, this shows that $r_3(t) \leq 2^{2^{ct}}$ for some constant c. On the other hand, by considering a random two-colouring of the edges of $K_N^{(k)}$, Erdős, Hajnal and Rado [98] showed that there is a positive constant c' such that $r_3(t) \geq 2^{c't^2}$. However, they conjectured that the upper bound is closer to the truth and Erdős later offered a \$500 reward for a proof.

Conjecture 2.6 *There exists a positive constant c' such that*

$$r_3(t) \geq 2^{2^{c't}}.$$

Fifty years after the work of Erdős, Hajnal and Rado, the bounds for $r_3(t)$ still differ by an exponential. Similarly, for $k \geq 4$, there is a difference of one exponential between the known upper and lower bounds for $r_k(t)$, our best bounds being

$$t_{k-1}(c't^2) \leq r_k(t) \leq t_k(ct),$$

where the tower function $t_k(x)$ is defined by $t_1(x) = x$ and $t_{i+1}(x) = 2^{t_i(x)}$. The upper bound here is a straightforward consequence of Theorem 2.5, while the lower bound follows from an ingenious construction of Erdős and Hajnal known as the stepping-up lemma (see, e.g., Chapter 4.7 in [130]). This allows one to construct lower bound colourings for uniformity $k+1$ from colourings for uniformity k, effectively gaining an extra exponential each time it is applied. Unfortunately, the smallest k for which it works is $k = 3$. However, if we could prove that $r_3(t)$ is double exponential in t, this would automatically close the gap between the upper and lower bounds for $r_k(t)$ for all uniformities k.

For more than two colours, the problem becomes easier and Erdős and Hajnal (see [130]) were able to construct a 4-colouring of the triples of a set of double-exponential size which does not contain a monochromatic clique of order t. By a standard extension of the Erdős–Rado upper bound to more than two colours, this result is sharp.

Theorem 2.7 *There exists a positive constant c' such that*

$$r_3(t; 4) \geq 2^{2^{c't}}.$$

We will now sketch this construction, since it is a good illustration of how the stepping-up lemma works. Let $m = 2^{(t-1)/2}$ and suppose we are given a red/blue-colouring χ of the edges of K_m with no monochromatic clique of order $t-1$ (in Section 2.1, we showed that such a colouring exists). Let $N = 2^m$ and consider the set of all binary strings of length m, where each string corresponds to the binary representation of an integer between 0 and $N-1$. For any two strings x and y, let $\delta(x, y)$ be the largest index in which they differ. Note that if $x < y < z$ (as numbers), then we have that $\delta(x, y) \neq \delta(y, z)$ and $\delta(x, z)$ is the maximum of $\delta(x, y)$ and $\delta(y, z)$. More generally, if $x_1 < \cdots < x_t$, then $\delta(x_1, x_t) = \max_i \delta(x_i, x_{i+1})$. Given vertices $x < y < z$ with $\delta_1 = \delta(x, y)$ and $\delta_2 = \delta(y, z)$, we let the colour of (x, y, z) be

- A if $\delta_1 < \delta_2$ and $\chi(\delta_1, \delta_2) = $ red;

- B if $\delta_1 < \delta_2$ and $\chi(\delta_1, \delta_2) = $ blue;

- C if $\delta_1 > \delta_2$ and $\chi(\delta_1, \delta_2) = $ red;

- D if $\delta_1 > \delta_2$ and $\chi(\delta_1, \delta_2) = $ blue.

Suppose now that $x_1 < \cdots < x_t$ is a monochromatic set in colour A (the other cases are similar) and let $\delta_i = \delta(x_i, x_{i+1})$. We claim that $\delta_1, \ldots, \delta_{t-1}$ form a red clique in the original colouring of K_m, which is a contradiction. Indeed, since (x_i, x_{i+1}, x_{i+2}) has colour A, we must have that $\delta_i < \delta_{i+1}$ for all i. Therefore, $\delta_1 < \cdots < \delta_{t-1}$ and $\delta(x_{i+1}, x_{j+1}) = \delta(x_j, x_{j+1}) = \delta_j$ for all $i < j$. Since the colour of the triple (x_i, x_{i+1}, x_{j+1}) is determined by the colour of (δ_i, δ_j), this now tells us that $\chi(\delta_i, \delta_j)$ is red for all $i < j$, as required.

For the intermediate case of three colours, Erdős and Hajnal [94] made a small improvement on the lower bound of $2^{c't^2}$, showing that $r_3(t; 3) \geq 2^{c't^2 \log^2 t}$. Extending the stepping-up approach described above, the authors [55] improved this bound as follows, giving a strong indication that $r_3(t; 3)$ is indeed double exponential.

Theorem 2.8 *There exists a positive constant c' such that*

$$r_3(t; 3) \geq 2^{t^{c' \log t}}.$$

Though Erdős [41, 85] believed that $r_3(t)$ is closer to $2^{2^{c't}}$, he and Hajnal [94] discovered the following interesting fact which they thought might indicate the opposite. They proved that there are positive constants c and ϵ such that every two-colouring of the triples of an N-element set contains a subset S of order $s \geq c(\log N)^{1/2}$ such that at least $(1/2+\epsilon)\binom{s}{3}$ triples of S have the same colour. That is, the density of each colour deviates from $1/2$ by at least some fixed positive constant.

In the graph case, a random colouring of the edges of K_N has the property that every subset of order $\omega(\log N)$ has roughly the same number of edges in both colours. That is, the Ramsey problem and the discrepancy problem have similar quantitative behaviour. Because of this, Erdős [87] remarked that he would begin to doubt that $r_3(t)$ is double exponential in t if one could prove that any two-colouring of the triples of an N-set contains some set of order $s = c(\epsilon)(\log N)^\delta$ for which at least $(1 - \epsilon)\binom{s}{3}$ triples have the same colour, where $\delta > 0$ is an absolute constant and $\epsilon > 0$ is arbitrary. Erdős and Hajnal proposed [94] that such a statement may even be true with $\delta = 1/2$, which would be tight up to the constant factor c. The following result, due to the authors [57], shows that this is indeed the case.

Theorem 2.9 *For each $\epsilon > 0$, there is $c = c(\epsilon) > 0$ such that every two-colouring of the triples of an N-element set contains a subset S of order $s = c\sqrt{\log N}$ such that at least $(1-\epsilon)\binom{s}{3}$ triples of S have the same colour.*

Unlike Erdős, we do not feel that this result suggests that the growth of $r_3(t)$ is smaller than double exponential. Indeed, this theorem also holds for any fixed number of colours q but, for $q \geq 4$, the hypergraph Ramsey number does grow as a double exponential. That is, the q-colour analogue of Theorem 2.9 shows that the largest almost monochromatic subset in a q-colouring of the triples of an N-element set is much larger than the largest monochromatic subset. This is in striking constrast to graphs, where we have already remarked that the two quantities have the same order of magnitude.

It would be very interesting to extend Theorem 2.9 to higher uniformities. In [55], the authors proved that for all k, q and $\epsilon > 0$ there is $\delta = \delta(k, q, \epsilon) > 0$ such that every q-colouring of the k-tuples of an N-element set contains a subset of order $s = (\log N)^\delta$ which contains at least $(1 - \epsilon)\binom{s}{k}$ k-tuples of the same colour. Unfortunately, δ here depends on ϵ. On the other hand, this result could hold with $\delta = 1/(k - 1)$ (which is the case for $k = 3$).

Problem 2.10 *Is it true that for any $k \geq 4$ and $\epsilon > 0$ there exists $c = c(k, \epsilon) > 0$ such that every two-colouring of the k-tuples of an N-element*

60 · D. Conlon, J. Fox and B. Sudakov

set contains a subset S of order $s = c(\log N)^{1/(k-1)}$ such that at least $(1-\epsilon)\binom{s}{k}$ k-tuples of S have the same colour?

Another wide open problem is that of estimating off-diagonal Ramsey numbers for hypergraphs. Progress on this question was slow and for several decades the best known bound was that obtained by Erdős and Rado [99]. Combining their estimate from Theorem 2.5 with the best upper bound on $r(s-1,t-1)$ shows that for fixed s,

$$r_3(s,t) \le 2^{\binom{r(s-1,t-1)}{2}} + 1 \le 2^{ct^{2s-4}/\log^{2s-6} t}.$$

Recently, the authors [55] discovered an interesting connection between the problem of bounding $r_3(s,t)$ and a new game-theoretic parameter. To describe this parameter, we start with the classical approach of Erdős and Rado and then indicate how it can be improved.

Let $p = r(s-1,t-1)$, $N = 2^{\binom{p}{2}}+1$ and consider a red/blue-colouring c of all triples on the vertex set $[N] = \{1,2,\ldots,N\}$. We will show how to find vertices $v_1,\ldots,v_p,v_{p+1}$ such that, for each $i < j$, all triples (v_i,v_j,v_k) with $k > j$ have the same colour, which we denote by $\chi(i,j)$. This will solve the problem, since, by the definition of p, the colouring χ of $v_1,\ldots,v_p$ contains either a red K_{s-1} or a blue K_{t-1}, which together with v_{p+1} would give a monochromatic set of triples of the correct order in the original colouring. We will pick the vertices v_i in rounds. Suppose that we already have vertices $v_1,\ldots,v_m$ with the required property as well as a set of vertices S_m such that for every v_i,v_j and every $w \in S_m$ the colour of the triple (v_i,v_j,w) is given by $\chi(i,j)$ and so does not depend on w. Pick $v_{m+1} \in S_m$ arbitrarily. For all other w in S_m, consider the colour vector $(c_1,\ldots,c_m)$ such that $c_i = c(v_i,v_{m+1},w)$, which are the only new triples we need worry about. Let S_{m+1} be the largest subset of S_m such that every vertex in this subset has the same colour vector $(c_1,\ldots,c_m)$. Clearly, this set has order at least $2^{-m}(|S_m|-1)$. Notice that $v_1,\ldots,v_{m+1}$ and S_{m+1} have the desired properties. We may therefore continue the algorithm, noting that we have lost a factor of 2^m in the size of the remaining set of vertices, i.e., a factor of 2 for every edge coloured by χ.

To improve this approach, we note that the colouring χ does not need to colour every pair of vertices. This idea is captured nicely by the notion of vertex on-line Ramsey number. Consider the following game, played by two players, Builder and Painter: at step $m+1$ a new vertex v_{m+1} is revealed; then, for every existing vertex v_j, $j = 1,\cdots,m$, the Builder decides, in order, whether to draw the edge v_jv_{m+1}; if he does expose such an edge, the Painter has to colour it either red or blue immediately. The *vertex on-line Ramsey number* $\tilde{r}_v(k,l)$ is then defined as the minimum number of edges that Builder has to draw in order to force Painter to create a red K_k or a blue K_l. Using an approach similar to that described in the previous paragraph, one can bound the Ramsey number $r_3(s,t)$ roughly by an exponential in $\tilde{r}_v(s-1,t-1)$. By estimating $\tilde{r}_v(s-1,t-1)$, this observation, together with some additional ideas, allowed the authors to improve the Erdős–Rado estimate for off-diagonal hypergraph Ramsey numbers as follows.

Theorem 2.11 *For every natural number $s \geq 4$, there exists a positive constant c such that*

$$r_3(s,t) \leq 2^{ct^{s-2}\log t}.$$

A similar improvement for off-diagonal Ramsey numbers of higher uniformity follows from combining this result with Theorem 2.5.

How accurate is this estimate? For the first non-trivial case, when $s = 4$, the problem was first considered by Erdős and Hajnal [92] in 1972. Using the following clever construction, they showed that $r_3(4,t)$ is exponential in t. Consider a random tournament with vertex set $[N]$. This is a complete graph on N vertices whose edges are oriented uniformly at random. Colour a triple in $[N]$ red if it forms a cyclic triangle and blue otherwise. Since it is well known and easy to show that every tournament on four vertices contains at most two cyclic triangles and a random tournament on N vertices with high probability does not contain a transitive subtournament of order $c \log N$, the resulting colouring has neither a red subset of order 4 nor a blue subset of order $c \log N$. In the same paper [92], Erdős and Hajnal conjectured that $\frac{\log r_3(4,t)}{t} \to \infty$. This was recently confirmed in [55], where the authors obtained a more general result which in particular implies that $r_3(4,t) \geq 2^{c't\log t}$. This should be compared with the upper bound $r_3(4,t) \leq 2^{ct^2\log t}$ obtained above.

2.3 Sparse graphs

After the complete graph, the next most classical topic in graph Ramsey theory concerns the Ramsey numbers of sparse graphs, i.e., graphs with certain constraints on the degrees of the vertices. Burr and Erdős [30] initiated the study of these Ramsey numbers in 1975 and this topic has since placed a central role in graph Ramsey theory, leading to the development of many important techniques with broader applicability.

In their foundational paper, Burr and Erdős [30] conjectured that for every positive integer Δ there is a constant $c(\Delta)$ such that every graph H with n vertices and maximum degree Δ satisfies $r(H) \leq c(\Delta)n$. This conjecture was proved by Chvátal, Rödl, Szemerédi and Trotter [44] as an early application of Szemerédi's regularity lemma [207]. We will now sketch their proof, first reviewing the statement of the regularity lemma.

Roughly speaking, the regularity lemma says that the vertex set of any graph may be partitioned into a small number of parts such that the bipartite subgraph between almost every pair of parts is random-like. More formally, we say that a pair of disjoint vertex subsets (A, B) in a graph G is ϵ-*regular* if, for every $A' \subseteq A$ and $B' \subseteq B$ with $|A'| \geq \epsilon|A|$ and $|B'| \geq \epsilon|B|$, the density $d(A', B')$ of edges between A' and B' satisfies $|d(A', B') - d(A, B)| \leq \epsilon$. That is, the density between any two large subsets of A and B is close to the density between A and B. The regularity lemma then says that for every $\epsilon > 0$ there exists $M = M(\epsilon)$ such that the vertex set of any graph G may be partitioned into $m \leq M$ parts $V_1, \ldots, V_m$ such that $||V_i| - |V_j|| \leq 1$ for all $1 \leq i, j \leq m$ and all but $\epsilon\binom{m}{2}$ pairs (V_i, V_j) are ϵ-regular.

Suppose now that $N = c(\Delta)n$ and the edges of K_N have been two-coloured. To begin, we apply the regularity lemma with approximation parameter $\epsilon = 4^{-\Delta}$ (since the colours are complementary, we may apply the regularity lemma to either the red or the blue subgraph, obtaining a regular partition for both). This gives a partition of the vertex set into $m \leq M$ parts of roughly equal size, where M depends only on Δ, such that all but $\epsilon\binom{m}{2}$ pairs of parts are ϵ-regular. By applying Turán's theorem, we may find 4^{Δ} parts such that every pair of parts is ϵ-regular. Since $r(\Delta+1) \leq 4^{\Delta}$, an application of Ramsey's theorem then implies that there are $\Delta+1$ parts $V_1, \ldots, V_{\Delta+1}$ such that every pair is ϵ-regular and the graph between each pair has density at least $1/2$ in one particular colour, say red. As $\chi(H) \leq \Delta+1$, we can partition the vertex set of H into independent sets $U_1, \ldots, U_{\Delta+1}$. The regularity between the sets $V_1, \ldots, V_{\Delta+1}$ now allows us to greedily construct a red copy of H, embedding one vertex at a time and mapping U_i into V_i for each i. Throughout the embedding process, we must ensure that for any vertex u of U_i which is not yet embedded the set of potential vertices in V_i into which one may embed u is large (at step t, we guarantee that it has order at least $4^{-d(t,u)}|V_i| - t$, where $d(t, u) \leq \Delta$ is the number of neighbours of u among the first t embedded vertices). Though an elegant application of the regularity lemma, this method gives a poor bound on $c(\Delta)$, namely, a tower of 2s with height exponential in Δ.

Since this theorem was first proved, the problem of determining the correct order of magnitude for $c(\Delta)$ as a function of Δ has received considerable attention from various researchers. The first progress was made by Eaton [76], who showed that $c(\Delta) \leq 2^{2^{c\Delta}}$ for some fixed c, the key observation being that the proof above does not need the full strength of the regularity lemma. Instead, one only needs to find 4^{Δ} large vertex subsets such that the graph between each pair is ϵ-regular. This may be achieved using a weak regularity lemma due to Duke, Lefmann and Rödl [75].

A novel approach of Graham, Rödl and Rucinski [128] was the first to give a linear upper bound on Ramsey numbers of bounded-degree graphs without using any form of the regularity lemma. Their proof also gave good quantitative control, showing that one may take $c(\Delta) \leq 2^{c\Delta \log^2 \Delta}$. As in the regularity proof, they try to greedily construct a red copy of H one vertex at a time, at each step ensuring that the set of potential vertices into which one might embed any remaining vertex is large. If this process fails, we will find two large vertex subsets such that the red graph between them has very low density. Put differently, this means that the blue graph between these vertex sets has very high density. We now iterate this procedure within each of the two subsets, trying to embed greedily in red and, if this fails, finding two large vertex subsets with high blue density between them. After $\log 8\Delta$ iterations, we will either have found the required red copy of H or we will have 8Δ subsets of equal size with high blue density between all pairs of sets. If the constants are chosen appropriately, the union of these sets will have blue density at least $1 - \frac{1}{4\Delta}$ and at least $4n$ vertices. One can then greedily embed a blue copy of H one vertex at a time.

Recently, the authors [58] improved this bound to $c(\Delta) \leq 2^{c\Delta \log \Delta}$.

Theorem 2.12 *There exists a constant c such that any graph H on n vertices with maximum degree Δ satisfies*

$$r(H) \leq 2^{c\Delta \log \Delta} n.$$

In the approach of Graham, Rödl and Ruciński, the two colours play asymmetrical roles. Either we find a set where the red graph has some reasonable density between any two large subsets or a set which is almost complete in blue. In either case, a greedy embedding gives the required monochromatic copy of H. The approach we take in [58] is more symmetrical. The basic idea is that once we find a pair of vertex subsets (V_1, V_2) such that the graph between them is almost complete in blue, we split H into two parts U_1 and U_2, each of which induces a subgraph of maximum degree at most $\Delta/2$, and try to embed blue copies of $H[U_i]$ into V_i for $i = 1, 2$, using the high blue density between V_1 and V_2 to ensure that this gives a blue embedding of H. The gain comes from the fact that when we iterate the maximum degree of the graph we wish to embed shrinks. Unfortunately, while this gives some of the intuition behind the proof, the details are rather more involved.

Graham, Rödl and Rucinski [129] observed that for bipartite graphs H on n vertices with maximum degree Δ their technique could be used to prove a bound of the form $r(H) \leq 2^{c\Delta \log \Delta} n$. Indeed, if greedily embedding a red copy of H fails, then there will be two large vertex subsets V_1 and V_2 such that the graph between them is almost complete in blue. A blue copy of H can then be greedily embedded between these sets. In the other direction, they showed that there is a positive constant c' such that for each Δ and n sufficiently large there is a bipartite graph H on n vertices with maximum degree Δ for which $r(H) \geq 2^{c'\Delta} n$. Conlon [46] and, independently, Fox and Sudakov [116] showed that this bound is essentially tight, that is, there is a constant c such that $r(H) \leq 2^{c\Delta} n$ for every bipartite graph H on n vertices with maximum degree Δ. Both proofs are quite similar, each relying on an application of dependent random choice and a hypergraph embedding lemma.

Dependent random choice is a powerful probabilistic technique which has recently led to a number of advances in extremal graph theory, additive combinatorics, Ramsey theory and combinatorial geometry. Early variants of this technique were developed by Gowers [125], Kostochka and Rödl [147] and Sudakov [202]. In many applications, including that under discussion, the technique is used to prove the useful fact that every dense graph contains a large subset U in which almost every set of d vertices has many common neighbours. To prove this fact, we let R be a random set of vertices from our graph and take U to be the set of all common neighbours of R. Intuitively, it is clear that if some subset of U of order d has only a few common neighbours, then it is unlikely that all the members of R could have been chosen from this set of neighbours. It is therefore unlikely that U contains any many subsets of this type. For more information about dependent random choice and its applications, we refer the interested reader to the recent survey [118].

Using the Lovász local lemma, the authors [62] recently improved on the hypergraph embedding lemmas used in their earlier proofs to obtain a bound of the form $r(H) \leq c2^\Delta n$ for every bipartite graph H on n vertices with maximum degree Δ. Like the earlier results, this follows from a more general density result which shows that the denser of the two colour classes will contain the required monochromatic copy of H.

By repeated application of the dependent random choice technique and an appropriate adaptation of the embedding technique, Fox and Sudakov [116] also proved that $r(H) \leq 2^{4\chi\Delta} n$ for all graphs H on n vertices with chromatic number χ and maximum degree Δ. However, the dependency on χ is unlikely to be necessary here.

Conjecture 2.13 *There is a constant c such that every graph H on n vertices with maximum degree Δ satisfies $r(H) \leq 2^{c\Delta} n$.*

One particular family of bipartite graphs that has received significant attention in Ramsey theory are hypercubes. The *hypercube* Q_n is the n-regular graph on vertex set $\{0,1\}^n$ where two vertices are connected by an edge if and only if they differ in exactly one coordinate. Burr and Erdős [30] conjectured that $r(Q_n)$ is linear in $|Q_n|$.

Conjecture 2.14
$$r(Q_n) = O(2^n).$$

After several improvements over the trivial bound $r(Q_n) \leq r(|Q_n|) \leq 4^{|Q_n|} = 2^{2^{n+1}}$ by Beck [15], Graham, Rödl and Ruciński [129], Shi [194, 195] and Fox and Sudakov [116], the authors [62] obtained the best known upper bound of $r(H) = O(2^{2n})$, which is quadratic in the number of vertices. This follows immediately from the general upper bound on Ramsey numbers of bipartite graphs with given maximum degree stated earlier.

Another natural notion of sparseness which has been studied extensively in the literature is that of degeneracy. A graph is said to be *d-degenerate* if every subgraph has a vertex of degree at most d. Equivalently, a graph is d-degenerate if there is an ordering of the vertices such that each vertex has at most d neighbours that precede it in the ordering. The *degeneracy* of a graph is the smallest d such that the graph is d-degenerate. Burr and Erdős [30] conjectured that every graph with bounded degeneracy has linear Ramsey number.

Conjecture 2.15 *For every natural number d, there is a constant c(d) such that every d-degenerate graph H on n vertices satisfies $r(H) \leq c(d)n$.*

This conjecture is one of the most important open problems in graph Ramsey theory. The first significant progress on the conjecture was made by Kostochka and Sudakov [149], who proved an almost linear upper bound. That is, for fixed d, they showed that every d-degenerate graph H on n vertices satisfies $r(H) = n^{1+o(1)}$. This result was later refined by Fox and Sudakov [117], who showed that every d-degenerate graph H on n vertices satisfies $r(H) \leq e^{c(d)\sqrt{\log n}}n$.

Partial progress of a different sort was made by Chen and Schelp [37], who considered a notion of sparseness which is intermediate between having bounded degree and having bounded degeneracy. We say that a graph is *p-arrangeable* if there is an ordering $v_1, v_2, \ldots, v_n$ of its vertices such that for each vertex v_i, its neighbours to the right of v_i have together at most p neighbours to the left of v_i (including v_i). The *arrangeability* of a graph is the smallest p such that the graph is p-arrangeable. Extending the result of Chvátal et al. [44], Chen and Schelp [37] proved that for every

p there is a constant $c(p)$ such that every p-arrangeable graph on n vertices has Ramsey number at most $c(p)n$. Graphs with bounded arrangeability include planar graphs and graphs embeddable on a fixed surface. More generally, Rödl and Thomas [185] proved that graphs which do not contain a subdivision of a fixed graph have bounded arrangeability and hence have linear Ramsey number. Another application was given by Fox and Sudakov [117], who proved that for fixed d the Erdős–Renyi random graph $G(n, d/n)$ almost surely has arrangeability on the order of d^2 and hence almost surely has linear Ramsey number.

In general, the Ramsey number of a graph appears to be intimately connected to its degeneracy. Indeed, if $d(H)$ is the degeneracy of H, a random colouring easily implies that $r(H) \geq 2^{d(H)/2}$. Since it is also clear that $r(H) \geq n$ for any n-vertex graph, we see that $\log r(H) = \Omega(d(H) + \log n)$. We conjecture that this bound is tight up to the constant. It is even plausible that $r(H) \leq 2^{O(d)}n$ for every d-degenerate graph H on n vertices. Since the degeneracy of a graph is easily computable, this would give a very satisfying approximation for the Ramsey number of a general graph.

Conjecture 2.16 *For every n-vertex graph H,*

$$\log r(H) = \Theta \left(d(H) + \log n \right).$$

For graphs of bounded chromatic number, Conjecture 2.16 follows from a bound on Ramsey numbers due to Fox and Sudakov (Theorem 2.1 in [117]). Moreover, another result from the same paper (Theorem 3.1 in [117]) shows that Conjecture 2.16 always holds up to a factor of $\log^2 d(H)$.

In graph Ramsey theory, it is natural to expect there should be no significant qualitative difference between the bounds for two colours and the bounds for any fixed number of colours. However, there are many well-known problems where this intuition has yet to be verified, the classic example being the bounds for hypergraph Ramsey numbers. Another important example is furnished by the results of this section. Indeed, the proof technique of Graham, Rödl and Ruciński can be extended to work for more than two colours, but only gives the estimate $r(H; q) \leq 2^{\Delta^{q-1+o(1)}}n$ for the Ramsey number of graphs H with n vertices and maximum degree Δ. While dependent random choice does better, giving a bound of the form $r(H; q) \leq 2^{O_q(\Delta^2)}n$, we believe that for a fixed number of colours, the exponent of Δ should still be 1. In particular, we conjecture that the following bound holds.

Conjecture 2.17 *For every graph H on n vertices with maximum degree*

Δ, *the 3-colour Ramsey number of H satisfies*

$$r(H, H, H) \leq 2^{\Delta^{1+o(1)}} n,$$

where the $o(1)$ is a function of Δ which tends to 0 as Δ tends to infinity.

With the development of the hypergraph regularity method [126, 164, 183], the result that bounded-degree graphs have linear Ramsey numbers was extended to 3-uniform hypergraphs by Cooley, Fountoulakis, Kühn and Osthus [66] and Nagle, Olsen, Rödl and Schacht [163] and to k-uniform hypergraphs by Cooley et al. [67]. That is, for each k and Δ there is $c(\Delta, k)$ such that every k-uniform hypergraph H on n vertices with maximum degree Δ satisfies $r(H) \leq c(\Delta, k)n$. However, because they use the hypergraph regularity lemma, their proof only gives an enormous Ackermann-type upper bound on $c(\Delta, k)$. In [54], the authors gave another shorter proof of this theorem which gives the right type of behaviour for $c(\Delta, k)$. The proof relies on an appropriate generalisation of the dependent random choice technique to hypergraphs. As in Section 2.2, we write $t_1(x) = x$ and $t_{i+1}(x) = 2^{t_i(x)}$.

Theorem 2.18 *For any natural numbers $k \geq 3$ and $q \geq 2$, there exists a constant $c = c(k, q)$ such that the q-colour Ramsey number of any k-uniform hypergraph H on n vertices with maximum degree Δ satisfies*

$$r_3(H; q) \leq 2^{2^{c\Delta \log \Delta}} n \text{ and, for } k \geq 4, \ r_k(H; q) \leq t_k(c\Delta)n.$$

We say that a hypergraph is *d-degenerate* if every subgraph has a vertex of degree at most d. Equivalently, a hypergraph is d-degenerate if there is an ordering of the vertices $v_1, v_2, \ldots, v_n$ such that each vertex v_i is the final vertex in at most d edges in this ordering. Kostochka and Rödl [148] showed that the hypergraph analogue of the Burr–Erdős conjecture is false for uniformity $k \geq 4$. In particular, they constructed a 4-uniform hypergraph on n vertices which is 1-degenerate but has Ramsey number at least $2^{\Omega(n^{1/3})}$.

2.4 Graphs with a given number of edges

In 1973, Erdős and Graham [89] conjectured that among all connected graphs with $m = \binom{n}{2}$ edges, the complete graph has the largest Ramsey number. As this question seems unapproachable, Erdős [84] asked whether one could at least show that the Ramsey number of any graph with m edges is not substantially larger than that of the complete graph with the same size. Since the number of vertices in a complete graph with m edges is a constant multiple of $\sqrt{m}$, he conjectured that there exists a constant c such that $r(H) \leq 2^{c\sqrt{m}}$ for any graph H with m edges and no isolated vertices.

The first progress on this conjecture was made by Alon, Krivelevich and Sudakov [5], who showed that there exists a constant c such that $r(H) \leq 2^{c\sqrt{m}\log m}$ for any graph H with m edges and no isolated vertices. They also proved the conjecture in the special case where H is bipartite. Another proof of the same bound, though starting from a different angle, was later given by Conlon [49]. This approach, which focused on estimating the Ramsey number of graphs with a given density, allowed one to show that graphs on n vertices with $o(n^2)$ edges have Ramsey number $2^{o(n)}$. Soon after this work, Erdős' conjecture was completely resolved by Sudakov [205], so that it may now be stated as a theorem.

Theorem 2.19 *There exists a constant c such that any graph H with m edges and no isolated vertices satisfies*

$$r(H) \leq 2^{c\sqrt{m}}.$$

The proof of this theorem relies upon several ingredients, including the machinery of Graham, Rödl and Ruciński [128] mentioned in the previous section and a result of Erdős and Szemerédi [103] which says that if a graph has low density then it contains a larger clique or independent set than would be guaranteed by Ramsey's theorem alone.[7] However, these techniques are very specific to two colours, so the following problem remains wide open.

Problem 2.20 *Show that for any $q \geq 3$ there exists c_q such that $r(H;q) \leq 2^{c_q\sqrt{m}}$ for any graph H with m edges and no isolated vertices.*

If no vertex in the graph H has unusually high degree, it is often possible to improve on Theorem 2.19. For example, the following result [49, 53] implies that if a graph with n vertices and m edges has degeneracy at most $10m/n$, say, then the Ramsey number is at most an exponential in $\frac{m}{n}\log^2(\frac{n^2}{m})$. For $m = o(n^2)$, this is significantly smaller than $\sqrt{m}$.

---

[7]The Erdős–Szemerédi theorem is the starting point for another interesting topic which we have not had space to discuss, namely, the problem of determining what properties a graph with no clique or independent set of order $c\log n$ must satisfy. The Erdős–Szemerédi theorem shows that any such graph must have density bounded away from both 0 and 1 and there are numerous further papers (see, for example, [3, 27, 114] and their references) showing that these graphs must exhibit random-like behaviour.

Theorem 2.21 *There exists a constant c such that any graph H on n vertices with degeneracy at most d satisfies*

$$r(H) \leq 2^{cd \log^2 (2n/d)}.$$

The analogous question for hypergraphs was studied by the authors in [54]. Though the same rationale that led Erdős to conjecture Theorem 2.19 naturally leads one to conjecture that $r_3(H) \leq 2^{2^{cm^{1/3}}}$ for all 3-uniform hypergraphs H with m edges and no isolated vertices, it turns out that there are connected 3-uniform hypergraphs H with m edges for which $r_3(H; 4) \geq 2^{2^{c' \sqrt{m}}}$. This is also close to being sharp, since $r_3(H; q) \leq 2^{2^{c_q \sqrt{m} \log m}}$ for any 3-uniform hypergraph H with m edges and no isolated vertices and any $q \geq 2$. For higher uniformities, $k \geq 4$, one can do slightly better. Writing $t_1(x) = x$ and $t_{i+1}(x) = 2^{t_i(x)}$ as in Section 2.2, the authors showed that $r_k(H; q) \leq t_k(c_{k,q} \sqrt{m})$ for any k-uniform hypergraph H with m edges and no isolated vertices and any $q \geq 2$. It would be interesting to improve the bound in the 3-uniform case to bring it in line with higher uniformities.

Problem 2.22 *Show that for any $q \geq 2$ there exists c_q such that $r_3(H; q) \leq 2^{2^{c_q \sqrt{m}}}$ for any 3-uniform hypergraph H with m edges and no isolated vertices.*

This would likely follow if the bound for the Ramsey number of 3-uniform hypergraphs with n vertices and maximum degree Δ given in Theorem 2.18 could be improved to $2^{2^{c\Delta}} n$.

2.5   Ramsey goodness

If one tries to prove a lower bound for the off-diagonal Ramsey number $r(G, H)$, one simple construction, usually attributed to Chvátal and Harary [43], is to take $\chi(H) - 1$ red cliques, each of order $|G| - 1$, and to colour all edges between these sets in blue. If G is connected, this colouring clearly contains no red copy of G and no blue copy of H and so $r(G, H) \geq (|G| - 1)(\chi(H) - 1) + 1$. If we write $\sigma(H)$ for the order of the smallest colour class in any $\chi(H)$-colouring of the vertices of H, we see, provided $|G| \geq \sigma(H)$, that we may add a further red clique of order $\sigma(H) - 1$ to our construction. This additional observation, due to Burr [28], allows us to improve our lower bound to

$$r(G, H) \geq (|G| - 1)(\chi(H) - 1) + \sigma(H),$$

provided $|G| \geq \sigma(H)$. Following Burr and Erdős [28, 31], we will say that a graph G is H-*good* if this inequality is an equality, that is, if $r(G, H) = (|G| - 1)(\chi(H) - 1) + \sigma(H)$. Given a family of graphs $\mathcal{G}$, we say that $\mathcal{G}$ is H-*good* if equality holds for all sufficiently large graphs $G \in \mathcal{G}$. In the particular case where $H = K_s$, we say that a graph or family of graphs is s-*good*.

The classical result on Ramsey goodness, which predates the definition, is the theorem of Chvátal [42] showing that all trees are s-good for every s. However, the family of trees is not H-good for every graph H. For example [32], there is a constant $c < \frac{1}{2}$ such that $r(K_{1,t}, K_{2,2}) \geq t + \sqrt{t} - t^c$ for t sufficiently large, whereas $(|K_{1,t}| - 1)(\chi(K_{2,2}) - 1) + \sigma(K_{2,2}) = t + 2$.

In an effort to determine what properties contribute to being good, Burr and Erdős [29, 31] conjectured that if Δ is fixed then the family of graphs with maximum degree at most Δ is s-good for every s. However, this conjecture was disproved by Brandt [26], who showed that if a graph is a good expander then it cannot be 3-good. In particular, his result implies that for $\Delta \geq \Delta_0$ almost every Δ-regular graph on a sufficiently large number of vertices is not 3-good.

On the other hand, graphs with poor expansion properties are often good. The first such result, due to Burr and Erdős [31], states that for any fixed ℓ the family of connected graphs with bandwidth at most ℓ is s-good for any s, where the *bandwidth* of a graph G is the smallest number ℓ for which there exists an ordering $v_1, v_2, \ldots, v_n$ of the vertices of G such that every edge $v_i v_j$ satisfies $|i - j| \leq \ell$. This result was recently extended by Allen, Brightwell and Skokan [2], who showed that the set of connected graphs with bandwidth at most ℓ is H-good for every H. Their result even allows the bandwidth ℓ to grow at a reasonable rate with the order of the graph G. If G is known to have bounded maximum degree, their results are particularly strong, their main theorem in this case being the following.

Theorem 2.23 *For any Δ and any fixed graph H, there exists $c > 0$ such that if G is a connected graph on n vertices with maximum degree Δ and bandwidth at most cn then G is H-good.*

Another result of this type, proved by Nikiforov and Rousseau [170], shows that graphs with small separators are s-good. Recall that the degeneracy $d(G)$ of a graph G is the smallest natural number d such that every induced subgraph of G has a vertex of degree at most d. Furthermore, we say that a graph G has a (t, η)-*separator* if there exists a vertex subset $T \subseteq V(G)$ such that $|T| \leq t$ and every connected component of $V(G) \backslash T$ has order at most $\eta |V(G)|$. The result of Nikiforov and Rousseau is now as follows.

Theorem 2.24 *For any* $s \geq 3$, $d \geq 1$ *and* $0 < \gamma < 1$, *there exists* $\eta > 0$ *such that the class* $\mathcal{G}$ *of connected d-degenerate graphs* G *with a* $(|V(G)|^{1-\gamma}, \eta)$-*separator is s-good.*

Nikiforov and Rousseau used this result to resolve a number of outstanding questions of Burr and Erdős [31] regarding Ramsey goodness. For example, they showed that the 1-subdivision of K_n, the graph formed by adding an extra vertex to each edge of K_n, is s-good for n sufficiently large. Moreover, using this result, it was shown in [50] that the family of connected planar graphs is s-good for every s. This is a special case of a more general result. We say that a graph H is a *minor* of G if H can be obtained from a subgraph of G by contracting edges. By an *H-minor* of G, we mean a minor of G which is isomorphic to H. For a graph H, let $\mathcal{G}_H$ be the family of connected graphs which do not contain an H-minor. Since the family of planar graphs consists precisely of those graphs which do not contain K_5 or $K_{3,3}$ as a minor, our claim about planar graphs is an immediate corollary of the following result. The proof is an easy corollary of Theorem 2.24, a result of Mader [158] which bounds the average degree of H-minor-free graphs and a separator theorem for H-minor-free graphs due to Alon, Seymour and Thomas [10].

Theorem 2.25 *For every fixed graph* H, *the class* $\mathcal{G}_H$ *of graphs* G *which do not contain an H-minor is s-good for every* $s \geq 3$.

One of the original problems of Burr and Erdős that was left open after the work of Nikiforov and Rousseau was to determine whether the family of hypercubes is s-good for every s. Recall that the hypercube Q_n is the graph on vertex set $\{0, 1\}^n$ where two vertices are connected by an edge if and only if they differ in exactly one coordinate. Since Q_n has 2^n vertices, the problem asks whether $r(Q_n, K_s) = (s-1)(2^n - 1) + 1$ for n sufficiently large. The first progress on this question was made by Conlon, Fox, Lee and Sudakov [50], who obtained an upper bound of the form $c_s 2^n$, the main tool in the proof being a novel technique for embedding hypercubes. Using a variant of this embedding technique and a number of additional ingredients, the original question was subsequently resolved by Fiz Pontiveros, Griffiths, Morris, Saxton and Skokan [105, 106].

Theorem 2.26 *The family of hypercubes is s-good for every* $s \geq 3$.

2.6   Ramsey multiplicity

For any fixed graph H, Ramsey's theorem tells us that when N is sufficiently large, any two-colouring of the edges of K_N contains a monochromatic copy of H. But how many monochromatic copies of H will this two-colouring contain? To be more precise, we let $m_H(G)$ be the number of copies of one graph H in another graph G and define

$$ m_H(N) = \min\{m_H(G) + m_H(\overline{G}) : |G| = N\}, $$

that is, $m_H(N)$ is the minimum number of monochromatic copies of H that occur in any two-colouring of K_N. For the clique K_t, we simply write $m_t(N)$. We now define the *Ramsey multiplicity constant*[8] to be

$$ c_H = \lim_{N \to \infty} \frac{m_H(N)}{m_H(K_N)}. $$

That is, we consider the minimum proportion of copies of H which are monochromatic, where the minimum is taken over all two-colourings of K_N, and then take the limit as N tends to infinity. Since one may show that the fractions $m_H(N)/m_H(K_N)$ are increasing in N and bounded above by 1, this limit is well defined. For cliques, we simply write $c_t := c_{K_t} = \lim_{N \to \infty} m_t(N)/\binom{N}{t}$. We also write $c_{H,q}$ and $c_{t,q}$ for the analogous functions with q rather than two colours.

The earliest result on Ramsey multiplicity is the famous result of Goodman [124], which says that $c_3 \geq \frac{1}{4}$. This result is sharp, as may be seen by considering a random two-colouring of the edges of K_N. Erdős [80] conjectured that a similar phenomenon should hold for larger cliques, that is, that the Ramsey multiplicity should be asymptotically minimised by the graph $G_{N,1/2}$. Quantitatively, this would imply that $c_t \geq 2^{1-\binom{t}{2}}$. This conjecture was later generalised by Burr and Rosta [34], who conjectured that $c_H \geq 2^{1-e(H)}$ for all graphs H. Following standard practice, we will call a graph *common* if it satisfies the Burr–Rosta conjecture.

The Burr–Rosta conjecture was disproved by Sidorenko [196], who showed that a triangle with a pendant edge is not common. Soon after, Thomason [211] disproved Erdős' conjecture by showing that K_4 is not common. Indeed, he showed that $c_4 < \frac{1}{33}$, where Erdős' conjecture would have implied that $c_4 \geq \frac{1}{32}$. More generally, Jagger, Šťovíček and

---

[8]We note that sometimes the term Ramsey multiplicity is used for the quantity $m_H(r(H))$, that is, the minimum number of copies of H that must appear once one copy of H appears. For example, it is well known that every two-colouring of K_6 contains not just one but at least two monochromatic copies of K_3. In general, this quantity is rather intractable and we will not discuss it further.

Thomason [139] showed that any graph which contains K_4 is not common. They also asked whether the conjecture holds for the 5-wheel, the graph formed by taking a cycle of length 5 and adding a central vertex connected to each of the vertices in the cycle. Determining whether this graph satisfies the Burr–Rosta conjecture was of particular interest because it is the smallest graph of chromatic number 4 which does not contain K_4. Using flag algebras [175], this question was answered positively by Hatami, Hladký, Král', Norine and Razborov [136].

Theorem 2.27 *The 5-wheel is common.*

Therefore, there exist 4-chromatic common graphs. The following question, whether there exist common graphs of any chromatic number, was stated explicitly in [136]. For example, is it the case that the graphs arising in Mycielski's famous construction of triangle-free graphs with arbitrarily high chromatic number are common?

Problem 2.28 *Do there exist common graphs of all chromatic numbers?*

For bipartite graphs (that is, graphs of chromatic number two), the question of whether the graph is common is closely related to a famous conjecture of Sidorenko [197] and Erdős–Simonovits [101]. This conjecture states that if H is a bipartite graph then the random graph with density p has in expectation asymptotically the minimum number of copies of H over all graphs of the same order and edge density. In particular, if this conjecture is true for a given bipartite graph H then so is the Burr–Rosta conjecture. Since Sidorenko's conjecture is now known to hold for a number of large classes of graphs, we will not attempt an exhaustive summary here, instead referring the reader to some of the recent papers on the subject [56, 142, 155].

In general, the problem of estimating the constants c_H seems to be difficult. For complete graphs, the upper bound $c_t \le 2^{1-\binom{t}{2}}$ has only ever been improved by small constant factors, while the best lower bound, due to Conlon [48], is $c_t \ge C^{-(1+o(1))t^2}$, where $C \approx 2.18$ is an explicitly defined constant. The argument that gives this bound may be seen as a multiplicity analogue of the usual Erdős–Szekeres argument that bounds Ramsey numbers. We accordingly expect that it will be difficult to improve. For fixed t, the flag algebra method offers some hope. For example, it is now known [169, 201] that $c_4 > \frac{1}{35}$. A more striking recent success of this method, by Cummings, Král', Pfender, Sperfeld, Treglown and Young [68], is an exact determination of $c_{3,3} = \frac{1}{25}$.

A strong quantitative counterexample to the Burr–Rosta conjecture was found by Fox [108]. Indeed, suppose that H is connected and split the vertex set of K_N into $\chi(H)-1$ vertex sets, each of order $\frac{N}{\chi(H)-1}$, colouring the edges between any two sets blue and those within each set red. Since there are only $\chi(H)-1$ sets, there cannot be a blue copy of H. As every red copy of H must lie completely within one of the $\chi(H)-1$ vertex sets, a simple calculation then shows that $c_H \leq (\chi(H)-1)^{1-v(H)}$. Consider now the graph H consisting of a clique with $t = \sqrt{m}$ vertices and an appended path with $m - \binom{t}{2} \geq \frac{m}{2}$ edges. Since $\chi(H) = \sqrt{m}$ and $v(H) \geq \frac{m}{2}$, we see that $c_H \leq m^{-(1-o(1))m/4}$. Since $e(H) = m$, this gives a strong disproof of the conjecture that $c_H \geq 2^{1-m}$. However, the following conjecture [108] still remains plausible.

Conjecture 2.29 *For any $\epsilon > 0$, there exists m_0 such that if H is a graph with at least m_0 edges, then*

$$c_H \geq 2^{-e(H)^{1+\epsilon}}.$$

When $q \geq 3$, the Ramsey multiplicity constants $c_{H,q}$ behave very differently. To see this, consider a two-colouring, in red and blue, of the complete graph on $r(t)-1$ vertices which contains no monochromatic copy of K_t. We now form a three-colouring of K_N by blowing up each vertex in this two-colouring to have order $\frac{N}{r(t)-1}$ and placing a green clique in each vertex set. This colouring contains no red or blue copies of K_t. Therefore, if H is the graph defined above, that is, a clique with $t = \sqrt{m}$ vertices and an appended path with $m - \binom{t}{2} \geq \frac{m}{2}$ edges, it is easy to check that $c_{H,3} \leq (r(t)-1)^{1-v(H)} \leq 2^{-(1-o(1))m^{3/2}/4}$, where we used that $r(t) \geq 2^{t/2}$. In particular, Conjecture 2.29 is false for more than two colours. We hope to discuss this topic further in a forthcoming paper [62].

3   Variants

There are a huge number of interesting variants of the usual Ramsey function. In this section, we will consider only a few of these, focusing on those that we believe to be of the greatest importance.

3.1   Induced Ramsey numbers

A graph H is said to be an *induced subgraph* of H if $V(H) \subset V(G)$ and two vertices of H are adjacent if and only if they are adjacent in G. The *induced Ramsey number* $r_{\text{ind}}(H)$ is the smallest natural number N for which there is a graph G on N vertices such that every two-colouring of the edges of G contains an induced monochromatic copy of H. The existence of these numbers was proved independently by Deuber [70], Erdős, Hajnal and Pósa [97] and Rödl [176], though the bounds these proofs give on $r_{\mathrm{ind}}(H)$ are enormous. However, Erdős [81] conjectured the existence of a constant c such that every graph H with n vertices satisfies $r_{\mathrm{ind}}(H) \leq 2^{cn}$. If true, this would clearly be best possible.

In a problem paper, Erdős [84] stated that he and Hajnal had proved a bound of the form $r_{\mathrm{ind}}(H) \leq 2^{2^{n^{1+o(1)}}}$. This remained the state of the art for some years until Kohayakawa, Prömel and Rödl [144] proved that there is a constant c such that every graph H on n vertices satisfies $r_{\mathrm{ind}}(H) \leq 2^{cn \log^2 n}$. Using similar ideas to those used in the proof of Theorem 2.12, the authors [58] recently improved this bound, removing one of the logarithmic factors from the exponent.

Theorem 3.1 *There exists a constant c such that every graph H with n vertices satisfies*
$$r_{ind}(H) \leq 2^{cn \log n}.$$

The graph G used by Kohayakawa, Prömel and Rödl in their proof is a random graph constructed with projective planes. This graph is specifically designed so as to contain many copies of the target graph H. Subsequently, Fox and Sudakov [114] showed how to prove the same bounds as Kohayakawa, Prömel and Rödl using explicit pseudorandom graphs. The approach in [58] also uses pseudorandom graphs.

A graph is said to be pseudorandom if it imitates some of the properties of a random graph. One such property, introduced by Thomason [208, 209], is that of having approximately the same density between any pair of large disjoint vertex sets. More formally, we say that a graph $G = (V, E)$ is (p, λ)-*jumbled* if, for all subsets A, B of V, the number of edges $e(A, B)$ between A and B satisfies
$$|e(A, B) - p|A||B|| \leq \lambda\sqrt{|A||B|}.$$

The *binomial random graph* $G(N, p)$, where each edge in an N-vertex graph is chosen independently with probability p, is itself a (p, λ)-jumbled graph with $\lambda = O(\sqrt{pN})$. An example of an explicit $(\frac{1}{2}, \sqrt{N})$-jumbled graph is the Paley graph P_N. This is the graph with vertex set $\mathbb{Z}_N$, where N is a prime which is congruent to 1 modulo 4 and two vertices x and y are adjacent if and only if $x - y$ is a quadratic residue. For further examples, we refer the reader to [151]. We may now state the result that lies behind Theorem 3.1.

Theorem 3.2 *There exists a constant c such that, for any $n \in \mathbb{N}$ and any $(\frac{1}{2}, \lambda)$-jumbled graph G on N vertices with $\lambda \leq 2^{-cn \log n} N$, every graph on* *n vertices occurs as an induced monochromatic copy in all two-colourings of the edges of G. Moreover, all of these induced monochromatic copies can be found in the same colour.*

For graphs of bounded maximum degree, Trotter conjectured that the induced Ramsey number is at most polynomial in the number of vertices. That is, for each Δ there should be $d(\Delta)$ such that $r_{\text{ind}}(H) \leq n^{d(\Delta)}$ for any n-vertex graph H with maximum degree Δ. This was proved by Łuczak and Rödl [157], who gave an enormous upper bound for $d(\Delta)$, namely, a tower of twos of height $O(\Delta^2)$. More recently, Fox and Sudakov [114] proved the much more reasonable bound $d(\Delta) = O(\Delta \log \Delta)$. This was improved by Conlon, Fox and Zhao [63] as follows.

Theorem 3.3 *For every natural number Δ, there exists a constant c such that $r_{ind}(H) \leq cn^{2\Delta+8}$ for every n-vertex graph H of maximum degree Δ.*

Again, this is a special case of a much more general result. Like Theorem 3.2, it says that if a graph on N vertices is (p, λ)-jumbled for λ sufficiently small in terms of p and N, then the graph has strong Ramsey properties.[9]

Theorem 3.4 *For every natural number Δ, there exists a constant c such that, for any $n \in \mathbb{N}$ and any $(\frac{1}{n}, \lambda)$-jumbled graph G on N vertices with $\lambda \leq cn^{-\Delta - \frac{9}{2}}N$, every graph on n vertices with maximum degree Δ occurs as an induced monochromatic copy in all two-colourings of the edges of G. Moreover, all of these induced monochromatic copies can be found in the same colour.*

In particular, this gives the stronger result that there are graphs G on $cn^{2\Delta+8}$ vertices such that in every two-colouring of the edges of G there is a colour which contains induced monochromatic copies of every graph on n vertices with maximum degree Δ. The exponent of n in this result is best possible up to a multiplicative factor, since, even for the much weaker condition that G contains an induced copy of all graphs on n vertices with maximum degree Δ, G must contain $\Omega(n^{\Delta/2})$ vertices [35].

Theorems 3.3 and 3.4 easily extend to more than two colours. This is not the case for Theorems 3.1 and 3.2, where the following problem remains open. As usual, $r_{\text{ind}}(H; q)$ denotes the q-colour analogue of the induced Ramsey number.

---

[9]We note that this is itself a simple corollary of the main result in [63], which gives a counting lemma for subgraphs of sparse pseudorandom graphs and thereby a mechanism for transferring combinatorial theorems such as Ramsey's theorem to the sparse context. For further details, we refer the interested reader to [63].

Problem 3.5 *Show that if H is a graph on n vertices and $q \geq 3$ is a natural number, then $r_{ind}(H; q) \leq 2^{n^{1+o(1)}}$.*

It also remains to decide whether Theorem 3.3 can be improved to show that the induced Ramsey number of every graph with n vertices and maximum degree Δ is at most a polynomial in n whose exponent is independent of Δ.

Problem 3.6 *Does there exist a constant d such that $r_{ind}(H) \leq c(\Delta)n^d$ for all graphs with n vertices and maximum degree Δ?*

3.2 Folkman numbers

In the late sixties, Erdős and Hajnal [91] asked whether, for any positive integers $t \geq 3$ and $q \geq 2$, there exists a graph G which is K_{t+1}-free but such that any q-colouring of the edges of G contains a monochromatic copy of K_t. For two colours, this problem was solved in the affirmative by Folkman [107]. However, his method did not generalise to more than two colours and it was several years before Nešetřil and Rödl [166] found another proof which worked for any number of colours.

Once we know that these graphs exist, it is natural to try and estimate their size. To do this, we define the *Folkman number* $f(t)$ to be the smallest natural number N for which there exists a K_{t+1}-free graph G on N vertices such that every two-colouring of the edges of G contains a monochromatic copy of K_t. The lower bound for $f(t)$ is essentially the same as for the usual Ramsey function, that is, $f(t) \geq 2^{c't}$. On the other hand, the proofs mentioned above (and some subsequent ones [167, 180]) use induction schemes which result in the required graphs G having enormous numbers of vertices.

Because of the difficulties involved in proving reasonable bounds for these numbers, a substantial amount of effort has gone into understanding the bounds for $f(3)$. In particular, Erdős asked for a proof that $f(3)$ is smaller than 10^{10}. This was subsequently given by Spencer [200], building on work of Frankl and Rödl [119], but has since been improved further [73, 156]. The current best bound, due to Lange, Radziszowski and Xu [153], stands at $f(3) \leq 786$.

The work of Frankl and Rödl [119] and Spencer [200] relied upon analysing the Ramsey properties of random graphs. Recall that the binomial random graph $G_{n,p}$ is a graph on n vertices where each of the $\binom{n}{2}$ possible edges is chosen independently with probability p. Building on the work of Frankl and Rödl, Rödl and Ruciński [179, 180] determined the threshold for Ramsey's theorem to hold in a binomial random graph and used it to give another proof of Folkman's theorem. To state their theorem, let us say that a graph G is (H, q)-*Ramsey* if any q-colouring of the edges of G contains a monochromatic copy of H.

Theorem 3.7 *For any graph H that is not a forest consisting of stars and paths of length 3 and any positive integer $q \geq 2$, there exist positive constants c and C such that*

$$\lim_{n \to \infty} \mathbb{P}[G_{n,p} \text{ is } (H, q)\text{-Ramsey}] = \begin{cases} 0 & \text{if } p < cn^{-1/m_2(H)}, \\ 1 & \text{if } p > Cn^{-1/m_2(H)}, \end{cases}$$

where

$$m_2(H) = \max \left\{ \frac{e(H') - 1}{v(H') - 2} : H' \subseteq H \text{ and } v(H') \geq 3 \right\}.$$

Very recently, it was noted [65, 181] that some new methods for proving this theorem yield significantly stronger bounds for Folkman numbers. As we have already remarked, the connection between these two topics is not a new one. However, in recent years, a number of very general methods have been developed for proving combinatorial theorems in random sets [13, 64, 121, 188, 189] and some of these methods return good quantitative estimates. In particular, the following result was proved by Rödl, Ruciński and Schacht [181]. The proof relies heavily on the hypergraph container method of Balogh, Morris and Samotij [13] and Saxton and Thomason [188] and an observation of Nenadov and Steger [165] that allows one to apply this machinery to Ramsey problems.

Theorem 3.8 *There exists a constant c such that*

$$f(t) \leq 2^{ct^4 \log t}.$$

Their method also returns a comparable bound for the q-colour analogue $f(t; q)$. Given how close these bounds now lie to the lower bound, we are willing to conjecture that, like the usual Ramsey number, the Folkman number is at most exponential in t.

Conjecture 3.9 *There exists a constant c such that*

$$f(t) \leq 2^{ct}.$$

3.3 The Erdős–Hajnal conjecture

There are several results and conjectures saying that graphs which do not contain a fixed induced subgraph are highly structured. The most famous conjecture of this type is due to Erdős and Hajnal [94] and asks whether any such graph must contain very large cliques or independent sets.[10]

Conjecture 3.10 *For every graph H, there exists a positive constant $c(H)$ such that any graph on n vertices which does not contain an induced copy of H has a clique or an independent set of order at least $n^{c(H)}$.*

This is in stark contrast with general graphs, since the probabilistic argument that gives the standard lower bound on Ramsey numbers shows that almost all graphs on n vertices contain no clique or independent set of order $2 \log n$. Therefore, the Erdős–Hajnal conjecture may be seen as saying that the bound on Ramsey numbers can be improved from exponential to polynomial when one restricts to colourings that have a fixed forbidden subcolouring.

The Erdős–Hajnal conjecture has been solved in some special cases. For example, the bounds for off-diagonal Ramsey numbers imply that it holds when H is itself a clique or an independent set. Moreover, Alon, Pach and Solymosi [8] observed that if the conjecture is true for two graphs H_1 and H_2, then it also holds for the graph H formed by blowing up a vertex of H_1 and replacing it with a copy of H_2. These results easily allow one to prove that the conjecture holds for all graphs on at most four vertices with the exception of P_4, the path with 3 edges. However, this case follows from noting that any graph which contains no induced P_4 is perfect. The conjecture remains open for a number of graphs on five vertices, including the cycle C_5 and the path P_5. However, Chudnovsky and Safra [39] recently proved the conjecture for the graph on five vertices known as the bull, consisting of a triangle with two pendant edges. We refer the reader to the survey by Chudnovsky [38] for further information on this and related results.

The best general bound, due to Erdős and Hajnal [94], is as follows.

Theorem 3.11 *For every graph H, there exists a positive constant $c(H)$ such that any graph on n vertices which does not contain an induced copy of H has a clique or an independent set of order at least $e^{c(H)\sqrt{\log n}}$.*

---

[10]Although their 1989 paper [94] is usually cited as the origin of this problem, the Erdős–Hajnal conjecture already appeared in a paper from 1977 [93].

Despite much attention, this bound has not been improved. However, an off-diagonal generalisation was proved by Fox and Sudakov [116] using dependent random choice. This says that for any graph H there exists a positive constant $c(H)$ such that for every induced-H-free graph G on n vertices and any positive integers n_1 and n_2 satisfying $(\log n_1)(\log n_2) \leq c(H) \log n$, G contains either a clique of order n_1 or an independent set of order n_2.

Another result of this type, due to Promel and Rödl [172], states that for each C there is $c > 0$ such that every graph on n vertices contains every graph on at most $c \log n$ vertices as an induced subgraph or has a clique or independent set of order at least $C \log n$. That is, every graph contains all small graphs as induced subgraphs or has an unusually large clique or independent set. Fox and Sudakov [114] proved a result which implies both the Erdős–Hajnal result and the Promel–Rödl result. It states that there are absolute constants $c, c' > 0$ such that for all positive integers n and k every graph on n vertices contains every graph on at most k vertices as an induced subgraph or has a clique or independent set of order $c2^{c'\sqrt{\frac{\log n}{k}}} \log n$. When k is constant, this gives the Erdős–Hajnal bound and when k is a small multiple of $\log n$, we obtain the Promel–Rödl result.

It is also interesting to see what happens if one forbids not just one but many graphs as induced subgraphs. A family $\mathcal{F}$ of graphs is *hereditary* if it is closed under taking induced subgraphs. We say that it is *proper* if it does not contain all graphs. A family $\mathcal{F}$ of graphs has the *Erdős–Hajnal property* if there is $c = c(\mathcal{F}) > 0$ such that every graph $G \in \mathcal{F}$ has a clique or an independent set of order $|G|^c$. The Erdős–Hajnal conjecture is easily seen to be equivalent to the statement that every proper hereditary family of graphs has the Erdős–Hajnal property.

A family $\mathcal{F}$ of graphs has the *strong Erdős–Hajnal property* if there is $c' = c'(\mathcal{F}) > 0$ such that for every graph $G \in \mathcal{F}$ on at least two vertices, G or its complement $\bar{G}$ contains a complete bipartite subgraph with parts of order $c'|G|$. A simple induction argument (see [110]) shows that if a hereditary family of graphs has the strong Erdős–Hajnal property, then it also has the Erdős–Hajnal property. However, not every proper hereditary family of graphs has the strong Erdős–Hajnal property. For example, it is easy to see that the family of triangle-free graphs does not have the strong Erdős–Hajnal property. Even so, the strong Erdős–Hajnal property has been a useful way to deduce the Erdős–Hajnal property for some families of graphs. A good example is the recent result of Bousquet, Lagoutte and Thomassé [25] which states that for each positive integer t the family of graphs that excludes both the path P_t on t vertices and its complement as induced subgraphs has the strong Erdős–Hajnal property (using different techniques, Chudnovsky and Seymour [40] had earlier proved that this family has the Erdős–Hajnal property when $t = 6$). Bonamy, Bousquet and Thomassé [24] later extended the result of [25], proving that for each $t \geq 3$ the family of graphs that excludes all cycles on at least t vertices and their complements as induced subgraphs has the strong Erdős–Hajnal property.

This approach also applies quite well in combinatorial geometry, where a common problem is to show that arrangements of geometric objects have large crossing or disjoint patterns. This is usually proved by showing that the auxiliary *intersection graph*, with a vertex for each object and an edge between two vertices if the corresponding objects intersect, has a large clique or independent set. Larman, Matoušek, Pach and Törőcsik [154] proved that the family of intersection graphs of convex sets in the plane has the Erdős–Hajnal property. This was later strengthened by Fox, Pach and Tóth [113], who proved that this family has the strong Erdős–Hajnal property. Alon, Pach, Pinchasi, Radoičić and Sharir [7] proved that the family of semi-algebraic graphs of bounded description complexity has the strong Erdős–Hajnal property. This implies the existence of large patterns in many graphs that arise naturally in discrete geometry.

String graphs are intersection graphs of curves in the plane. It is still an open problem to decide whether every family of n curves in the plane contains a subfamily of size n^c whose elements are either pairwise intersecting or pairwise disjoint, i.e., whether the family S of string graphs has the Erdős–Hajnal property. The best known bound is $n^{c/\log\log n}$, due to Fox and Pach [111]. This follows by first proving that every string graph on $n \geq 2$ vertices contains a complete or empty bipartite subgraph with parts of order $\Omega(n/\log n)$. This latter result is tight up to the constant factor, so the family of string graphs does not have the strong Erdős–Hajnal property. On the other hand, Fox, Pach and Tóth [113] proved that the family S_k of intersection graphs of curves where each pair of curves intersects at most k times does have the strong Erdős–Hajnal property.

We have already noted that the strong Erdős–Hajnal property does not always hold for induced-H-free graphs. However, Erdős, Hajnal and Pach [96] proved that a bipartite analogue of the Erdős-Hajnal conjecture does hold. That is, for every graph H there is a positive constant $c(H)$ such that every induced-H-free graph on $n \geq 2$ vertices contains a complete or empty bipartite graph with parts of order $n^{c(H)}$. Using dependent random choice, Fox and Sudakov [116] proved a strengthening of this result, showing that every such graph contains a complete bipartite graph with parts of order $n^{c(H)}$ or an independent set of order $n^{c(H)}$.

In a slightly different direction, Rödl [177] showed that any graph with a forbidden induced subgraph contains a linear-sized subset which is close to being complete or empty. That is, for every graph H and every $\epsilon > 0$, there is $\delta > 0$ such that every induced-H-free graph on n vertices contains an induced subgraph on at least δn vertices with edge density at most ϵ or at least $1 - \epsilon$. Rödl's proof uses Szemerédi's regularity lemma and consequently gives a tower-type bound on δ^{-1}. Fox and Sudakov [114] proved the much better bound $\delta \geq 2^{-c|H|(\log 1/\epsilon)^2}$, which easily implies Theorem 3.11 as a corollary. They also conjectured that a polynomial dependency holds, which would in turn imply the Erdős–Hajnal conjecture.

Conjecture 3.12 *For every graph H, there is a positive constant $c(H)$ such that for every $\epsilon > 0$ every induced-H-free graph on n vertices contains an induced subgraph on $\epsilon^{c(H)}n$ vertices with density at most ϵ or at least $1 - \epsilon$.*

One of the key steps in proving Theorem 3.11 is to find, in an induced-H-free graph on n vertices, two disjoint subsets of order at least $\epsilon^c n$ for some $c = c(H) > 0$ such that the edge density between them is at most ϵ or at least $1 - \epsilon$. We wonder whether this can be improved so that one part is of linear size.

Problem 3.13 *Is it true that for every graph H there is $c = c(H) > 0$ such that for every $\epsilon > 0$ every induced-H-free graph on n vertices contains two disjoint subsets of orders cn and $\epsilon^c n$ such that the edge density between them is at most ϵ or at least $1 - \epsilon$?*

A positive answer to this question would improve the bound on the Erdős–Hajnal conjecture to $e^{c\sqrt{\log n \log \log n}}$. However, we do not even know the answer when H is a triangle. A positive answer in this case would imply the following conjecture.

Conjecture 3.14 *There is a positive constant c such that every triangle-free graph on $n \geq 2$ vertices contains disjoint subsets of orders cn and n^c with no edges between them.*

Restated, this conjecture says that there exists a positive constant c such that the Ramsey number of a triangle versus a complete bipartite graph with parts of orders cn and n^c is at most n.

There is also a multicolour generalisation of the Erdős–Hajnal conjecture.

Conjecture 3.15 *For every q-edge-coloured complete graph K, there exists a positive constant $c(K)$ such that every q-edge-colouring of the complete graph on n vertices which does not contain a copy of K has an induced subgraph on $n^{c(K)}$ vertices which uses at most $q - 1$ colours.*

The Erdős–Hajnal conjecture clearly corresponds to the case $q = 2$, as we can take the edges of our graph as one colour and the non-edges as the other colour. For $q = 3$, Fox, Grinshpun and Pach [109] proved that every rainbow-triangle-free 3-edge-colouring of the complete graph on n vertices contains a two-coloured subset with at least $cn^{1/3} \log^2 n$ vertices. This bound is tight up to the constant factor and answers a question of Hajnal [134], the construction that demonstrates tightness being the lexicographic product of three two-colourings of the complete graph on $n^{1/3}$ vertices, one for each pair of colours and each having no monochromatic clique of order $\log n$.

Alon, Pach and Solymosi [8] observed that the Erdős–Hajnal conjecture is equivalent to the following variant for tournaments. For every tournament T, there is a positive constant $c(T)$ such that every tournament on n vertices which does not contain T as a subtournament has a transitive subtournament of order $n^{c(T)}$. Recently, Berger, Choromanski and Chudnovsky [19] proved that this conjecture holds for every tournament T on at most five vertices, as well as for an infinite family of tournaments that cannot be obtained through the tournament analogue of the substitution procedure of Alon, Pach and Solymosi.

Analogues of the Erdős–Hajnal conjecture have also been studied for hypergraphs. The authors [59] proved that for $k \geq 4$ no analogue of the standard Erdős–Hajnal conjecture can hold in k-uniform hypergraphs. That is, there are k-uniform hypergraphs H and sequences of induced-H-free hypergraphs which do not contain cliques or independent sets of order appreciably larger than is guaranteed by Ramsey's theorem. The proof uses the fact that the stepping-up construction of Erdős and Hajnal has forbidden induced subgraphs.

Nevertheless, one can still show that 3-uniform hypergraphs with forbidden induced subgraphs contain some unusually large configurations. It is well known that every 3-uniform hypergraph on n vertices contains a complete or empty tripartite subgraph with parts of order $c(\log n)^{1/2}$ and a random 3-uniform hypergraph shows that this bound is tight up to the constant factor. Rödl and Schacht [182] proved that this bound can be improved by any constant factor for sufficiently large induced-H-free hypergraphs. This result was subsequently improved by the authors [59], who showed that for every 3-uniform hypergraph H there exists a positive constant $\delta(H)$ such that, for n sufficiently large, every induced-H-free 3-

uniform hypergraph on n vertices contains a complete or empty tripartite subgraph with parts of order $(\log n)^{1/2+\delta(H)}$. We believe that this bound can be improved further. If true, the following conjecture would be best possible.

Conjecture 3.16 *For every 3-uniform hypergraph H, any induced-H-free hypergraph on n vertices contains a complete or empty tripartite subgraph with parts of order $(\log n)^{1-o(1)}$.*

3.4 Size Ramsey numbers

Given a graph H, the *size Ramsey number* $\hat{r}(H)$ is defined to be the smallest m for which there exists a graph G with m edges such that G is Ramsey with respect to H, that is, such that any two-colouring of the edges of G contains a monochromatic copy of H. This concept was introduced by Erdős, Faudree, Rousseau and Schelp [88]. Since the complete graph on $r(H)$ vertices is Ramsey with respect to H, it is clear that $\hat{r}(H) \leq \binom{r(H)}{2}$. Moreover, as observed by Chvátal (see [88]), this inequality is tight when H is a complete graph. This follows easily from noting that any graph which is Ramsey with respect to K_t must have chromatic number at least $r(t)$.

The most famous result in this area is the following rather surprising theorem of Beck [16], which says that the size Ramsey number of a path is linear in the number of vertices. Here P_n is the path with n vertices.

Theorem 3.17 *There exists a constant c such that $\hat{r}(P_n) \leq cn$.*

This result, which answered a question of Erdős, Faudree, Rousseau and Schelp [88] (see also [83]), was later extended to trees of bounded maximum degree [122] and to cycles [137]. For a more general result on the size Ramsey number of trees, we refer the reader to the recent work of Dellamonica [69].

Beck [17] raised the question of whether this result could be generalised to graphs of bounded maximum degree. That is, he asked whether for any Δ there exists a constant c, depending only on Δ, such that any graph on n vertices with maximum degree Δ has size Ramsey number at most cn. This question was answered in the negative by Rödl and Szemerédi [184], who proved that there are already graphs of maximum degree 3 with superlinear size Ramsey number.

Theorem 3.18 *There are positive constants c and α and, for every n, a graph H with n vertices and maximum degree 3 such that*

$$\hat{r}(H) \geq cn(\log n)^{\alpha}.$$

On the other hand, a result of Kohayakawa, Rödl, Schacht and Szemerédi [145] shows that the size Ramsey number of graphs with bounded maximum degree is subquadratic.

Theorem 3.19 *For every natural number* Δ, *there exists a constant* c_Δ *such that any graph* H *on* n *vertices with maximum degree* Δ *satisfies*

$$\hat{r}(H) \leq c_\Delta n^{2-1/\Delta}(\log n)^{1/\Delta}.$$

We are not sure where the truth lies, though it seems likely that Theorem 3.18 can be improved by a polynomial factor. This was formally conjectured by Rödl and Szemerédi [184].

Conjecture 3.20 *For every natural number* $\Delta \geq 3$, *there exists a constant* $\epsilon > 0$ *such that for all sufficiently large* n *there is a graph* H *on* n *vertices with maximum degree* Δ *for which* $\hat{r}(H) \geq n^{1+\epsilon}$.

More generally, given a real-valued graph parameter f, we may define the f-*Ramsey number* $r_f(H)$ of H to be the minimum value of $f(G)$, taken over all graphs G which are Ramsey with respect to H. The usual Ramsey number is the case where $f(G) = v(G)$, while the size Ramsey number is the case where $f(G) = e(G)$. However, there have also been studies of other variants, such as the *chromatic Ramsey number* $r_\chi(H)$, where $f(G) = \chi(G)$, and the *degree Ramsey number* $r_\Delta(H)$, where $f(G) = \Delta(G)$. We will point out one result concerning the first parameter and a problem concerning the second.

The chromatic Ramsey number was introduced by Burr, Erdős and Lovász [33], who observed that any graph H with chromatic number t has $r_\chi(H) \geq (t-1)^2 + 1$ and conjectured that there are graphs of chromatic number t for which this bound is sharp. In their paper, they outlined a proof of this conjecture based on the still unproven Hedetniemi conjecture, which concerns the chromatic number of the tensor product of graphs. Recently, Zhu [214] proved a fractional version of the Hedetniemi conjecture, which, by an observation of Paul and Tardif [171], was sufficient to establish the conjecture.

Theorem 3.21 *For every natural number* t, *there exists a graph* H *of chromatic number* t *such that*

$$r_\chi(H) = (t-1)^2 + 1.$$

The outstanding open problem concerning the degree Ramsey number is the following, which seems to have been first noted by Kinnersley, Milans and West [143].

Problem 3.22 *Is it true that for every* $\Delta \geq 3$, *there exists a natural number* Δ' *such that* $r_\Delta(H) \leq \Delta'$ *for every graph H of maximum degree* Δ?

We suspect that the answer is no, but the problem appears to be difficult. For $\Delta = 2$, the answer is yes (see, for example, [138]).

An on-line variant of the size Ramsey number was introduced by Beck [18] and, independently, by Kurek and Ruciński [152]. It is best described as a game between two players, known as Builder and Painter. Builder draws a sequence of edges and, as each edge appears, Painter must colour it in either red or blue. Builder's goal is to force Painter to draw a monochromatic copy of some fixed graph H. The smallest number of turns needed by Builder to force Painter to draw a monochromatic copy of H is known as the *on-line Ramsey number* of H and denoted $\tilde{r}(H)$. As usual, we write $\tilde{r}(t)$ for $\tilde{r}(K_t)$.

The basic question in this area, attributed to Rödl (see [152]), is to show that $\lim_{t\to\infty} \tilde{r}(t)/\hat{r}(t) = 0$. Put differently, we would like to show that $\tilde{r}(t) = o(\binom{r(t)}{2})$. This conjecture remains open (and is probably difficult), but the following result, due to Conlon [47], shows that the on-line Ramsey number $\tilde{r}(t)$ is exponentially smaller than the size Ramsey number $\hat{r}(t)$ for infinitely many values of t.

Theorem 3.23 *There exists a constant* $c > 1$ *such that for infinitely many* t,

$$\tilde{r}(t) \leq c^{-t}\binom{r(t)}{2}.$$

On-line analogues of f-Ramsey numbers were considered by Grytczuk, Hałuszczak and Kierstead [132]. The most impressive result in this direction, proved by Grytczuk, Hałuszczak, Kierstead and Konjevod over two papers [132, 140], says that Builder may force Painter to draw a monochromatic copy of any graph with chromatic number t while only exposing a graph of chromatic number t herself. We also note that the on-line analogue of Problem 3.22 was studied in [36] but again seems likely to have a negative answer for $\Delta \geq 3$ (though we refer the interested reader to [61] for a positive answer to the analogous question when maximum degree is replaced by degeneracy).

3.5 Generalised Ramsey numbers

In this section, we will consider two generalisations of the usual Ramsey function, both of which have been referred to in the literature as generalised Ramsey numbers.

3.5.1 The Erdős–Gyárfás function
Let p and q be positive integers with $2 \leq q \leq \binom{p}{2}$. An edge colouring of the complete graph K_n is said to be a (p,q)-colouring if every K_p receives at least q different colours. The function $f(n,p,q)$ is defined to be the minimum number of colours that are needed for K_n to have a (p,q)-colouring. This function generalises the usual Ramsey function, as may be seen by noting that $f(n,p,2)$ is the minimum number of colours needed to guarantee that no K_p is monochromatic. In particular, if we invert the bounds $2^s \leq r(3;s) \leq es!$, we get

$$c' \frac{\log n}{\log \log n} \leq f(n,3,2) \leq c \log n.$$

This function was first introduced by Erdős and Shelah [81, 82] and studied in depth by Erdős and Gyárfás [90], who proved a number of interesting results, demonstrating how the function falls off from being equal to $\binom{n}{2}$ when $q = \binom{p}{2}$ and $p \geq 4$ to being at most logarithmic when $q = 2$. They also determined ranges of p and q where the function $f(n,p,q)$ is linear in n, where it is quadratic in n and where it is asymptotically equal to $\binom{n}{2}$. Many of these results were subsequently strengthened by Sárközy and Selkow [186, 187].

One simple observation of Erdős and Gyárfás is that $f(n,p,p)$ is always polynomial in n. To see this, it is sufficient to show that a colouring with fewer than $n^{1/(p-2)} - 1$ colours contains a K_p with at most $p-1$ colours. For $p = 3$, this follows since one only needs that some vertex has at least two neighbours in the same colour. For $p = 4$, we have that any vertex will have at least $n^{1/2}$ neighbours in some fixed colour. But then there are fewer than $n^{1/2} - 1$ colours on this neighbourhood of order at least $n^{1/2}$, so the $p = 3$ case implies that it contains a triangle with at most two colours. The general case follows similarly.

Erdős and Gyárfás [90] asked whether this result is best possible, that is, whether $q = p$ is the smallest value of q for which $f(n,p,q)$ is polynomial in n. For $p = 3$, this is certainly true, since we know that $f(n,3,2) \leq c \log n$. However, for general p, they were only able to show that $f(n,p,\lceil \log p \rceil)$ is subpolynomial. This left the question of determining whether $f(n,p,p-1)$ is subpolynomial wide open, even for $p = 4$.

The first progress on this question was made by Mubayi [162], who found a $(4,3)$-colouring of K_n with only $e^{c\sqrt{\log n}}$ colours, thus showing that $f(n,4,3) \leq e^{c\sqrt{\log n}}$. Later, Eichhorn and Mubayi [77] showed that this colouring is also a $(5,4)$-colouring and, more generally, a $(p, 2\lceil \log p \rceil - 2)$-colouring for all $p \geq 5$. It will be instructive to describe this colouring (or rather a slight variant).

Given n, let t be the smallest integer such that $n \leq 2^{t^2}$ and $m = 2^t$.

We consider the vertex set $[n]$ as a subset of $[m]^t$. For two vertices $x = (x_1, \ldots, x_t)$ and $y = (y_1, \ldots, y_t)$, let

$$c_M(x, y) = \Big(\{x_i, y_i\}, a_1, \ldots, a_t \Big),$$

where i is the minimum index in which x and y differ and $a_j = 0$ or 1 depending on whether $x_j = y_j$ or not. Since $2^{(t-1)^2} < n$, the total number of colours used is at most

$$m^2 \cdot 2^t = 2^{3t} < 2^{3(1+\sqrt{\log n})} \leq 2^{6\sqrt{\log n}}.$$

Hence, c_M uses at most $2^{6\sqrt{\log n}}$ colours to colour the edge set of the complete graph K_n. The proof that c_M is a $(4,3)$-colouring is a straightforward case analysis which we leave as an exercise. We have already noted that it is also a $(5,4)$-colouring. However, as observed in [51], it cannot be a $(p, p-1)$-colouring for all p.

Nevertheless, in a recent paper, Conlon, Fox, Lee and Sudakov [51] found a way to extend this construction and answer the question of Erdős and Gyárfás for all p. Stated in a quantitative form (though one which we expect to be very far from best possible), this result is as follows.

Theorem 3.24 *For any natural number $p \geq 4$, there exists a constant c_p such that*
$$f(n, p, p-1) \leq 2^{c_p(\log n)^{1-1/(p-2)}}.$$

Our quantitative understanding of these functions is poor, even for $f(n, 4, 3)$. Improving a result of Kostochka and Mubayi [146], Fox and Sudakov [115] showed that $f(n, 4, 3) \geq c' \log n$ for some positive constant c'. Though a substantial improvement on the trivial bound $f(n, 4, 3) \geq f(n, 4, 2) \geq c' \log n / \log \log n$, it still remains very far from the upper bound of $e^{c\sqrt{\log n}}$. We suspect that the upper bound may be closer to the truth. An answer to the following question would be a small step in the right direction.

Problem 3.25 *Show that $f(n, 4, 3) = \omega(\log n)$.*

For $p \geq k+1$ and $2 \leq q \leq \binom{p}{k}$, we define the natural hypergraph generalisation $f_k(n, p, q)$ as the minimum number of colours that are needed for $K_n^{(k)}$ to have a (p, q)-colouring, where here a (p, q)-colouring means that every $K_p^{(k)}$ receives at least q distinct colours. As in the graph case, it is comparatively straightforward to show that $f_k(n, p, \binom{p-1}{k-1} + 1)$ is polynomial in n for all $p \geq k+1$. With Lee [52], we conjecture the following.

Conjecture 3.26 $f_k(n, p, \binom{p-1}{k-1}))$ *is subpolynomial for all* $p \geq k + 1$.

Theorem 3.24 addresses the $k = 2$ case, while the cases where $k = 3$ and $p = 4$ and 5 were addressed in [52]. These cases already require additional ideas beyond those used to resolve the graph case. The case where $k = 3$ and $p = 4$ is of particular interest, because it is closely related to Shelah's famous primitive recursive bound for the Hales–Jewett theorem [192].

Shelah's proof relied in a crucial way on a lemma now known as the Shelah cube lemma. The simplest case of this lemma concerns the *grid graph* $\Gamma_{m,n}$, the graph on vertex set $[m] \times [n]$ where two distinct vertices (i, j) and (i', j') are adjacent if and only if either $i = i'$ or $j = j'$. That is, $\Gamma_{m,n}$ is the Cartesian product $K_m \times K_n$. A *rectangle* in $\Gamma_{m,n}$ is a copy of $K_2 \times K_2$, that is, an induced subgraph over a vertex subset of the form $\{(i, j), (i', j), (i, j'), (i', j')\}$ for some integers $1 \leq i < i' \leq m$ and $1 \leq j < j' \leq n$. We will denote this rectangle by (i, j, i', j'). For an edge-coloured grid graph, an *alternating rectangle* is a rectangle (i, j, i', j') such that the colour of the edges $\{(i, j), (i', j)\}$ and $\{(i, j'), (i', j')\}$ are equal and the colour of the edges $\{(i, j), (i, j')\}$ and $\{(i', j), (i', j')\}$ are equal, that is, opposite sides of the rectangle receive the same colour. The basic case of Shelah's lemma, which we refer to as the grid Ramsey problem, asks for an estimate on $G(r)$, the smallest n such that every r-colouring of the edges of $\Gamma_{n,n}$ contains an alternating rectangle.

It is easy to show that $G(r) \leq r^{\binom{r+1}{2}} + 1$. Indeed, let $n = r^{\binom{r+1}{2}} + 1$ and suppose that an r-colouring of $\Gamma_{r+1,n}$ is given. Since each column is a copy of K_{r+1}, there are at most $r^{\binom{r+1}{2}}$ ways to colour the edges of a fixed column with r colours. Since $n > r^{\binom{r+1}{2}}$, the pigeonhole principle implies that there are two columns which are identically coloured. Let these columns be the j-th column and the j'-th column and consider the edges that connect these two columns. Since there are $r + 1$ rows, the pigeonhole principle implies that there are i and i' such that the edges $\{(i, j), (i, j')\}$ and $\{(i', j), (i', j')\}$ have the same colour. Since the edges $\{(i, j), (i', j)\}$ and $\{(i, j'), (i', j')\}$ also have the same colour, the rectangle (i, j, i', j') is alternating.

This argument is very asymmetrical and yet the resulting bound on $G(r)$ remains essentially the best known. The only improvement, due to Gyárfás [133], is $G(r) \leq r^{\binom{r+1}{2}} - r^{\binom{r-1}{2}+1} + 1$. Though it seems likely that $G(r)$ is significantly smaller than this, the following problem already appears to be difficult.

Problem 3.27 *Show that* $G(r) = o(r^{\binom{r+1}{2}})$.

In the second edition of their book on Ramsey theory [130], Graham, Rothschild and Spencer suggested that $G(r)$ may even be polynomial in r. This was recently disproved by Conlon, Fox, Lee and Sudakov [52], who showed the following.

Theorem 3.28 *There exists a positive constant c such that*

$$G(r) > 2^{c(\log r)^{5/2}/\sqrt{\log\log r}}.$$

To see how this relates back to estimating $f_3(n, 4, 3)$, we let $g(n)$ be the inverse function of $G(r)$, defined as the minimum integer s for which there exists an s-colouring of the edges of $\Gamma_{n,n}$ with no alternating rectangle. Letting $K^{(3)}(n, n)$ be the 3-uniform hypergraph with vertex set $A \cup B$, where $|A| = |B| = n$, and edge set consisting of all those triples which intersect both A and B, we claim that $g(n)$ is within a factor of two of the minimum integer r for which there exists an r-colouring of the edges of $K^{(3)}(n, n)$ such that any copy of $K_4^{(3)}$ has at least three colours on its edges.

To prove this claim, we define a bijection between the edges of $\Gamma_{n,n}$ and the edges of $K^{(3)}(n, n)$ such that the rectangles of $\Gamma_{n,n}$ are in one-to-one correspondence with the copies of $K_4^{(3)}$ in $K^{(3)}(n, n)$. For $i \in A$ and $j, j' \in B$, we map the edge (i, j, j') of $K^{(3)}(n, n)$ to the edge $\{(i, j), (i, j')\}$ of $\Gamma_{n,n}$ and, for $i, i' \in A$ and $j \in B$, we map the edge (i, i', j) of $K^{(3)}(n, n)$ to the edge $\{(i, j), (i', j)\}$ of $\Gamma_{n,n}$. Given a colouring of $K^{(3)}(n, n)$ where every $K_4^{(3)}$ receives at least three colours, this correspondence gives a colouring of $\Gamma_{n,n}$ where every rectangle receives at least three colours, showing that $g(n) \leq r$. Similarly, given a colouring of $\Gamma_{n,n}$ with no alternating rectangles, we may double the number of colours to ensure that the set of colours used for row edges is disjoint from the set used for column edges. This gives a colouring where every $K_4^{(3)}$ receives at least three colours, so $r \leq 2g(n)$.

Therefore, essentially the only difference between $g(n)$ and $f_3(2n, 4, 3)$ is that the base hypergraph for $g(n)$ is $K^{(3)}(n, n)$ rather than $K_{2n}^{(3)}$. This observation allows us to show that

$$g(n) \leq f_3(2n, 4, 3) \leq 2\lceil \log n \rceil^2 g(n).$$

In particular, this allows us to establish a subpolynomial upper bound for $f_3(n, 4, 3)$.

More generally, Shelah's work on the Hales–Jewett theorem requires an estimate for the function $f_{2d-1}(n, 2d, d+1)$. If the growth rate of these functions was bounded below by, say, $c_d' \log \log \log n$, then it might be possible to give a tower-type bound for Hales–Jewett numbers. However, we expect that this is not the case.

Problem 3.29 *Show that for all s, there exist d and n_0 such that*

$$f_{2d-1}(n, 2d, d+1) \leq \underbrace{\log \log \ldots \log \log}_{s} n$$

for all $n \geq n_0$.

We conclude this section with one further problem which arose in studying $f(n, p, q)$ and its generalisations. Mubayi's colouring c_M was originally designed to have the property that the union of any two colour classes contains no K_4. However, in [52], it was shown to have the stronger property that the union of any two colour classes has chromatic number at most three. We suspect that this property can be generalised.

Problem 3.30 *Let $p \geq 5$ be an integer. Does there exist an edge colouring of K_n with $n^{o(1)}$ colours such that the union of every $p-1$ colour classes has chromatic number at most p?*

For $p = 4$, Mubayi's colouring again has the desired property, though it is known that it cannot work for all p. However, it may be that the colourings used in the proof of Theorem 3.24 suffice.

3.5.2 The Erdős–Rogers function

Given an integer $s \geq 2$, a set of vertices U in a graph G is said to be *s-independent* if $G[U]$ contains no copy of K_s. When $s = 2$, this simply means that U is an independent set in G. We write $\alpha_s(G)$ for the order of the largest s-independent subset in a graph G.

The problem of estimating Ramsey numbers can be rephrased as a problem about determining the minimum independence number over all K_t-free graphs with a given number of vertices. In 1962, Erdős and Rogers [100] initiated the study of the more general question obtained by replacing the notion of independence number with the s-independence number. Suppose $2 \leq s \leq t < n$ are integers. Erdős and Rogers defined

$$f_{s,t}(n) = \min \alpha_s(G),$$

where the minimum is taken over all K_t-free graphs G on n vertices. In particular, for $s = 2$, we have $f_{2,t}(n) < \ell$ if and only if the Ramsey number $r(\ell, t)$ satisfies $r(\ell, t) > n$.

The first lower bound for $f_{s,t}$ was given by Bollobás and Hind [23], who proved that $f_{s,t}(n) \geq n^{1/(t-s+1)}$. Their proof is by induction on t. When

$t = s$, the bound holds trivially, since the graph contains no K_s. Now suppose that G is an n-vertex graph with no K_t and let v be a vertex of maximum degree. If $|N(v)| \geq n^{\frac{t-s}{t-s+1}}$, then we can apply induction to the subgraph of G induced by this set, since this subgraph is clearly K_{t-1}-free. Otherwise, by Brooks' theorem, the independence number of G is at least $n/|N(v)| \geq n^{1/(t-s+1)}$. The bound in this argument can be improved by a polylogarithmic factor using a result of Shearer [191] on the independence number of K_t-free graphs. As was pointed out by Bollobás and Hind [23], this proof usually finds an independent set rather than an s-independent set. Another approach, which better utilises the fact that we are looking for an s-independent set, was proposed by Sudakov [203].

To illustrate this approach, we show that $f_{3,5}(n) \geq cn^{2/5}$ for some constant $c > 0$, improving on the bound of $n^{1/3}$ given above. Let G be a K_5-free graph on n vertices and assume that it does not contain a 3-independent subset of order $n^{2/5}$. For every edge (u, v) of G, the set of common neighbours $N(u, v)$ is triangle-free. Therefore, we may assume that it has order less than $n^{2/5}$. Moreover, for any vertex v, its set of neighbours $N(v)$ is K_4-free. But, by the Bollobás–Hind bound, $N(v)$ contains a triangle-free subset of order $|N(v)|^{1/2}$. Therefore, if there is a vertex v of degree at least $n^{4/5}$, there will be a triangle-free subset of order $|N(v)|^{1/2} \geq n^{2/5}$. Hence, we may assume that all degrees in G are less than $n^{4/5}$. This implies that every vertex in G is contained in at most $n^{4/5} \cdot n^{2/5} = n^{6/5}$ triangles.

We now consider the auxiliary 3-uniform hypergraph H on the same vertex set as G whose edges are the triangles in G. Crucially, an independent set in H is a 3-independent set in G. The number m of edges in H satisfies $m \leq n \cdot n^{6/5} = n^{11/5}$. Therefore, using a well-known bound on the independence number of 3-uniform hypergraphs, we conclude that $\alpha_3(G) = \alpha(H) \geq cn^{3/2}/\sqrt{m} \geq cn^{2/5}$. This bound can be further improved by combining the above argument with a variant of dependent random choice. Using this approach, Sudakov [204] showed that $f_{3,5}(n)$ is at least $n^{5/12}$ times a polylogarithmic factor. For $t > s + 1$, he also proved that $f_{s,t}(n) = \Omega(n^{a_t})$, where $a_t(s)$ is roughly $s/2t + O_s(t^{-2})$. More precisely, he showed the following.

Theorem 3.31 *For any $s \geq 3$ and $t > s + 1$, $f_{s,t}(n) = \Omega(n^{a_t})$, where*

$$\frac{1}{a_t} = 1 + \frac{1}{s-1} \sum_{i=1}^{s-1} \frac{1}{a_{t-i}}, \quad a_{s+1} = \frac{3s-4}{5s-6} \quad and \quad a_3 = \cdots = a_s = 1.$$

The study of upper bounds for $f_{s,t}(n)$ goes back to the original paper of Erdős and Rogers [100]. They considered the case where s and $t =$

$s + 1$ are fixed and n tends to infinity, proving that there exists a positive constant $\epsilon(s)$ such that $f_{s,s+1}(n) \leq n^{1-\epsilon(s)}$. That is, they found a K_{s+1}-free graph of order n such that every induced subgraph of order $n^{1-\epsilon(s)}$ contains a copy of K_s. About thirty years later, Bollobás and Hind [23] improved the estimate for $\epsilon(s)$. This bound was then improved again by Krivelevich [150], who showed that

$$f_{s,t}(n) \leq cn^{\frac{s}{t+1}}(\log n)^{\frac{1}{s-1}},$$

where c is some constant depending only on s and t. Note that this upper bound is roughly the square of the lower bound from [204]. We also note that all of the constructions mentioned above rely on applications of the probabilistic method, but explicit constructions showing that $f_{s,s+1}(n) \leq n^{1-\epsilon(s)}$ were obtained by Alon and Krivelevich [4].

One of the most intriguing problems in this area concerned the case where $t = s + 1$. For many years, the best bounds for this question were very far apart, the lower bound being roughly $n^{1/2}$ and the upper bound being $n^{1-\epsilon(s)}$, with $\epsilon(s)$ tending to zero as s tends to infinity. Both Krivelevich [150] and Sudakov [204] asked whether the upper bound is closer to the correct order of magnitude for $f_{s,s+1}(n)$. Quite surprisingly, this was recently disproved in a sequence of three papers. First, Dudek and Rödl [74] proved that $f_{s,s+1}(n) = O(n^{2/3})$. Then Wolfovitz [213], building on their work but adding further ideas, managed to show that the lower bound for $f_{3,4}(n)$ is correct up to logarithmic factors. Finally, Dudek, Retter and Rödl [72], extending the approach from [213], proved that $f_{s,s+1}(n) = n^{1/2+o(1)}$. More explicitly, they proved the following.

Theorem 3.32 *For every $s \geq 3$, there exists a constant c_s such that*

$$f_{s,s+1}(n) \leq c_s(\log n)^{4s^2}\sqrt{n}.$$

It would be interesting to close the gap between this and the best lower bound, observed by Dudek and Mubayi [71], which stands at

$$f_{s,s+1}(n) \geq c_s'\left(\frac{n\log n}{\log\log n}\right)^{1/2}.$$

We will now sketch the neat construction from [74], showing that $f_{3,4}(n) = O(n^{2/3})$. Let p be a prime, $n = p^3 + p^2 + p + 1$ and let $L_1, \ldots, L_n$ be the lines of a generalised quadrangle. The reader not familiar with this concept may consult [123]. For our purposes, it will be sufficient to note that this is a collection of points and lines with the following two properties:

- every line is a subset of $[n]$ of order $p+1$ and every vertex in $[n]$ lies on $p+1$ lines;

- any two vertices belong to at most one line and every three lines with non-empty pairwise intersection have one point in common (i.e., every triangle of lines is degenerate).

We construct a random graph G on $[n]$ as follows. Partition the vertex set of every line L_i into three parts $L_{i,j}, 1 \leq j \leq 3$, uniformly at random. Take a complete 3-partite graph on these parts and let G be the union of all such graphs for $1 \leq i \leq n$. Note that the second property above implies that the vertices of every triangle in G belong to some line. This easily implies that G is K_4-free. Consider now an arbitrary subset X of G of order $6p^2$ and let $x_i = |L_i \cap X|$. If X contains no triangles, then, for every i, there is an index j such that the set $L_{i,j} \cap X$ is empty. The probability that this happens for a fixed i is at most $3(2/3)^{x_i}$. Therefore, since these events are independent for different lines, the probability that X is triangle-free is at most $3^n(2/3)^{\sum x_i}$. Since every vertex lies on $p+1$ lines, we have that $\sum x_i = (p+1)|X| > 5n$. Since the number of subsets X is at most 2^n and $2^n 3^n (2/3)^{5n} \ll 1$, we conclude that with probability close to one every subset of G of order at least $10n^{2/3} > 6p^2$ contains a triangle.

There are many open problems remaining regarding the Erdős–Rogers function. For example, it follows from the work of Sudakov [204] and Dudek, Retter and Rödl [72] that for any $\epsilon > 0$ there exists s_0 such that if $s \geq s_0$, then

$$c'n^{1/2-\epsilon} \leq f_{s,s+2}(n) \leq cn^{1/2}$$

for some positive constants c' and c. It remains to decide if the upper bound can be improved for fixed values of s. The following question was posed by Dudek, Retter and Rödl [72].

Problem 3.33 *For any $s \geq 3$, is it true that $f_{s,s+2}(n) = o(\sqrt{n})$?*

The hypergraph generalisation of the Erdős–Rogers function was first studied by Dudek and Mubayi [71]. For $s \leq t$, let $f_{s,t}^{(k)}(n)$ be given by

$$f_{s,t}^{(k)}(n) = \min\{\max\{|W| : W \subseteq V(G) \text{ and } G[W] \text{ contains no } K_s^{(k)}\}\},$$

where the minimum is taken over all $K_t^{(k)}$-free k-uniform hypergraphs G on n vertices. Dudek and Mubayi proved the following.

Theorem 3.34 *For any $s \geq 3$ and $t \geq s+1$,*

$$f_{s-1,t-1}(\lfloor \sqrt{\log n} \rfloor) \leq f_{s,t}^{(3)}(n) \leq c_s \log n.$$

In particular, for $t = s + 1$, this gives constants c_1 and c_2 depending only on s such that

$$c_1 (\log n)^{1/4} \left(\frac{\log \log n}{\log \log \log n} \right)^{1/2} \leq f_{s,s+1}^{(3)}(n) \leq c_2 \log n.$$

The lower bound was subsequently improved by the authors [61], using ideas on hypergraph Ramsey numbers developed in [55].

Theorem 3.35 *For any natural number* $s \geq 3$, *there exists a positive constant* c *such that*

$$f_{s,s+1}^{(3)}(n) \geq c \left(\frac{\log n}{\log \log \log n} \right)^{1/3}.$$

This result easily extends to higher uniformities to give $f_{s,s+1}^{(k)}(n) \geq (\log_{(k-2)} n)^{1/3 - o(1)}$, where $\log_{(0)} x = x$ and $\log_{(i+1)} x = \log(\log_{(i)} x)$. This improves an analogous result of Dudek and Mubayi [71] with a $1/4$ in the exponent but is far from their upper bound $f_{s,s+1}^{(k)}(n) \leq c_{s,k}(\log n)^{1/(k-2)}$. It would be interesting to close the gap between the upper and lower bounds. In particular, we have the following problem.

Problem 3.36 *Is it the case that*

$$f_{s,s+1}^{(4)}(n) = (\log n)^{o(1)}?$$

3.6 Monochromatic cliques with additional structure

There are a number of variants of the classical Ramsey question which ask for further structure on the monochromatic cliques being found. The classic example of such a theorem is the Paris–Harrington theorem [135], which says that for any t, k and q, there exists an N such that any q-colouring of the edges of the complete k-uniform hypergraph on the set $\{1, 2, \ldots, N\}$ contains a monochromatic $K_s^{(k)}$ with vertices $a_1 < \cdots < a_s$ for which $s \geq \max\{t, a_1\}$. That is, the clique is at least as large as its minimal element. This theorem, which follows easily from a compactness argument, is famous for being a natural statement which is not provable in Peano arithmetic (though we note that for graphs and two colours, the function is quite well behaved and grows as a double exponential in t [160]). In this section, we will discuss two decidedly less pathological strengthenings of Ramsey's theorem.

3.6.1   Weighted cliques   In the early 1980s, Erdős considered the following variant of Ramsey's theorem. For a finite set S of integers greater than one, define its weight $w(S)$ by

$$w(S) = \sum_{s \in S} \frac{1}{\log s},$$

where, as usual, log is assumed to be base 2. For a red/blue-colouring c of the edges of the complete graph on $[2,n] = \{2,\ldots,n\}$, let $f(c)$ be the maximum weight $w(S)$ taken over all sets $S \subset [2,n]$ which form a monochromatic clique in the colouring c. For each integer $n \geq 2$, let $f(n)$ be the minimum of $f(c)$ over all red/blue-colourings c of the edges of the complete graph on $\{2,\ldots,n\}$.

Erdős [83] conjectured that $f(n)$ tends to infinity, choosing this particular weight function because the standard bound $r(t) \leq 2^{2t}$ only allows one to show that $f(n) \geq \frac{\log n}{2} \cdot \frac{1}{\log n} = \frac{1}{2}$. Erdős' conjecture was verified by Rödl [178], who proved that there exist positive constants c and c' such that

$$c' \frac{\log\log\log\log n}{\log\log\log\log\log n} \leq f(n) \leq c \log\log\log n.$$

To prove Rödl's upper bound, we cover the interval $[2,n]$ by $s = \lfloor \log\log n \rfloor + 1$ intervals, where the ith interval is $[2^{2^{i-1}}, 2^{2^i})$. Using the bound $r(t) \geq 2^{t/2}$, we can colour the edges of the complete graph on the ith interval so that the maximum monochromatic clique in this interval has order 2^{i+1}. Since the log of any element in this interval is at least 2^{i-1}, the maximum weight of any monochromatic clique is at most 4. If we again use the lower bound on $r(t)$, we see that there is a red/blue-colouring of the edges of the complete graph on vertex set $\{1, 2, \ldots, s\}$ whose largest monochromatic clique is of order $O(\log s)$. Colour the edges of the complete bipartite graph between the ith and jth interval by the colour of edge (i,j) in this colouring. We get a red/blue-colouring of the edges of the complete graph on $[2,n]$ such that any monochromatic clique in this colouring has a non-empty intersection with at most $O(\log s)$ intervals. Since every interval can contribute at most 4 to the weight of this clique, the total weight of any monochromatic clique is $O(\log s) = O(\log\log\log n)$.

Answering a further question of Erdős, the authors [60] showed that this upper bound is tight up to a constant factor. The key idea behind the proof is to try to force the type of situation that arises in the upper bound construction. In practice, this means that we split our graph into intervals $I_1, \ldots, I_s$ of the form $[2^{2^{i-1}}, 2^{2^i})$ and, for each $i = 1, \ldots, s$, we find a subset $I_i' \subset I_i$ such that I_i' is the union of a red and a blue clique and all edges between I_i' and I_j' are monochromatic for each $1 \leq i < j \leq s$.

In broad outline, this was also the method used by Rödl to prove his lower bound but our proof uses two additional ingredients, dependent random choice and a certain weighted version of Ramsey's theorem.

Theorem 3.37 *For n sufficiently large, every two-colouring of the edges of the complete graph on the interval $\{2, \ldots, n\}$ contains a monochromatic clique with vertex set S such that*

$$\sum_{s \in S} \frac{1}{\log s} \geq 2^{-8} \log \log \log n.$$

Hence, $f(n) = \Theta(\log \log \log n)$.

It also makes sense to consider the function $f_q(n)$, defined now as the minimum over all q-colourings of the edges of the complete graph on $\{2, 3, \ldots, n\}$ of the maximum weight of a monochromatic clique. However, as observed by Rödl, the analogue of Erdős' conjecture for three colours does not hold. To see this, we again cover the interval $[2, n]$ by $s = \lfloor \log \log n \rfloor + 1$ intervals of the form $[2^{2^{i-1}}, 2^{2^i})$. The edges inside these intervals are coloured red and blue as in the previous construction, while the edges between the intervals are coloured green. But then the maximum weight of any red or blue clique is at most 4 and the maximum weight of any green clique is at most $\sum_{i \geq 1} 2^{-i+1} = 2$.

We may also ask whether there are other weight functions for which an analogue of Rödl's result holds. If $w(i)$ is a weight function defined on all positive integers $n \geq a$, we let $f(n, w)$ be the minimum over all red/blue-colourings of the edges of the complete graph on $[a, n]$ of the maximum weight of a monochromatic clique. In particular, if $w_1(i) = 1/\log i$ and $a = 2$, then $f(n, w_1) = f(n)$.

The next interesting case is when $w_2(i) = 1/\log i \log \log \log i$, since, for any function $u(i)$ which tends to infinity with i, Theorem 3.37 implies that $f(n, u') \to \infty$, where $u'(i) = u(i)/\log i \log \log \log i$. To derive a lower bound for $f(n, w_2)$, we colour the interval $I_i = [2^{2^{i-1}}, 2^{2^i})$ so that the largest clique has order at most 2^{i+1}. Then the contribution of the ith interval will be $O(1/\log i)$. If we now treat I_i as though it were a vertex of weight $1/\log i$, we may blow up Rödl's colouring and colour monochromatically between the I_i so that the weight of any monochromatic clique is $O(\log \log \log s) = O(\log \log \log \log \log n)$. This bound is also sharp [60], that is, $f(n, w_2) = \Theta(\log \log \log \log \log n)$.

More generally, we have the following theorem, which essentially determines the boundary below which $f(n, \cdot)$ converges. Here $\log_{(i)}(x)$ is again the iterated logarithm given by $\log_{(0)} x = x$ and $\log_{(i+1)} x = \log(\log_{(i)} x)$.

Theorem 3.38 *Let* $w_s(i) = 1/\prod_{j=1}^{s} \log_{(2j-1)} i$. *Then*

$$f(n, w_s) = \Theta(\log_{(2s+1)} n).$$

However, if $w_s'(i) = w_s(i)/(\log_{(2s-1)} i)^\epsilon$ *for any fixed* $\epsilon > 0$, $f(n, w_s')$ *converges.*

3.6.2 Cliques of fixed order type

Motivated by an application in model theory, Väänänen [168] asked whether, for any positive integers t and q and any permutation π of $[t-1] = \{1, 2, \ldots, t-1\}$, there is a positive integer R such that every q-colouring of the edges of the complete graph on vertex set $[R]$ contains a monochromatic K_t with vertices $a_1 < \cdots < a_t$ satisfying

$$a_{\pi(1)+1} - a_{\pi(1)} > a_{\pi(2)+1} - a_{\pi(2)} > \cdots > a_{\pi(t-1)+1} - a_{\pi(t-1)}.$$

That is, we want the set of consecutive differences $\{a_{i+1} - a_i : 1 \leq i \leq t-1\}$ to have a prescribed order. The least such positive integer R is denoted by $R_\pi(t; q)$ and we let $R(t; q) = \max_\pi R_\pi(t; q)$, where the maximum is over all permutations π of $[t-1]$.

Väänänen's question was answered positively by Alon [168] and, independently, by Erdős, Hajnal and Pach [95]. Alon's proof uses the Gallai–Witt theorem and so gives a weak bound on $R(t; q)$, whereas the proof of Erdős, Hajnal and Pach uses a compactness argument and gives no bound at all. Later, Alon, Shelah and Stacey all found proofs giving tower-type bounds for $R(t; q)$, but these were never published, since a double-exponential upper bound $R(t; q) \leq 2^{(q(t+1)^3)^{qt}}$ was then found by Shelah [193].

A natural conjecture, made by Alon (see [193]), is that for any q there exists a constant c_q such that $R(t; q) \leq 2^{c_q t}$. For the trivial permutation, this was confirmed by Alon and Spencer. For a general permutation, the best known bound, due to the authors [60], is as follows. Once again, dependent random choice plays a key role in the proof.

Theorem 3.39 *For any positive integers t and q and any permutation π of $[t-1]$, every q-colouring of the edges of the complete graph on vertex set $[R]$ with $R = 2^{t^{20q}}$ contains a monochromatic K_t with vertices $a_1 < \cdots < a_t$ satisfying*

$$a_{\pi(1)+1} - a_{\pi(1)} > a_{\pi(2)+1} - a_{\pi(2)} > \cdots > a_{\pi(t-1)+1} - a_{\pi(t-1)}.$$

That is, $R(t; q) \leq 2^{t^{20q}}$.

There are several variants of Väänänen's question which have negative answers. For example, the natural hypergraph analogue fails. To see this, we colour an edge (a_1, a_2, a_3) with $a_1 < a_2 < a_3$ red if $a_3 - a_2 \geq a_2 - a_1$ and blue otherwise. Hence, if the subgraph with vertices $a_1 < \cdots < a_t$ is monochromatic, the sequence $a_2 - a_1, \ldots, a_t - a_{t-1}$ must be monotone increasing or decreasing, depending on whether the subgraph is coloured red or blue.

3.7 Ordered Ramsey numbers

An *ordered graph* on n vertices is a graph whose vertices have been labelled with the vertex set $[n] = \{1, 2, \ldots, n\}$. We say that an ordered graph G on vertex set $[N]$ contains another ordered graph H on vertex set $[n]$ if there exists a map $\phi : [n] \to [N]$ such that $\phi(i) < \phi(j)$ for all $i < j$ and $(\phi(i), \phi(j))$ is an edge of G whenever (i, j) is an edge of H. Given an ordered graph H, we define the *ordered Ramsey number* $r_<(H)$ to be the smallest N such that every two-colouring of the complete graph on vertex set $[N]$ contains a monochromatic ordered copy of H.

As a first observation, we note the elementary inequalities,

$$r(H) \leq r_<(H) \leq r(K_{v(H)}).$$

In particular, $r_<(K_t) = r(K_t)$. However, for sparse graphs, the ordered Ramsey number may differ substantially from the usual Ramsey number. This was first observed by Conlon, Fox, Lee and Sudakov [53] and by Balko, Cibulka, Král and Kynčl [12], who proved the following result.

Theorem 3.40 *There exists a positive constant c such that, for every even n, there exists an ordered matching M on n vertices for which*

$$r_<(M) \geq n^{c \log n / \log \log n}.$$

In [53], it was proved that this lower bound holds for almost all orderings of a matching. This differs considerably from the usual Ramsey number, where it is trivial to show that $r(M)$ is linear in the number of vertices. It is also close to best possible, since, for all matchings M, $r_<(M) \leq n^{\lceil \log n \rceil}$.

For general graphs, it was proved in [53] that the ordered Ramsey number cannot be too much larger than the usual Ramsey number. Recall, from Section 2.3, that a graph is d-degenerate if there is an ordering of the vertices, say $v_1, v_2, \ldots, v_n$, such that every vertex v_i has at most d neighbours v_j preceding it in the ordering, that is, such that $j < i$. We stress that in the following theorems the degenerate ordering need not agree with the given ordering.

Theorem 3.41 *There exists a constant c such that for any ordered graph H on n vertices with degeneracy d,*

$$r_<(H) \leq r(H)^{c\gamma(H)},$$

where $\gamma(H) = \min\{\log^2(2n/d), d\log(2n/d)\}.$

An important role in ordered Ramsey theory is played by the concept of interval chromatic number. The *interval chromatic number* $\chi_<(H)$ of an ordered graph H is defined to be the minimum number of intervals into which the vertex set of H may be partitioned so that each interval forms an independent set in the graph. This is similar to the usual chromatic number but with arbitrary vertex sets replaced by intervals. For an ordered graph H with bounded degeneracy and bounded interval chromatic number, the ordered Ramsey number is at most polynomial in the number of vertices. This is the content of the following theorem from [53] (we note that a weaker version was also proved in [12]).

Theorem 3.42 *There exists a constant c such that any ordered graph H on n vertices with degeneracy at most d and interval chromatic number at most χ satisfies*

$$r_<(H) \leq n^{cd\log\chi}.$$

If H is an ordered graph with vertices $\{1, 2, \ldots, n\}$, we define the *bandwidth* of H to be the smallest ℓ such that $|i - j| \leq \ell$ for all edges $ij \in E(H)$. Answering a question of Lee and the authors [53], Balko, Cibulka, Král and Kynčl [12] showed that the ordered Ramsey number of ordered graphs with bounded bandwidth is at most polynomial in the number of vertices.

Theorem 3.43 *For any positive integer ℓ, there exists a constant c_ℓ such that any ordered graph on n vertices with bandwidth at most ℓ satisfies*

$$r_<(H) \leq n^{c_\ell}.$$

In [12], it is shown that for n sufficiently large in terms of ℓ one may take $c_\ell = O(\ell)$. It is plausible that the correct value of c_ℓ is significantly smaller than this.

A large number of questions about ordered Ramsey numbers remain open. Here we will discuss just one such problem, referring the reader to [53] for a more complete discussion. As usual, we define $r_<(G, H)$ to be the smallest N such that any red/blue-colouring of the edges of the complete graph on $[N]$ contains a red ordered copy of G or a blue ordered copy of H. Given an ordered matching M on n vertices, it is easy to see that

$$r_<(K_3, M) \le r(3, n) = O\left(\frac{n^2}{\log n}\right).$$

In the other direction, it is known [53] that there exists a positive constant c such that, for all even n, there exists an ordered matching M on n vertices for which $r_<(K_3, M) \ge c(\frac{n}{\log n})^{4/3}$. It remains to determine which bound is closer to the truth. In particular, we have the following problem.

Problem 3.44 *Does there exist an $\epsilon > 0$ such that for any matching M on n vertices $r(K_3, M) = O(n^{2-\epsilon})$?*

Finally, we note that for hypergraphs the difference between Ramsey numbers and their ordered counterparts is even more pronounced. If we write $P_n^{(k)}$ for the monotone k-uniform tight path on $\{1, 2, \ldots, n\}$, where $\{i, i+1, \ldots, i+k-1\}$ is an edge for $1 \le i \le n-k+1$, then results of Fox, Pach, Sudakov and Suk [112] and Moshkovitz and Shapira [161] (see also [159]) show that for $k \ge 3$ the ordered Ramsey number $r_<(P_n^{(k)})$ grows as a $(k-2)$-fold exponential in n. This is in stark contrast with the unordered problem, where $r(P_n^{(k)})$ is known to grow linearly in n for all k.

4   Concluding remarks

Given the length of this survey, it is perhaps unnecessary to add any further remarks. However, we would like to highlight two further problems which we believe to be of particular importance but which did not fit neatly in any of the sections above.

The first problem we wish to mention, proposed by Erdős, Fajtlowicz and Staton [41, 86], asks for an estimate on the order of the largest regular induced subgraph in a graph on n vertices. Ramsey's theorem tells us that any graph on n vertices contains a clique or an independent set of order at least $\frac{1}{2} \log n$. Since cliques and independent sets are both regular, this shows that there is always a regular induced subgraph of order at least $\frac{1}{2} \log n$. The infamous conjecture of Erdős, Fajtlowicz and Staton, which we now state, asks whether this simple bound can be improved.

Conjecture 4.1 *Any graph on n vertices contains a regular induced subgraph with $\omega(\log n)$ vertices.*

By using an inhomogeneous random graph, Bollobás showed that for any $\epsilon > 0$ and n sufficiently large depending on ϵ there are graphs on n vertices for which the largest regular induced subgraph has order at most $n^{1/2+\epsilon}$. This result was sharpened slightly by Alon, Krivelevich and Sudakov [6], who showed that there is a constant c and graphs on n vertices with no regular induced subgraph of order at least $cn^{1/2}\log^{1/4}n$. Any polynomial improvement on this upper bound would be of considerable interest.

The second problem is that of constructing explicit Ramsey graphs. Erdős' famous probabilistic lower bound argument, discussed at length in Section 2.1, shows that almost all colourings of the complete graph on $\sqrt{2}^t$ vertices do not contain a monochromatic copy of K_t. While this proves that the Ramsey number $r(t)$ is greater than $\sqrt{2}^t$, it does not give any constructive procedure for producing a colouring which exhibits this fact.

For many years, the best explicit example of a Ramsey graph was the following remarkable construction due to Frankl and Wilson [120]. Let p be a prime and let $r = p^2 - 1$. Let G be the graph whose vertices are all subsets of order r from the set $[m] = \{1, 2, \ldots, m\}$ and where two vertices are adjacent if and only if their corresponding sets have intersection of size congruent to $-1 \pmod{p}$. This is a graph with $\binom{m}{r}$ vertices and may be shown to contain no clique or independent set of order larger than $\binom{m}{p-1}$. Taking $m = p^3$ and $t = \binom{p^3}{p-1}$, this gives a graph on $t^{c\log t/\log\log t}$ vertices with no clique or independent set of order t.

Recently, Barak, Rao, Shaltiel and Wigderson [14] found a construction which improves on the Frankl–Wilson bound, giving graphs on

$$2^{2^{(\log\log t)^{1+\epsilon}}}$$

vertices with no clique or independent set of order t, where $\epsilon > 0$ is a fixed constant. Unfortunately, their construction does not have any simple description. Instead, it is constructive in the sense that given the labels of any two vertices in the graph, it is possible to decide whether they are connected in polynomial time. It would be very interesting to know whether the same bound, or any significant improvement over the Frankl–Wilson bound, could be achieved by graphs with a simpler description. It still seems that we are a long way from resolving Erdős' problem [41] of constructing explicit graphs exhibiting $r(t) > (1 + \epsilon)^t$, but for those who do not believe that hard work is its own reward, Erdős has offered the princely sum of \$100 as an enticement.

Acknowledgements. The authors would like to thank the anonymous referee for a number of useful comments.

References

[1] M. Ajtai, J. Komlós, and E. Szemerédi, A note on Ramsey numbers, *J. Combin. Theory Ser. A* **29** (1980), 354–360.

[2] P. Allen, G. Brightwell and J. Skokan, Ramsey-goodness – and otherwise, *Combinatorica* **33** (2013), 125–160.

[3] N. Alon, J. Balogh, A. Kostochka and W. Samotij, Sizes of induced subgraphs of Ramsey graphs, *Combin. Probab. Comput.* **18** (2009), 459–476.

[4] N. Alon and M. Krivelevich, Constructive bounds for a Ramsey-type problem, *Graphs Combin.* **13** (1997), 217–225.

[5] N. Alon, M. Krivelevich and B. Sudakov, Turán numbers of bipartite graphs and related Ramsey-type questions, *Combin. Probab. Comput.* **12** (2003), 477–494.

[6] N. Alon, M. Krivelevich and B. Sudakov, Large nearly regular induced subgraphs, *SIAM J. Discrete Math.* **22** (2008), 1325–1337.

[7] N. Alon, J. Pach, R. Pinchasi, R. Radoičić and M. Sharir, Crossing patterns of semi-algebraic sets, *J. Combin. Theory Ser. A* **111** (2005), 310–326.

[8] N. Alon, J. Pach and J. Solymosi, Ramsey-type theorems with forbidden subgraphs, *Combinatorica* **21** (2001), 155–170.

[9] N. Alon and V. Rödl, Sharp bounds for some multicolor Ramsey numbers, *Combinatorica* **25** (2005), 125–141.

[10] N. Alon, P. Seymour and R. Thomas, A separator theorem for non-planar graphs, *J. Amer. Math. Soc.* **3** (1990), 801–808.

[11] N. Alon and J. H. Spencer, **The Probabilistic Method,** 3rd edition, Wiley, 2007.

[12] M. Balko, J. Cibulka, K. Král and J. Kynčl, Ramsey numbers of ordered graphs, *preprint*.

[13] J. Balogh, R. Morris and W. Samotij, Independent sets in hypergraphs, to appear in *J. Amer. Math. Soc.*

[14] B. Barak, A. Rao, R. Shaltiel and A. Wigderson, 2-source dispersers for $n^{o(1)}$ entropy, and Ramsey graphs beating the Frankl–Wilson construction, *Ann. of Math.* **176** (2012), 1483–1543.

[15] J. Beck, An upper bound for diagonal Ramsey numbers, *Studia Sci. Math. Hungar.* **18** (1983), 401–406.

[16] J. Beck, On size Ramsey number of paths, trees and cycles I, *J. Graph Theory* **7** (1983), 115–130.

[17] J. Beck, On size Ramsey number of paths, trees and cycles II, in Mathematics of Ramsey theory, Algorithms Combin., Vol. 5, 34–45, Springer, Berlin, 1990.

[18] J. Beck, Achievement games and the probabilistic method, in Combinatorics, Paul Erdős is Eighty, Vol. 1, 51–78, Bolyai Soc. Math. Stud., János Bolyai Math. Soc., Budapest, 1993.

[19] E. Berger, K. Choromanski and M. Chudnovsky, Forcing large transitive subtournaments, to appear in *J. Combin. Theory Ser. B.*

[20] T. Bohman, The triangle-free process, *Adv. Math.* **221** (2009), 1653–1677.

[21] T. Bohman and P. Keevash, The early evolution of the H-free process, *Invent. Math.* **181** (2010), 291–336.

[22] T. Bohman and P. Keevash, Dynamic concentration of the triangle-free process, *preprint*.

[23] B. Bollobás and H. R. Hind, Graphs without large triangle free subgraphs, *Discrete Math.* **87** (1991), 119–131.

[24] M. Bonamy, N. Bousquet and S. Thomassé, The Erdős–Hajnal conjecture for long holes and anti-holes, *preprint*.

[25] N. Bousquet, A. Lagoutte and S. Thomassé, The Erdős–Hajnal conjecture for paths and antipaths, to appear in *J. Combin. Theory Ser. B.*

[26] S. Brandt, Expanding graphs and Ramsey numbers, available at Freie Universität, Berlin preprint server, ftp://ftp.math.fu-berlin.de/pub/math/publ/pre/1996/pr-a-96-24.ps.

[27] B. Bukh and B. Sudakov, Induced subgraphs of Ramsey graphs with many distinct degrees, *J. Combin. Theory Ser. B* **97** (2007), 612–619.

[28] S. A. Burr, Ramsey numbers involving graphs with long suspended paths, *J. London Math. Soc.* **24** (1981), 405–413.

[29] S. A. Burr, What can we hope to accomplish in generalized Ramsey theory?, *Discrete Math.* **67** (1987), 215–225.

[30] S. A. Burr and P. Erdős, On the magnitude of generalized Ramsey numbers for graphs, in Infinite and Finite Sets, Vol. 1 (Keszthely, 1973), 214–240, Colloq. Math. Soc. János Bolyai, Vol. 10, North-Holland, Amsterdam, 1975.

[31] S. A. Burr and P. Erdős, Generalizations of a Ramsey-theoretic result of Chvátal, *J. Graph Theory* **7** (1983), 39–51.

[32] S. A. Burr, P. Erdős, R. J. Faudree, C. C. Rousseau and R. H. Schelp, Some complete bipartite graph-tree Ramsey numbers, *Ann. Discrete Math.* **41** (1989), 79–90.

[33] S. A. Burr, P. Erdős and L. Lovász, On graphs of Ramsey type, *Ars Combin.* **1** (1976), 167–190.

[34] S. A. Burr and V. Rosta, On the Ramsey multiplicity of graphs – problems and recent results, *J. Graph Theory* **4** (1980), 347–361.

[35] S. Butler, Induced-universal graphs for graphs with bounded maximum degree, *Graphs Combin.* **25** (2009), 461–468.

[36] J. Butterfield, T. Grauman, W. B. Kinnersley, K. G. Milans, C. Stocker and D. B. West, On-line Ramsey theory for bounded degree graphs, *Electron. J. Combin.* **18** (2011), P136.

[37] G. Chen and R. H. Schelp, Graphs with linearly bounded Ramsey numbers, *J. Combin. Theory Ser. B* **57** (1993), 138–149.

[38] M. Chudnovsky, The Erdős–Hajnal conjecture – a survey, *J. Graph Theory* **75** (2014), 178–190.

[39] M. Chudnovsky and S. Safra, The Erdős–Hajnal conjecture for bull-free graphs, *J. Combin. Theory Ser. B* **98** (2008), 1301–1310.

[40] M. Chudnovsky and P. Seymour, Excluding paths and antipaths, to appear in *Combinatorica*.

[41] F. Chung and R. L. Graham, **Erdős on Graphs. His Legacy of Unsolved Problems**, A K Peters, Ltd., Wellesley, MA, 1998.

[42] V. Chvátal, Tree-complete graph Ramsey numbers, *J. Graph Theory* **1** (1977), 93.

[43] V. Chvátal and F. Harary, Generalized Ramsey theory for graphs. III. Small off-diagonal numbers, *Pacific J. Math.* **41** (1972), 335–345.

[44] V. Chvátal, V. Rödl, E. Szemerédi and W. T. Trotter Jr, The Ramsey number of a graph with bounded maximum degree, *J. Combin. Theory Ser. B* **34** (1983), 239–243.

[45] D. Conlon, A new upper bound for diagonal Ramsey numbers, *Ann. of Math.* **170** (2009), 941–960.

[46] D. Conlon, Hypergraph packing and sparse bipartite Ramsey numbers, *Combin. Probab. Comput.* **18** (2009), 913–923.

[47] D. Conlon, On-line Ramsey numbers, *SIAM J. Discrete Math.* **23** (2009), 1954–1963.

[48] D. Conlon, On the Ramsey multiplicity of complete graphs, *Combinatorica* **32** (2012), 171–186.

[49] D. Conlon, The Ramsey number of dense graphs, *Bull. Lond. Math. Soc.* **45** (2013), 483–496.

[50] D. Conlon, J. Fox, C. Lee and B. Sudakov, Ramsey numbers of cubes versus cliques, to appear in *Combinatorica*.

[51] D. Conlon, J. Fox, C. Lee and B. Sudakov, The Erdős–Gyárfás problem on generalized Ramsey numbers, *Proc. Lond. Math. Soc.* **110** (2015), 1–18.

[52] D. Conlon, J. Fox, C. Lee and B. Sudakov, On the grid Ramsey problem and related questions, to appear in *Int. Math. Res. Not.*

[53] D. Conlon, J. Fox, C. Lee and B. Sudakov, Ordered Ramsey numbers, *submitted*.

[54] D. Conlon, J. Fox and B. Sudakov, Ramsey numbers of sparse hypergraphs, *Random Structures Algorithms* **35** (2009), 1–14.

[55] D. Conlon, J. Fox and B. Sudakov, Hypergraph Ramsey numbers, *J. Amer. Math. Soc.* **23** (2010), 247–266.

[56] D. Conlon, J. Fox and B. Sudakov, An approximate version of Sidorenko's conjecture, *Geom. Funct. Anal.* **20** (2010), 1354–1366.

[57] D. Conlon, J. Fox and B. Sudakov, Large almost monochromatic subsets in hypergraphs, *Israel J. Math.* **181** (2011), 423–432.

[58] D. Conlon, J. Fox and B. Sudakov, On two problems in graph Ramsey theory, *Combinatorica* **32** (2012), 513–535.

[59] D. Conlon, J. Fox and B. Sudakov, Erdős–Hajnal-type theorems in hypergraphs, *J. Combin. Theory Ser. B* **102** (2012), 1142–1154.

[60] D. Conlon, J. Fox and B. Sudakov, Two extensions of Ramsey's theorem, *Duke Math. J.* **162** (2013), 2903–2927.

[61] D. Conlon, J. Fox and B. Sudakov, Short proofs of some extremal results, *Combin. Probab. Comput.* **23** (2014), 8–28.

[62] D. Conlon, J. Fox and B. Sudakov, Short proofs of some extremal results II, *in preparation.*

[63] D. Conlon, J. Fox and Y. Zhao, Extremal results in sparse pseudorandom graphs, *Adv. Math.* **256** (2014), 206–290.

[64] D. Conlon and W. T. Gowers, Combinatorial theorems in sparse random sets, *submitted.*

[65] D. Conlon and W. T. Gowers, An upper bound for Folkman numbers, *preprint.*

[66] O. Cooley, N. Fountoulakis, D. Kühn and D. Osthus, 3-uniform hypergraphs of bounded degree have linear Ramsey numbers, *J. Combin. Theory Ser. B* **98** (2008), 484–505.

[67] O. Cooley, N. Fountoulakis, D. Kühn and D. Osthus, Embeddings and Ramsey numbers of sparse k-uniform hypergraphs, *Combinatorica* **28** (2009), 263–297.

[68] J. Cummings, D. Král', F. Pfender, K. Sperfeld, A. Treglown and M. Young, Monochromatic triangles in three-coloured graphs, *J. Combin. Theory Ser. B* **103** (2013), 489–503.

[69] D. Dellamonica, The size-Ramsey number of trees, *Random Structures Algorithms* **40** (2012), 49–73.

[70] W. Deuber, A generalization of Ramsey's theorem, in Infinite and Finite Sets, Vol. 1 (Keszthely, 1973), 323–332, Colloq. Math. Soc. János Bolyai, Vol. 10, North-Holland, Amsterdam, 1975.

[71] A. Dudek and D. Mubayi, On generalized Ramsey numbers for 3-uniform hypergraphs, *J. Graph Theory* **76** (2014), 217–223.

108   D. Conlon, J. Fox and B. Sudakov

[72] A. Dudek, T. Retter and V. Rödl, On generalized Ramsey numbers of Erdős and Rogers, *J. Combin. Theory Ser. B* **109** (2014), 213–227.

[73] A. Dudek and V. Rödl, On the Folkman number $f(2,3,4)$, *Exp. Math.* **17** (2008), 63–67.

[74] A. Dudek and V. Rödl, On K_s-free subgraphs in K_{s+k}-free graphs and vertex Folkman numbers, *Combinatorica* **31** (2011), 39–53.

[75] R. A. Duke, H. Lefmann, and V. Rödl, A fast approximation algorithm for computing the frequencies of subgraphs in a given graph, *SIAM J. Comput.* **24** (1995), 598–620.

[76] N. Eaton, Ramsey numbers for sparse graphs, *Discrete Math.* **185** (1998), 63–75.

[77] D. Eichhorn and D. Mubayi, Edge-coloring cliques with many colors on subcliques, *Combinatorica* **20** (2000), 441–444.

[78] P. Erdős, Some remarks on the theory of graphs, *Bull. Amer. Math. Soc.* **53** (1947), 292–294.

[79] P. Erdős, Graph theory and probability II, *Canad. J. Math.* **13** (1961), 346–352.

[80] P. Erdős, On the number of complete subgraphs contained in certain graphs, *Magyar Tud. Akad. Mat. Kutató Int. Közl.* **7** (1962), 459–464.

[81] P. Erdős, Problems and results on finite and infinite graphs, in Recent advances in graph theory (Proc. Second Czechoslovak Sympos., Prague, 1974), 183–192, Academia, Prague, 1975.

[82] P. Erdős, Solved and unsolved problems in combinatorics and combinatorial number theory, in Proceedings of the Twelfth Southeastern Conference on Combinatorics, Graph Theory and Computing, Vol. I (Baton Rouge, La., 1981), *Congr. Numer.* **32** (1981), 49–62.

[83] P. Erdős, On the combinatorial problems which I would most like to see solved, *Combinatorica* **1** (1981), 25–42.

[84] P. Erdős, On some problems in graph theory, combinatorial analysis and combinatorial number theory, in Graph theory and combinatorics (Cambridge, 1983), 1–17, Academic Press, London, 1984.

[85] P. Erdős, Problems and results on graphs and hypergraphs: similarities and differences, in Mathematics of Ramsey theory, 12–28, Algorithms Combin., 5, Springer, Berlin, 1990.

[86] P. Erdős, On some of my favourite problems in various branches of combinatorics, in Proceedings of the Fourth Czechoslovakian Symposium on Combinatorics, Graphs and Complexity (Prachatice, 1990), 69–79, Ann. Discrete Math., 51, North-Holland, Amsterdam, 1992.

[87] P. Erdős, Problems and results in discrete mathematics, *Discrete Math.* **136** (1994), 53–73.

[88] P. Erdős, R. J. Faudree, C. C. Rousseau and R. H. Schelp, The size Ramsey number, *Period. Math. Hungar.* **9** (1978), 145–161.

[89] P. Erdős and R. Graham, On partition theorems for finite graphs, in Infinite and Finite Sets, Vol. 1 (Keszthely, 1973), 515–527, Colloq. Math. Soc. János Bolyai, Vol. 10, North-Holland, Amsterdam, 1975.

[90] P. Erdős and A. Gyárfás, A variant of the classical Ramsey problem, *Combinatorica* **17** (1997), 459–467.

[91] P. Erdős and A. Hajnal, Research problems 2–3, *J. Combin. Theory* **2** (1967), 104–105.

[92] P. Erdős and A. Hajnal, On Ramsey like theorems, problems and results, in Combinatorics (Proc. Conf. Combinatorial Math., Math. Inst., Oxford, 1972), 123–140, Inst. Math. Appl., Southend-on-Sea, 1972.

[93] P. Erdős and A. Hajnal, On spanned subgraphs of graphs, in Contributions to graph theory and its applications (Internat. Colloq., Oberhof, 1977), 80–96, Tech. Hochschule Ilmenau, Ilmenau, 1977.

[94] P. Erdős and A. Hajnal, Ramsey-type theorems, *Discrete Appl. Math.* **25** (1989), 37–52.

[95] P. Erdős, A. Hajnal and J. Pach, On a metric generalization of Ramsey's theorem, *Israel J. Math.* **102** (1997), 283–295.

[96] P. Erdős, A. Hajnal and J. Pach, A Ramsey-type theorem for bipartite graphs, *Geombinatorics* **10** (2000), 64–68.

[97] P. Erdős, A. Hajnal and L. Pósa, Strong embeddings of graphs into colored graphs, in Infinite and Finite Sets, Vol. 1 (Keszthely, 1973), 585–595, Colloq. Math. Soc. János Bolyai, Vol. 10, North-Holland, Amsterdam, 1975.

[98] P. Erdős, A. Hajnal and R. Rado, Partition relations for cardinal numbers, *Acta Math. Acad. Sci. Hungar.* **16** (1965), 93–196.

110   D. Conlon, J. Fox and B. Sudakov

[99] P. Erdős and R. Rado, Combinatorial theorems on classifications of subsets of a given set, *Proc. London Math. Soc.* **3** (1952), 417–439.

[100] P. Erdős and C. A. Rogers, The construction of certain graphs, *Canad. J. Math.* **14** (1962), 702–707.

[101] P. Erdős and M. Simonovits, Cube-supersaturated graphs and related problems, in Progress in graph theory (Waterloo, Ont., 1982), 203–218, Academic Press, Toronto, ON, 1984.

[102] P. Erdős and G. Szekeres, A combinatorial problem in geometry, *Compos. Math.* **2** (1935), 463–470.

[103] P. Erdős and E. Szemerédi, On a Ramsey type theorem, *Period. Math. Hungar.* **2** (1972), 295–299.

[104] G. Fiz Pontiveros, S. Griffiths and R. Morris, The triangle-free process and $R(3,k)$, *preprint*.

[105] G. Fiz Pontiveros, S. Griffiths, R. Morris, D. Saxton and J. Skokan, On the Ramsey number of the triangle and the cube, to appear in *Combinatorica*.

[106] G. Fiz Pontiveros, S. Griffiths, R. Morris, D. Saxton and J. Skokan, The Ramsey number of the clique and the hypercube, *J. Lond. Math. Soc.* **89** (2014), 680–702.

[107] J. Folkman, Graphs with monochromatic complete subgraphs in every edge coloring, *SIAM J. Appl. Math.* **18** (1970), 19–24.

[108] J. Fox, There exist graphs with super-exponential Ramsey multiplicity constant, *J. Graph Theory* **57** (2008), 89–98.

[109] J. Fox, A. Grinshpun and J. Pach, The Erdős–Hajnal conjecture for rainbow triangles, *J. Combin. Theory Ser. B* **111** (2015), 75–125.

[110] J. Fox and J. Pach, Erdős–Hajnal-type results on intersection patterns of geometric objects, in Horizons of combinatorics, 79–103, Bolyai Soc. Math. Stud., 17, Springer, Berlin, 2008.

[111] J. Fox and J. Pach, Applications of a new separator theorem for string graphs, *Combin. Probab. Comput.* **23** (2014), 66–74.

[112] J. Fox, J. Pach, B. Sudakov and A. Suk, Erdős–Szekeres-type theorems for monotone paths and convex bodies, *Proc. Lond. Math. Soc.* **105** (2012), 953–982.

[113] J. Fox, J. Pach and Cs. D. Tóth, Intersection patterns of curves, *J. Lond. Math. Soc.* **83** (2011), 389–406.

[114] J. Fox and B. Sudakov, Induced Ramsey-type theorems, *Adv. Math.* **219** (2008), 1771–1800.

[115] J. Fox and B. Sudakov, Ramsey-type problem for an almost monochromatic K_4, *SIAM J. Discrete Math.* **23** (2008/09), 155–162.

[116] J. Fox and B. Sudakov, Density theorems for bipartite graphs and related Ramsey-type results, *Combinatorica* **29** (2009), 153–196.

[117] J. Fox and B. Sudakov, Two remarks on the Burr–Erdős conjecture, *European J. Combin.* **30** (2009), 1630–1645.

[118] J. Fox and B. Sudakov, Dependent Random Choice, *Random Structures Algorithms* **38** (2011), 68–99.

[119] P. Frankl and V. Rödl, Large triangle-free subgraphs in graphs without K_4, *Graphs Combin.* **2** (1986), 135–144.

[120] P. Frankl and R. M. Wilson, Intersection theorems with geometric consequences, *Combinatorica* **1** (1981), 357–368.

[121] E. Friedgut, V. Rödl and M. Schacht, Ramsey properties of discrete random structures, *Random Structures Algorithms* **37** (2010), 407–436.

[122] J. Friedman and N. Pippenger, Expanding graphs contain all small trees, *Combinatorica* **7** (1987), 71–76.

[123] C. Godsil and G. Royle, **Algebraic Graph Theory**, Springer, 2001.

[124] A. W. Goodman, On sets of acquaintances and strangers at any party, *Amer. Math. Monthly* **66** (1959), 778–783.

[125] W. T. Gowers, A new proof of Szemerédi's theorem for arithmetic progressions of length four, *Geom. Funct. Anal.* **8** (1998), 529–551.

[126] W. T. Gowers, Hypergraph regularity and the multidimensional Szemerédi theorem, *Ann. of Math.* **166** (2007), 897–946.

[127] R. L. Graham and V. Rödl, Numbers in Ramsey theory, in Surveys in Combinatorics 1987, 111–153, London Math. Soc. Lecture Note Ser., Vol. 123, Cambridge University Press, Cambridge, 1987.

[128] R. L. Graham, V. Rödl and A. Ruciński, On graphs with linear Ramsey numbers, *J. Graph Theory* **35** (2000), 176–192.

[129] R. L. Graham, V. Rödl and A. Ruciński, On bipartite graphs with linear Ramsey numbers, *Combinatorica* **21** (2001), 199–209.

[130] R. L. Graham, B. L. Rothschild and J. H. Spencer, **Ramsey theory**, 2nd edition, Wiley, 1990.

[131] B. Green and T. Tao, The primes contain arbitrarily long arithmetic progressions, *Ann. of Math.* **167** (2008), 481–547.

[132] J. A. Grytczuk, M. Hałuszczak and H. A. Kierstead, On-line Ramsey theory, *Electron. J. Combin.* **11** (2004), Research Paper 60, 10pp.

[133] A. Gyárfás, On a Ramsey type problem of Shelah, in Extremal problems for finite sets (Visegrád, 1991), 283–287, Bolyai Soc. Math. Stud., 3, János Bolyai Math. Soc., Budapest, 1994.

[134] A. Hajnal, Rainbow Ramsey theorems for colorings establishing negative partition relations, *Fund. Math.* **198** (2008), 255–262.

[135] L. Harrington and J. Paris, A mathematical incompleteness in Peano arithmetic, in Handbook of Mathematical Logic, 1133–1142, North-Holland, Amsterdam, 1977.

[136] H. Hatami, J. Hladký, D. Král', S. Norine and A. Razborov, Non-three-colorable common graphs exist, *Combin. Probab. Comput.* **21** (2012), 734–742.

[137] P. E. Haxell, Y. Kohayakawa and T. Łuczak, The induced size-Ramsey number of cycles, *Combin. Probab. Comput.* **4** (1995), 217–239.

[138] P. Horn, K. G. Milans and V. Rödl, Degree Ramsey numbers of closed blowups of trees, *Electron. J. Combin.* **21** (2014), Paper 2.5, 6pp.

[139] C. Jagger, P. Šťovíček and A. Thomason, Multiplicities of subgraphs, *Combinatorica* **16** (1996), 123–141.

[140] H. A. Kierstead and G. Konjevod, Coloring number and on-line Ramsey theory for graphs and hypergraphs, *Combinatorica*, **29** (2009), 49–64.

[141] J. H. Kim, The Ramsey number $R(3,t)$ has order of magnitude $t^2/\log t$, *Random Structures Algorithms* **7** (1995), 173–207.

[142] J. H. Kim, C. Lee and J. Lee, Two approaches to Sidorenko's conjecture, to appear in *Trans. Amer. Math. Soc.*

[143] W. B. Kinnersley, K. G. Milans and D. B. West, Degree Ramsey numbers of graphs, *Combin. Probab. Comput.* **21** (2012), 229–253.

[144] Y. Kohayakawa, H. Prömel and V. Rödl, Induced Ramsey numbers, *Combinatorica* **18** (1998), 373–404.

[145] Y. Kohayakawa, V. Rödl, M. Schacht and E. Szemerédi, Sparse partition universal graphs for graphs of bounded degree, *Adv. Math.* **226** (2011), 5041–5065.

[146] A. V. Kostochka and D. Mubayi, When is an almost monochromatic K_4 guaranteed? *Combin. Probab. Comput.* **17** (2008), 823–830.

[147] A. V. Kostochka and V. Rödl, On graphs with small Ramsey numbers, *J. Graph Theory* **37** (2001), 198–204.

[148] A. V. Kostochka and V. Rödl, On Ramsey numbers of uniform hypergraphs with given maximum degree, *J. Combin. Theory Ser. A* **113** (2006), 1555–1564.

[149] A. V. Kostochka and B. Sudakov, On Ramsey numbers of sparse graphs, *Combin. Probab. Comput.* **12** (2003), 627–641.

[150] M. Krivelevich, Bounding Ramsey numbers through large deviation inequalities, *Random Structures Algorithms* **7** (1995), 145–155.

[151] M. Krivelevich and B. Sudakov, Pseudo-random graphs, in More sets, graphs and numbers, 199–262, Bolyai Soc. Math. Stud., 15, Springer, Berlin, 2006.

[152] A. Kurek and A. Ruciński, Two variants of the size Ramsey number, *Discuss. Math. Graph Theory* **25** (2005), 141–149.

[153] A. R. Lange, S. P. Radziszowski and X. Xu, Use of MAX-CUT for Ramsey arrowing of triangles, *J. Combin. Math. Combin. Comput.* **88** (2014), 61–71.

[154] D. Larman, J. Matoušek, J. Pach and J. Törőcsik, A Ramsey-type result for convex sets, *Bull. London Math. Soc.* **26** (1994), 132–136.

[155] J. L. X. Li and B. Szegedy, On the logarithmic calculus and Sidorenko's conjecture, to appear in *Combinatorica*.

[156] L. Lu, Explicit construction of small Folkman graphs, *SIAM J. Discrete Math.* **21** (2007), 1053–1060.

[157] T. Łuczak and V. Rödl, On induced Ramsey numbers for graphs with bounded maximum degree, *J. Combin. Theory Ser. B* **66** (1996), 324–333.

[158] W. Mader, Homomorphiesätze für Graphen, *Math. Ann.* **178** (1968), 154–168.

[159] K. G. Milans, D. Stolee and D. B. West, Ordered Ramsey theory and track representations of graphs, to appear in *J. Combin.*

[160] G. Mills, Ramsey–Paris–Harrington numbers for graphs, *J. Combin. Theory Ser. A* **38** (1985), 30–37.

[161] G. Moshkovitz and A. Shapira, Ramsey theory, integer partitions and a new proof of the Erdős–Szekeres theorem, *Adv. Math.* **262** (2014), 1107–1129.

[162] D. Mubayi, Edge-coloring cliques with three colors on all 4-cliques, *Combinatorica* **18** (1998), 293–296.

[163] B. Nagle, S. Olsen, V. Rödl and M. Schacht, On the Ramsey number of sparse 3-graphs, *Graphs Combin.* **27** (2008), 205–228.

[164] B. Nagle, V. Rödl and M. Schacht, The counting lemma for regular k-uniform hypergraphs, *Random Structures Algorithms* **28** (2006), 113–179.

[165] R. Nenadov and A. Steger, A short proof of the random Ramsey theorem, to appear in *Combin. Probab. Comput.*

[166] J. Nešetřil and V. Rödl, The Ramsey property for graphs with forbidden complete subgraphs, *J. Combin. Theory Ser. B* **20** (1976), 243–249.

[167] J. Nešetřil and V. Rödl, Simple proof of the existence of restricted Ramsey graphs by means of a partite construction, *Combinatorica* **1** (1981), 199–202.

[168] J. Nešetřil and J. A. Väänänen, Combinatorics and quantifiers, *Comment. Math. Univ. Carolin.* **37** (1996), 433–443.

[169] S. Nieß, Counting monochromatic copies of K_4: a new lower bound for the Ramsey multiplicity problem, *preprint.*

[170] V. Nikiforov and C. C. Rousseau, Ramsey goodness and beyond, *Combinatorica* **29** (2009), 227–262.

[171] N. Paul and C. Tardif, The chromatic Ramsey number of odd wheels, *J. Graph Theory* **69** (2012), 198–205.

[172] H. Prömel and V. Rödl, Non-Ramsey graphs are $c \log n$-universal, *J. Combin. Theory Ser. A* **88** (1999), 379–384.

[173] S. Radziszowski, Small Ramsey numbers, *Electron. J. Combin.* (2014), DS1.

[174] F. P. Ramsey, On a problem of formal logic, *Proc. London Math. Soc.* **30** (1930), 264–286.

[175] A. A. Razborov, Flag algebras, *J. Symbolic Logic* **72** (2007), 1239–1282.

[176] V. Rödl, The dimension of a graph and generalized Ramsey theorems, Master's thesis, Charles University, 1973.

[177] V. Rödl, On universality of graphs with uniformly distributed edges, *Discrete Math.* **59** (1986), 125–134.

[178] V. Rödl, On homogeneous sets of positive integers, *J. Combin. Theory Ser. A* **102** (2003), 229–240.

[179] V. Rödl and A. Ruciński, Lower bounds on probability thresholds for Ramsey properties, in Combinatorics, Paul Erdős is eighty, Vol. 1, 317–346, Bolyai Soc. Math. Stud., János Bolyai Math. Soc., Budapest, 1993.

[180] V. Rödl and A. Ruciński, Threshold functions for Ramsey properties, *J. Amer. Math. Soc.* **8** (1995), 917–942.

[181] V. Rödl, A. Ruciński and M. Schacht, An exponential-type upper bound for Folkman numbers, *preprint*.

[182] V. Rödl and M. Schacht, Complete partite subgraphs in dense hypergraphs, *Random Structures Algorithms* **41** (2012), 557–573.

[183] V. Rödl and J. Skokan, Regularity lemma for uniform hypergraphs, *Random Structures Algorithms* **25** (2004), 1–42.

[184] V. Rödl and E. Szemerédi, On size Ramsey numbers of graphs with bounded maximum degree, *Combinatorica* **20** (2000), 257–262.

[185] V. Rödl and R. Thomas, Arrangeability and clique subdivisions, in The mathematics of Paul Erdős, II, 236–239, Algorithms Combin., 14, Springer, Berlin, 1997.

[186] G. N. Sárközy and S. M. Selkow, On edge colorings with at least q colors in every subset of p vertices, *Electron. J. Combin.* **8** (2001), Research Paper 9, 6pp.

[187] G. N. Sárközy and S. M. Selkow, An application of the regularity lemma in generalized Ramsey theory, *J. Graph Theory* **44** (2003), 39–49.

[188] D. Saxton and A. Thomason, Hypergraph containers, to appear in *Invent. Math.*

[189] M. Schacht, Extremal results for discrete random structures, *preprint*.

[190] J. Shearer, A note on the independence number of triangle-free graphs, *Discrete Math.* **46** (1983), 83–87.

[191] J. Shearer, On the independence number of sparse graphs, *Random Structures Algorithms* **7** (1995), 269–271.

[192] S. Shelah, Primitive recursive bounds for van der Waerden numbers, *J. Amer. Math. Soc.* **1** (1989), 683–697.

[193] S. Shelah, A finite partition theorem with double exponential bound, in The mathematics of Paul Erdős, II, 240–246, Algorithms Combin., 14, Springer, Berlin, 1997.

[194] L. Shi, Cube Ramsey numbers are polynomial, *Random Structures Algorithms* **19** (2001), 99–101.

[195] L. Shi, The tail is cut for Ramsey numbers of cubes, *Discrete Math.* **307** (2007), 290–292.

[196] A. F. Sidorenko, Cycles in graphs and functional inequalities, *Math. Notes* **46** (1989), 877–882.

[197] A. F. Sidorenko, A correlation inequality for bipartite graphs, *Graphs Combin.* **9** (1993), 201–204.

[198] J. H. Spencer, Ramsey's theorem – a new lower bound, *J. Combin. Theory Ser. A* **18** (1975), 108–115.

[199] J. H. Spencer, Asymptotic lower bounds for Ramsey functions, *Discrete Math.* **20** (1977/78), 69–76.

[200] J. H. Spencer, Three hundred million points suffice, *J. Combin. Theory Ser. A* **49** (1988), 210–217. See also the erratum by M. Hovey in *J. Combin. Theory Ser. A* **50** (1989), 323.

[201] K. Sperfeld, On the minimal monochromatic K_4-density, *preprint*.

[202] B. Sudakov, A few remarks on the Ramsey–Turán-type problems, *J. Combin. Theory Ser. B* **88** (2003), 99–106.

[203] B. Sudakov, A new lower bound for a Ramsey-type problem, *Combinatorica* **25** (2005), 487–498.

[204] B. Sudakov, Large K_r-free subgraphs in K_s-free graphs and some other Ramsey-type problems, *Random Structures Algorithms* **26** (2005), 253–265.

[205] B. Sudakov, A conjecture of Erdős on graph Ramsey numbers, *Adv. Math.* **227** (2011), 601–609.

[206] E. Szemerédi, On sets of integers containing no k elements in arithmetic progression, *Acta Arith.* **27** (1975), 199–245.

[207] E. Szemerédi, Regular partitions of graphs, in Problèmes Combinatoires et Théorie des Graphes (Orsay 1976), 399–401, Colloq. Internat. CNRS, 260, CNRS, Paris, 1978.

[208] A. Thomason, Pseudorandom graphs, in Random graphs '85 (Poznań, 1985), 307–331, North-Holland Math. Stud., Vol. 144, North-Holland, Amsterdam, 1987.

[209] A. Thomason, Random graphs, strongly regular graphs and pseudorandom graphs, in Surveys in Combinatorics 1987, 173–195, London Math. Soc. Lecture Note Ser., Vol. 123, Cambridge University Press, Cambridge, 1987.

[210] A. Thomason, An upper bound for some Ramsey numbers, *J. Graph Theory* **12** (1988), 509–517.

[211] A. Thomason, A disproof of a conjecture of Erdős in Ramsey theory, *J. London Math. Soc.* **39** (1989), 246–255.

[212] B. L. van der Waerden, Beweis einer Baudetschen Vermutung, *Nieuw. Arch. Wisk.* **15** (1927), 212–216.

[213] G. Wolfovitz, K_4-free graphs without large induced triangle-free subgraphs, *Combinatorica* **33** (2013), 623–631.

[214] X. Zhu, The fractional version of Hedetniemi's conjecture is true, *European J. Combin.* **32** (2011), 1168–1175.

Mathematical Institute
University of Oxford
Oxford OX2 6GG
United Kingdom
david.conlon@maths.ox.ac.uk

Department of Mathematics
Stanford University
Stanford, CA 94305
USA
fox@math.mit.edu

Department of Mathematics
ETH
8092 Zurich
Switzerland
benjamin.sudakov@math.ethz.ch

Controllability and matchings in random bipartite graphs

Paul Balister and Stefanie Gerke

Abstract

Motivated by an application in controllability we consider maximum matchings in random bipartite graphs $G = (A, B)$. First we analyse Karp–Sipser's algorithm to determine the asymptotic size of maximum matchings in random bipartite graphs with a fixed degree distribution. We then allow an adversary to delete one edge adjacent to every vertex in A in the more restricted model where each vertex in A chooses d neighbours uniformly at random from B.

1 Introduction

We are interested in finding large matchings in random bipartite graphs. The motivation comes in part from recent work by Liu, Slotine, and Barabási [16], in which they used a characterisation by Lin [15] of structural controllability to show how large matchings in random bipartite graphs play a crucial role in obtaining bounds on the number of nodes needed to control directed networks. We will give a short description of this connection in Section 2.

Matchings in bipartite graphs are a classical problem in graph theory. The famous theorem of Hall [10] states that a bipartite graph with vertex sets V_1 and V_2 contains a matching of size $|V_1|$ if and only if for every set $S \subseteq V_1$ we have $|S| \leq |\Gamma(S)|$ where $\Gamma(S)$ is the neighbourhood of S. One can use this characterisation to show that in a random bipartite graph $G(n, n, p)$ with vertex sets V_1 and V_2 of the same size n where each of the possible n^2 edges is present with probability p independent of the presences or absence of all other edges, with high probability (whp) there is a matching of size n if there is no isolated vertex. The random bipartite graph $G(n, n, p)$ has no isolated vertex with high probability if $np - \log n$ tends to infinity as n tends to infinity, see for example [11].

Finding matchings in general graphs is a well-studied problem and polynomial time algorithms are known to find maximum matchings, see for example [7, 17]. We will analyse a far simpler algorithm developed by Karp and Sipser [14] which we will introduce in Section 3.3. This algorithm is known to work well with high probability on sparse random graphs [1]. Bohman and Frieze [4] analysed this algorithm for the class of graphs which have no vertices of degree smaller than $\delta \geq 2$ or larger than Δ, and $\delta n_\delta, \ldots, \Delta n_\Delta$ forms a log-concave sequence. Here n_d is the number of vertices of degree d. We generalise this result in Section 3 to more general degree distributions and develop a simpler proof. Bohman and Frieze used the differential equation method to track the number of vertices of each degree whereas we use generating functions and will see that we only need to track two variables. Non-algorithmic proofs for slightly more precise results can be found in [5] and also in [18]. In particular the authors determine the size of a maximum matching in a random bipartite graph with a fixed degree distribution under some mild assumption on these distributions.

Motivated by the application in controllability we discuss briefly in Section 4 the presence of an adversary in a more restricted model. More precisely, we consider a random bipartite graph $G = (A \cup B, E)$ with $|A| = n$ and $|B| = (1 + \varepsilon)n$. Each vertex in A is adjacent to d neighbours chosen uniformly at random from B. We allow repetition, so this is a multigraph. An adversary is then able to remove a single edge adjacent to each vertex of A, with the aim of minimising the size of the largest matching. The complete proofs can be found in [2]. Melsted and Frieze [9] analysed the Karp–Sipser algorithm for the random bipartite graph where each vertex in a partition class of size n chooses d neighbours uniformly at random from a partition class of size αn for some $\alpha > 0$ but without an adversary.

2 Controllability

Roughly speaking, controllability is concerned with which elements of a network one needs to be able to manipulate — a control set — to be able to force the entire network into any desired state. More precisely, we have a vector $\mathbf{x} \in \mathbb{R}^n$ that evolves according to a system of equations $d\mathbf{x}(t)/dt = \mathbf{f}(\mathbf{x}) + \mathbf{y}(t)$ where $\mathbf{y} \in \mathbb{R}^n$ is a 'controlling' term with $y_i(t) = 0$ whenever i is not in the control set. We are interested in the smallest control set S such that by varying y_i for $i \in S$ suitably we can force $\mathbf{x}$ into any desired state in finite time. We consider the simpler case when the changes are linear. We have a vector $\mathbf{x}(t) = (x_1(t), x_2(t), \ldots x_n(t))$ which represents the current state of a system with n nodes at time t; the $n \times n$ matrix A showing the topology of the system signalling interactions between nodes, and the $n \times m$ matrix B that identifies the set of nodes controlling the system. Thus we are interested in $d\mathbf{x}(t)/dt = A\mathbf{x}(t) + B\mathbf{u}(t)$. The canonical controllability criterion according to Kalman [12] is that the $n \times nm$ matrix $C = (B, AB, A^2B, \ldots, A^{n-1}B)$ has full rank. In applications it is often impossible to measure the entries of A exactly or they may be time dependent (for example internet traffic). Structural Controllability circumvents this problem by allowing us to choose the non-zero entries in A and B such that $C = (B, AB, A^2B, \ldots, A^{n-1}B)$ has maximal rank. It can be shown [15, 19] that a system that is structurally controllable is controllable for almost all choices of the entries, except for some pathological cases of zero measure that occur when the system parameters satisfy certain accidental constraints. Structural controllability has applicability in many areas beyond classical control systems, including in large-scale networks found in national critical infrastructures such as energy and telecommunication networks, and where such networks interact, as e.g., in Smart Grid control systems [6].

From a graph-theoretic point of view structural controllability is attractive because one can interpret the problem as a directed graph problem: first one observes that if one can choose the entries freely one can always assume that the entries b_{ij} of B are 0 if $i \neq j$ and 1 if $i = j$. Hence one only needs to consider A which can be represented by a directed graph G with a directed arc between a vertex i and j if $a_{ij} \neq 0$, i.e., if variable x_i affects dx_j/dt. For example:

$$\frac{dx_1}{dt} = ax_1 + bx_2 + y_1$$
$$\frac{dx_2}{dt} = cx_1$$
$$\frac{dx_3}{dt} = dx_2$$

$S = \{1\}$

|Equations | Graph G | Controlling set|

There are fairly simple obstructions for a set S to control a network:

Inaccessibility: There is a vertex v such that there is no directed path from any vertex $u \in S$ to v.

Dilation: There is a subset of vertices X, $X \cap S = \emptyset$, such that the size of the in-neighbourhood is smaller than $|X|$.

Lin [15] showed that these necessary conditions are indeed sufficient, that is, if there are no inaccessible vertices and no dilation then the set S controls the network. He also showed that the number of nodes needed to control a system is equal to the minimum number of vertex-disjoint *cacti* that span G. To define a cactus we need to define stems and buds first. A *stem* is a simple directed path. The first vertex is called the root and the last vertex is called the top of the stem. A *bud* is a simple directed cycle with an additional edge that ends but does not begin in a vertex of the cycle. A cactus consists of a stem and a collection of disjoint buds such that the initial vertex of the additional edge of each bud belongs to the stem. We may also assume that this initial vertex is not the top vertex. The following is an example of a cactus.

Consider the undirected bipartite graph B obtained from a directed graph G by splitting each vertex v into v^+ and v^- and adding an edge between u^+ and v^- in B if there is a directed edge uv in G.

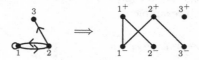

Note that a matching M in B corresponds to a union of $|V(G)| - |M|$ (possibly trivial) directed paths and some number of directed cycles in G. But in a sparse random graph there are whp only $o(n)$ directed cycles as the expected number of directed cycles of length k is constant for each fixed k. Hence if one wants to find the minimum number of disjoint cacti that span a sparse random graph up to an $o(n)$ term then one is interested in a maximum matching in B. Indeed, a vertex cover by t cacti gives rise to a vertex cover of t directed paths and $o(n)$ directed cycles, while a cover by t directed paths and $s = o(n)$ directed cycles gives rise to a cover by $t + s$ directed paths, which can be considered as $t + s$ cacti.

3 Maximum matchings in random bipartite graphs with a fixed degree distribution

3.1 The random graph models

Fix two sequences $(z_i)_{i=0}^{\infty}$ and $(\hat{z}_i)_{i=0}^{\infty}$ such that $z_i, \hat{z}_i \geq 0$ and

$$0 < \mu := \sum_{i=0}^{\infty} i z_i = \sum_{i=0}^{\infty} i \hat{z}_i < \infty. \qquad (3.1)$$

Co

We consider a sequence of random bipartite (multi-)graphs $B_n = (V_n \cup \hat{V}_n, E_n)$ with bipartite classes V_n and $\hat{V}_n$ such that

$$|V_n| = (1 + o(1))n \sum_{i=0}^{\infty} z_i, \qquad |\hat{V}_n| = (1 + o(1))n \sum_{i=0}^{\infty} \hat{z}_i,$$

by fixing the degrees of the vertices in V_n and $\hat{V}_n$ in such a way that as n tends to infinity the number of vertices of degree i in V_n is $(1 + o(1))nz_i$ and the number of vertices of degree i in $\hat{V}_n$ is $(1 + o(1))n\hat{z}_i$. Naturally, we require that the sum of the degrees in V_n and $\hat{V}_n$ are the same. For technical reasons we assume that the total number of edges is $(1+o(1))n\mu$. This assumption is mainly to avoid examples such as a sequence of double stars which have average degree 2 but all but two vertices have degree 1 and the size of the maximum matching is 2. Note that we do not require that $|V_n| = |\hat{V}_n|$ nor do we require that the distributions (z_i) and $(\hat{z}_i)$ are the same.

Having fixed the degrees $d(v)$ of the vertices $v \in V_n \cup \hat{V}_n$, we choose the edges randomly using the configuration model, that is, each vertex v is replaced by the appropriate number of *configuration points* $v_1, \ldots, v_{d(v)}$ and a perfect matching is chosen uniformly at random over all perfect matchings between the set of configuration points of V and the configuration points of $\hat{V}$. We then identify the configuration points corresponding to each vertex to obtain the bipartite multi-graph B_n with vertex classes V and $\hat{V}$.

We will also consider the corresponding non-bipartite version $G_n = (V_n, E_n)$ where we have a single degree distribution $(z_i)_{i=0}^{\infty}$, $|V_n| = (1+o(1))n \sum z_i$, and the total number of configuration points is chosen to be even and asymptotically $(1 + o(1))n\mu$, where $\mu = \sum i z_i < \infty$.

We are interested in the size of a maximum matching up to an $o(n)$ error. Because of the error term we may assume that the degrees are in fact bounded by an absolute constant Δ. To see this, note that adding or removing a single edge from a graph can affect the size of the maximum matching by at most 1. Since our distributions have finite mean we have that for all $\varepsilon > 0$, we can choose Δ sufficiently large such that

$$\sum_{i=0}^{\Delta} i z_i \geq \mu - \varepsilon.$$

Let $n_i^{(n)}$ be the number of vertices of degree i in V_n. Then, since $|E_n| = (1 + o(1))n\mu$ (in the bipartite case), we have that for sufficiently large n

$$\sum_{i=0}^{\Delta} i n_i^{(n)} \geq n(\mu - 2\varepsilon) \geq |E_n| - 3\varepsilon n.$$

Therefore there are at most $3\varepsilon n$ edges in B_n that are incident to vertices in V_n of degree more than Δ. By choosing Δ sufficiently large we can assume the same for $\hat{V}_n$. Hence removing at most $6\varepsilon n$ edges results in a graph of maximum degree at most Δ with a degree distribution that is asymptotically 6ε close to the original distribution. Moreover, it has a maximum matching that is within $6\varepsilon n$ of the original graph. A similar argument holds in the non-bipartite case of $G_n = (V_n, E_n)$. Thus we may assume for the remainder of the paper that the degrees are bounded by an absolute constant Δ which is independent of n. Similarly, by modifying the degrees of $o(n)$ vertices we may in fact assume that there are *exactly* $z_i n$ vertices in V_n and $\hat{z}_i n$ vertices in $\hat{V}_n$ of degree i. (Assuming of course that the z_i, $\hat{z}_i$ are rational, n is a multiple of their denominators, and $\sum i z_i n$ is even in the non-bipartite case.) To simplify notation we shall usually drop the subscript n and write the graphs as $B = (V \cup \hat{V}, E)$ or $G = (V, E)$.

3.2 Results

We consider the following generating functions. Let

$$f(x) = \sum_{i=0}^{\Delta} z_i x^i \quad \text{and} \quad \hat{f}(x) = \sum_{i=0}^{\Delta} \hat{z}_i x^i.$$

Note that $|V| = f(1)n$, $|\hat{V}| = \hat{f}(1)n$ and $\mu = f'(1) = \hat{f}'(1)$, so that $|E| = f'(1)n = \hat{f}'(1)n$.

Let w_1, w_2, $\hat{w}_1$, $\hat{w}_2$ be the smallest non-negative solutions to the following simultaneous equations

$$w_1 = \frac{f'(\hat{w}_2)}{f'(1)}, \quad w_2 = 1 - \frac{f'(1-\hat{w}_1)}{f'(1)}, \quad \hat{w}_1 = \frac{\hat{f}'(w_2)}{f'(1)}, \quad \hat{w}_2 = 1 - \frac{\hat{f}'(1-w_1)}{f'(1)}. \tag{3.2}$$

Define

$$\xi = f(1) + \hat{f}(1) - f(\hat{w}_2) - \hat{f}(1-w_1) - \hat{f}'(1-w_1)w_1,$$
$$\hat{\xi} = f(1) + \hat{f}(1) - \hat{f}(w_2) - f(1-\hat{w}_1) - f'(1-\hat{w}_1)\hat{w}_1. \tag{3.3}$$

Theorem 3.1 *Suppose that $\xi \leq \hat{\xi}$ and that*

$$\frac{\beta f''(\alpha)}{f'(\alpha+\beta) - f'(\alpha)} \cdot \frac{\hat{\beta}\hat{f}''(\hat{\alpha}+\hat{\beta})}{\hat{f}'(\hat{\alpha}+\hat{\beta}) - \hat{f}'(\hat{\alpha})} \leq 1 \tag{3.4}$$

along the trajectory $(\alpha, \beta, \hat{\alpha}, \hat{\beta})$ starting at $(\hat{w}_2, 1-\hat{w}_1-\hat{w}_2, w_2, 1-w_1-w_2)$ and evolving according to the differential equations given in (3.9) up until *the point at which $\phi = \sqrt{\delta}$. Then with probability tending to 1 as $n \to \infty$, the size of a maximum matching in $B = (V \cup \hat{V}, E)$ is*

$$(\xi - O(\sqrt{\delta}))n.$$

The conditions in Theorem 3.1 are somewhat awkward to check, however a much simpler statement is possible in the symmetric case when $z_i = \hat{z}_i$.

Theorem 3.2 *Suppose that $z_i = \hat{z}_i$ for all i, so that $f = \hat{f}$, $w_i = \hat{w}_i$, and $\xi = \hat{\xi}$, and suppose that $f''(x)^{-1/2}$ is convex on $[0, 1]$. Then with probability tending to 1 as $n \to \infty$, the size of a maximum matching in $B = (V \cup \hat{V}, E)$ is*

$$(\xi - o(1))n.$$

This last result also generalises to the non-bipartite case.

Theorem 3.3 *Let $f(x) = \sum_i z_i x^i$ be the generating function of $(z_i)_{i=0}^{\infty}$. Suppose that $f''(x)^{-1/2}$ is convex on $[0, 1]$. Define w_1 and w_2 to be the smallest non-negative solutions to the following simultaneous equations*

$$w_1 = \frac{f'(w_2)}{f'(1)}, \qquad w_2 = 1 - \frac{f'(1-w_1)}{f'(1)}.$$

Let $\xi = 2f(1) - f(w_2) - f(1 - w_1) - f'(1 - w_1)w_1$. Then with probability tending to 1 as $n \to \infty$, the size of a maximum matching in $G = (V, E)$ is

$$(\xi - o(1))n.$$

Let us remark that the condition that $f''(x)^{-1/2}$ is convex (or equivalently $2f''(x)f^{(4)}(x) \le 3f^{(3)}(x)^2$) holds for a wide variety of distributions, in particular it is implied by the log-concavity of dz_d which was assumed in [4]. We will discuss this condition later.

3.3 Karp–Sipser's algorithm

Given a graph G, Karp–Sipser's algorithm is a randomized algorithm that starts with an empty graph G' on the same vertex set, and chooses at each step an edge e of G, adds e to the graph G' and deletes e and all edges incident to e from G. If there are vertices of degree 1 in G then Karp–Sipser's algorithm chooses an edge incident to a vertex of degree 1 independently at random from all those edges. If there are no vertices of degree 1 then the Karp–Sipser's algorithm chooses an edge uniformly at random from all remaining edges of the graph.

For ease of exposition, we will delete the end vertices of e from the graph G when e is added to the matching. We shall say that a vertex *becomes isolated* if its degree is reduced to zero *without* it being included in the matching.

One key property of the Karp–Sipser algorithm is that it is optimal as long as there are degree 1 vertices, as given any degree 1 vertex v meeting the edge e, there is always some maximum matching containing the edge e. We split the Karp–Sipser algorithm into two phases. The first phase is all steps of the algorithm that occur before the first time when there are no degree 1 vertices left. The second phase is the remaining steps of the algorithm. As noted, the first phase is always optimal, so we can bound the difference between the size of the matching M_{KS} produced by the Karp–Sipser algorithm and the size of a maximum matching M_{max}. Indeed, suppose after the first phase the graph G has N remaining non-isolated vertices. If N_0 of these N vertices are not matched in the second phase of Karp–Sipser, then

$$|M_{\mathrm{KS}}| \leq |M_{\mathrm{max}}| \leq |M_{\mathrm{KS}}| + \lfloor N_0/2 \rfloor.$$

Indeed, the best we can hope for in the second phase is an additional $\lfloor N/2 \rfloor$ edges in our matching, so if we have $(N - N_0)/2$ additional edges then we are at most $\lfloor N_0/2 \rfloor$ short of a maximum matching. In the bipartite case we can say even more. Suppose there after the first phase the graph B has N remaining non-isolated vertices in the *smaller* bipartite class. If N_0 of these N vertices are not matched in the second phase of Karp–Sipser, then

$$|M_{\mathrm{KS}}| \leq |M_{\mathrm{max}}| \leq |M_{\mathrm{KS}}| + N_0,$$

as the maximum possible matching of the remaining edges would contain at most N edges.

Another key property of the Karp–Sipser algorithm is that at every step, if we condition on the degrees of the remaining vertices, the edges are distributed according to the configuration model. Indeed, we need only reveal the configuration as necessary as the algorithm proceeds. So, for example, when we choose a random edge we choose two configuration points uniformly at random and when we determine the neighbour of a particular configuration point, it is chosen uniformly at random from the collection of remaining configuration points. This gives the correct probability distribution of choices at each stage as in a uniformly chosen perfect matching, choosing a random edge is equivalent to picking two configuration points uniformly at random, and each configuration point is matched to another configuration point with a uniform distribution. Also, conditioning on the existence of an edge, the perfect matching of the remaining configuration points has a uniform distribution.

In Section 3.4 it will be convenient to modify the algorithm slightly so that when there are several vertices of degree 1 we do not necessarily choose one uniformly at random, but instead choose them according to some other scheme. The argument above still applies however, as long as the choice of degree 1 vertex does not depend on the edge it is incident to (i.e., on the choice of its matching configuration point). In Section 3.5 we will modify the algorithm further, allowing mistakes where an edge is chosen uniformly at random from the graph even when a small number of degree 1 vertices exist. Once again, the above argument is unaffected as long as the algorithm depends only on the degrees of the remaining vertices and not on the specific edges between them.

3.4   The first phase of Karp–Sipser

To analyse the first phase of the Karp–Sipser algorithm we will use branching-process techniques. It will turn out that the finite neighbourhood of most vertices is a tree and therefore branching processes are a good approximation. We will analyse the first phase of the algorithm in rounds. A round consist of marking all the degree 1 vertices and then processing each of these in some order. (The actual order is arbitrary, as long as it does not depend on the edges incident to these vertices.) Note that some of the degree 1 vertices may become isolated before they are processed in which case they are simply ignored in subsequent rounds. Note that strictly speaking this is a slight modification of Karp–Sipser, as we insist that all vertices of degree 1 in round t, say, are processed before any vertex of degree 1 that is generated by the removal of edges in round t.

Consider a vertex v and all vertices at distance at most t from v. The vertex v is called t-good or good if its t-neighbourhood is a tree. Otherwise it is bad. The next lemma shows that most vertices are good when $t = o(\log n)$.

Lemma 3.4 *Assume that $t = o(\log n)$. Then with high probability the number of bad vertices is $o(n)$.*

Proof We just give the proof for the bipartite case as non-bipartite case is similar. Fix a vertex v_1 and consider its t-neighbourhood. If v_1 is bad then this neighbourhood must contain a cycle. Pick a shortest such cycle and a shortest path connecting this cycle to v_1. This way we obtain a path $v_1, v_2, \ldots, v_\ell$ such that $\ell \le 2t + 1$ and v_ℓ is adjacent to some v_j, $j < \ell - 1$, or doubly adjacent to $v_{\ell-1}$. We count the expected number of such configurations. The probability of one such configuration is bounded from above by

$$\left(\prod_{i=1}^{\ell-1} \frac{d(v_i)d(v_{i+1})}{|E| - i + 1}\right) \frac{d(v_\ell)d(v_j)}{|E| - \ell + 1} \le \frac{\Delta^{2\ell}}{(|E| - \ell)^\ell} \le \frac{\Delta^{2\ell}}{|E|^{\ell-1}(|E| - \ell^2)}.$$

Indeed, conditioning on the presence of particular edges from v_k to v_{k+1} for $k < i$, the probability of an edge from v_i to v_{i+1} is at most

$$\frac{d(v_i)d(v_{i+1})}{|E| - i + 1}$$

as for each of the $d(v_i)$ (actually $d(v_i) - 1$ for $i > 1$) configuration points corresponding to v_i, there are at most $d(v_{i+1})$ out of the remaining $|E| - i + 1$ configuration points in the other bipartite class that this point can be joined to, and each are equally likely.

The number of such configurations is at most $\ell|E|^{\ell-1}$ as there are at most $|E|$ choices for each of the the vertices $v_2, \ldots, v_\ell$ and ℓ choices for j. (Clearly isolated vertices are not valid choices for the v_i and so there at most $|E|$ choices for each vertex.) Hence the probability of v_1 being bad is at most $\ell\Delta^{2\ell}/(|E| - \ell^2) \le c^\ell/n$ for some $c > 0$ as $|E| \sim \mu n$ and $\ell = o(\log n)$. The result now follows from Markov's inequality as

$$\mathbb{P}[\text{number of bad vertices} \ge \alpha] \le c^\ell/\alpha.$$

$\square$

Note that the fate of a vertex v during the first k rounds of the first phase of Karp–Sipser's algorithm is completely determined by the $2k + 1$ neighbourhood of v. For the remainder of this section we assume that the vertex v is t-good.

Fix a good vertex v and consider the tree of depth $t + 1$ rooted at v. Inductively, we classify all vertices at distance at most $t+1$ other than v as either v-*lonely*, v-*popular* or v-*normal* with respect to t. At depth $t + 1$ all vertices are v-normal. Suppose we have classified all vertices at level $\ell + 1$. Then a vertex at level ℓ is v-lonely if all its children on level $\ell + 1$ are v-popular. This also applies to the case when it has no children on level $\ell+1$. A vertex is v-popular if at least one of its children is v-lonely. Otherwise a vertex is v-normal. The root v itself will be *lonely* with respect to t if all of its children are v-popular (including the case when it is isolated), it will be *popular* with respect to t if at least *two* of its children are v-lonely and *normal* with respect to t otherwise (see Figure 1).

Level $t+1$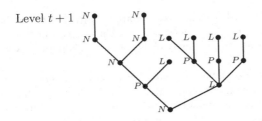

Figure 1: Classification of lonely, popular, and normal vertices.

Lemma 3.5 *Assume v is a t-good vertex and $t \geq 1$.*

(a) If v is lonely with respect to t then it will either be matched or become isolated within the first $\lfloor t/2 \rfloor$ rounds. If it is matched to u then u is popular with respect to $t - 1$.

(b) If v is popular with respect to t then it will be matched within the first $\lfloor t/2 \rfloor$ rounds. If it is matched to u then u is lonely with respect to $t + 1$.

(c) If v is normal with respect to t then it will not become isolated during the first $\lfloor t/2 \rfloor$ rounds. If it is matched to another vertex u during the first $\lfloor t/2 \rfloor$ rounds, then u is normal with respect to $t - 1$.

Proof Note first that whatever happens during the first $\lfloor t/2 \rfloor$ rounds to vertex v only depends on the tree rooted at v of depth $t + 1$. In particular whether a vertex becomes isolated or of degree 1 during the first $\lfloor t/2 \rfloor$ rounds only depends on this tree.

We now observe that as long as no chosen matching edge contains the root v, after i rounds, all matching edges within distance $t - 2i + 2$ of v connect a v-lonely vertex (with respect to t) with an v-popular parent (with respect to t). Indeed, after i rounds, if v is not matched then any v-popular vertex at distance at least $t - 2i$ from v will be matched with a child that is v-lonely, and any v-lonely vertex at distance at least $t + 1 - 2i$ from v will be either matched or become isolated. Note that no vertex at level t can be v-popular, so any v-lonely vertex at distance t or $t-1$ from v has no children. Hence, assuming v is not matched or isolated after $\lfloor t/2 \rfloor$ rounds, then all its v-popular children will have been matched, leaving it isolated if v is itself lonely. If v is lonely and matched, then it must have done so because it became degree 1. The vertex u that it is matched to is clearly popular with respect to $t - 1$ as it is v-popular. Thus u has a v-lonely child, which is u-lonely with respect to $t - 1$. On the other hand, all the children of v other than u were matched after $\lfloor t/2 \rfloor - 1$ rounds and so must be v-popular with respect to $t - 2$. Thus v is u-lonely with respect to $t - 1$. Hence u has two u-lonely children and so is popular with respect to $t - 1$.

Similarly, if v is popular, then at least two of its children are v-lonely. But all of their children are then matched within the first $\lfloor t/2 \rfloor - 1$ rounds, and so these children are of degree 1 before round $\lfloor t/2 \rfloor$. Thus v is matched in this round if not before. Since v has two children that are v-lonely, and v-lonely vertices can only be matched to their parent prior to v being matched or isolated, it is clear v does not become isolated or degree 1, but instead is matched to one of its children. This child u is clearly lonely with respect to $t + 1$ as all its neighbours (including v) are u-popular (v itself having a lonely child other than u).

Now assume v is normal. Then not all of its children are v-popular and therefore not all of its children can be removed during this process. Therefore v never becomes isolated. Now assume v is matched to u and assume for a contradiction that u is not normal with respect to $t-1$. Before v is matched, the only children it can lose are v-popular by the observation above. Moreover, these v-popular neighbours have degree at least 2 or are matched to an v-lonely child. Assume first that u is popular with respect to $t - 1$. Since v is normal there must be at least one neighbour w of v that is not v-popular with respect to t and therefore won't be removed before v is matched. This means v has degree at least 2 when it is matched. But u must have a u-lonely neighbour other than v that cannot be matched except to u. Hence u also has degree at least 2 or is matched to a u-lonely vertex which is not v. Therefore uv cannot be a matching edge.

Now assume that u is lonely with respect $t - 1$. Then u is v-lonely with respect to t. Also v is u-popular with respect to $t - 1$ which means that there is at least one child other than u which is u-lonely with respect to $t - 1$ and in particular it is v-lonely with respect to t as being v-lonely is monotone in t. Thus v has two v-lonely children with respect to t, a contradiction to the assumption that v is normal with respect to t. $\square$

We will show that the solutions w_1, w_2, $\hat{w}_1$, and $\hat{w}_2$ of (3.2) mentioned in the introduction represent the probability of a vertex in V or $\hat{V}$ being v-lonely or v-popular if they are close to a good vertex v. We use iterated versions of w_i and $\hat{w}_2$ to make this precise. In particular, we define $w_i^{(t)}$ and $\hat{w}_i^{(t)}$ using the following recursion.

Set
$$w_1^{(0)} = w_2^{(0)} = \hat{w}_1^{(0)} = \hat{w}_2^{(0)} = 0,$$

and for $t \geq 1$ inductively define

$$w_1^{(t)} = \frac{f'(\hat{w}_2^{(t-1)})}{f'(1)}, \quad w_2^{(t)} = 1 - \frac{f'(1-\hat{w}_1^{(t-1)})}{f'(1)},$$

and

$$\hat{w}_1^{(t)} = \frac{\hat{f}'(w_2^{(t-1)})}{\hat{f}'(1)}, \quad \hat{w}_2^{(t)} = 1 - \frac{\hat{f}'(1-w_1^{(t-1)})}{\hat{f}'(1)}. \tag{3.5}$$

Theorem 3.6 *Let $t = o(\log n)$ and fix a good vertex v. For all $i = 1, \ldots, t$ the following holds.*

(a) *The probability that a vertex u in V at distance $t + 1 - i$ is v-lonely with respect to t conditioned on its existence and the tree of depth t from v except for the branch rooted at u is $w_1^{(i)} + O(c^{t+i}/n)$ for some constant c.*

(b) *The probability that a vertex u in V at distance $t + 1 - i$ is v-popular with respect to t conditioned on its existence and the tree of depth t from v except for the branch rooted at u is $w_2^{(i)} + O(c^{t+i}/n)$ for some constant c.*

The corresponding statements hold for $u \in \hat{V}$ with $w_1^{(i)}$, $w_2^{(i)}$ replaced with $\hat{w}_1^{(i)}$, $\hat{w}_2^{(i)}$ respectively. Moreover, these results remain true even conditioned on the location of $O(c^t)$ edges outside of the tree of depth t from v.

Proof The statements are clearly true for $i = 0$ as all vertices at level $t + 1$ are v-normal with respect to t.

Assume that we explored the entire tree with root v of depth $t + 1$ except for one branch starting at level $t + 1 - i$. We know that there is an edge from level $t - i$ to a vertex u but we do not know anything further about this branch. Note that in the configuration model we have fixed at most Δ^{t+1} edges. The probability that u is v-lonely (conditioning on the remaining tree) is

$$\sum_{k=0}^{\Delta-1} \mathbb{P}(u \text{ has children } u_1 \ldots u_k) \prod_{j=1}^{k} \mathbb{P}(u_j \text{ is } v\text{-pop.} \mid u_1 \ldots u_{j-1} \text{ are } v\text{-pop.}). \tag{3.6}$$

The probability that at the beginning, we choose an edge adjacent to vertex of degree $k + 1$ is $(k+1)z_{k+1}/\mu$ and hence the probability that u has k children is

$$\frac{(k+1)z_{k+1}n - O(\Delta^t)}{|E| + O(\Delta^t)} = \frac{(k+1)z_{k+1}}{\mu} + O(\Delta^t/n).$$

Note that the $O(\Delta^t)$ term takes care of the edges that are already fixed. The probability p_j that a child u_j of u is v-popular with respect to t conditioned on the remaining tree including the previous children of u is by induction hypothesis $\hat{w}_2^{(i-1)} + O(c^{t+i-1}/n)$. Thus (3.6) gives

$$\sum_{k=0}^{\Delta-1} \left(\frac{(k+1)z_{k+1}}{\mu} + O(\Delta^t/n) \right) p_1 p_2 \ldots p_k$$

$$= \sum_{k=0}^{\Delta-1} \frac{(k+1)z_{k+1}}{\mu} p_1 \ldots p_k + O(\Delta^{t+1}/n)$$

$$= \sum_{k=0}^{\Delta-1} \frac{(k+1)z_{k+1}}{\mu} ((\hat{w}_2^{(i-1)})^k + k \cdot O(c^{t+i-1}/n)) + O(\Delta^{t+1}/n)$$

$$= \frac{f'(\hat{w}_2^{(i-1)})}{f'(1)} + \frac{f''(1)}{f'(1)} O(c^{t+i-1}/n) + O(\Delta^{t+1}/n).$$

Here we used that $p_j = \hat{w}_2^{(i-1)} + O(c^{t+i-1}/n)$ is between 0 and 1. Note that the first term is by definition $w_1^{(i)}$ and the error term is $O(c^{t+1}/n)$ provided $c > \max\{f''(1)/f'(1), \Delta\}$.

The other statements are proved analogously. $\qquad\square$

Theorem 3.7 *Let $t = t(n) \to \infty$ with $t = o(\log n)$. Then the following hold whp.*

(a) *The number of good vertices $v \in V$ that are lonely w.r.t. t is $(f(\hat{w}_2) + o(1))n$.*

(b) *The number of good vertices $v \in V$ that are popular w.r.t. t is $(f(1) - f(1 - \hat{w}_1) - f'(1 - \hat{w}_1)\hat{w}_1 + o(1))n$.*

(c) *The number of good vertices $v \in V$ that remain unmatched and non-isolated after t rounds is $(f(1 - \hat{w}_1) - f(\hat{w}_2) - f'(\hat{w}_2)\hat{w}_2 + o(1))n$.*

(d) *The number of vertices $v \in V$ that are of degree $k \geq 1$ after t rounds is $(z'_k + o(1))n$, where z'_k are defined by the generating function $\sum z'_k x^k = f(\alpha + \beta x) - f(\alpha) - \beta f'(\alpha)x$, and $\alpha = \hat{w}_2$, $\beta = 1 - \hat{w}_1 - \hat{w}_2$.*

(e) *The number of vertices that have degree 1 after t rounds is $o(n)$.*

The same statements apply to $\hat{V}$ with f replaced by $\hat{f}$ and $\hat{w}_i$ replaced by w_i.

Proof Note that $w_i^{(t)}$ and $\hat{w}_i^{(t)}$ are bounded increasing sequences and as $t \to \infty$, $w_i^{(t)} \to w_i$ and $\hat{w}_i^{(t)} \to \hat{w}_i$. Also, the properties of being v-lonely or v-popular with respect to t are increasing in t. A vertex $v \in V$ is lonely with respect to t if all its children are popular, so by Theorem 3.6, this occurs with probability $(\hat{w}_2^{(t)} + o(1))^{\deg(v)}$. The expected number of vertices in V that are lonely with respect to t is therefore $f(\hat{w}_2^{(t)} + o(1))n = (f(\hat{w}_2) + o(1))n$. To get concentration we use the second moment method. Let $d(u,v)$ denote the distance between u and v. By coupling a random instance of the bipartite graph with an instance in which the $t+1$-neighbourhood of v_2 is fixed, it is clear that

$$\left| \mathbb{P}[v_1 \text{ lonely} \mid v_2 \text{ lonely}] - \mathbb{P}[v_1 \text{ lonely}] \right| \le \mathbb{P}[d(v_1, v_2) < 2t + 1].$$

Hence, if X represents the number of lonely vertices in V,

$$\begin{aligned}
\mathrm{Var}(X) &= \mathbb{E}(X^2) - \mathbb{E}(X)^2 \\
&= \sum_{v_1, v_2} \left(\mathbb{P}[v_1, v_2 \text{ lonely}] - \mathbb{P}[v_1 \text{ lonely}] \mathbb{P}[v_2 \text{ lonely}] \right) \\
&= O\left(\sum_{v_1, v_2} \mathbb{P}[d(v_1, v_2) < 2t + 1] \right) = O(n\Delta^{2t})
\end{aligned}$$

Thus $|X - \mathbb{E}(X)| \le n^{1/2+\varepsilon}$ with high probability. Parts (b)-(d) are proved similarly. In (c) we note that as $t \to \infty$ and the events of being popular and lonely are monotone, it is enough to count vertices v that have no v-lonely children and at least two children that are not popular. In (d) we are counting the number of these vertices with exactly $k \ge 2$ children that are not v-popular. Statement (e) is an instant corollary of (d). $\square$

3.5 Evolution of the generating functions

We generalise Theorem 3.7(d) by showing that after any number of steps of the Karp–Sipser algorithm, in either phase 1 or phase 2, the generating function of the degree distribution of the remaining vertices is of a simple form. Recall that we delete vertices that are used in the matching.

We shall use the differential equation method of Wormald (see [20]). However, to directly apply this method in the form proved in [20], we shall modify the Karp–Sipser algorithm slightly. Fix a $\delta > 0$ and run the Karp–Sipser algorithm with the following modification. At each step, if the total number N_1 of degree 1 vertices is less than δn, but more than zero, then with probability $N_1/(\delta n)$ we run the original Karp–Sipser algorithm by picking uniformly at random an edge incident to a degree 1 vertex. However, with probability $1 - N_1/(\delta n)$ we instead pick an edge uniformly from the graph (i.e., we don't notice that there are degree 1 vertices present). The advantage of this modification is that the derivatives in the differential equation method will now be Lipschitz. (In [4] a different method we employed to avoid the discontinuity in the algorithm between $N_1 = 0$ and $N_1 > 0$, namely steps were grouped together so that $N_1 = 0$ always held after one of these multiple steps. Here however we are less concerned about the error terms so will use this simpler method.)

Let $n_d(t)$ and $\hat{n}_d(t)$ be the number of degree d vertices in V and $\hat{V}$ respectively after t steps of the algorithm, and let $\mathcal{F}_t$ be the filtration given by the edges revealed up to and including the tth step. Write $\Delta n_d(t) = n_d(t) - n_d(t-1)$ and define

$$\eta(t) = \frac{n_1(t)}{\max\{\delta n, n_1(t) + \hat{n}_1(t)\}}$$

and similarly for $\hat{\eta}(t)$. Then with probability $\eta(t)$ the algorithm picks an edge adjacent to a degree 1 vertex in V, with probability $\hat{\eta}(t)$ the algorithm picks an edge adjacent to a degree 1 vertex in $\hat{V}$, and with probability $1 - \eta(t) - \hat{\eta}(t)$ an edge is picked uniformly at random from the entire graph. When an edge is chosen adjacent to a degree 1 vertex in $\hat{V}$, or arbitrarily in $\hat{V}$, then the matching vertex in V is of degree d with probability $dn_d(t)/E(t)$, where $E(t) = \sum dn_d(t) = \sum d\hat{n}_d(t)$ is the number of remaining edges in the graph. From this is is possible to calculate the expected change in the $n_d(t)$ for $0 \leq d \leq \Delta$ as

$$\mathbb{E}[\Delta n_d(t) \mid \mathcal{F}_{t-1}] = -\eta \delta_{1d} - (1 - \eta)\frac{dn_d}{E}$$
$$+ (1 - \hat{\eta})\left\{\frac{(d+1)n_{d+1} - dn_d}{E} \sum_{k=1}^{\Delta} \frac{k\hat{n}_k}{E}(k-1) + O(\Delta^3/E)\right\}.$$

$$(3.7)$$

Here $\delta_{1d} = 1$ if $d = 1$ and $\delta_{1d} = 0$ if $d \neq 1$. The first two terms gives the change caused in $n_d(t)$ as a result of a degree d vertex in V being chosen in the matching. Either this vertex is chosen from one of the degree 1 vertices (with probability η), or it is the vertex containing a randomly chosen configuration point. The last term gives the change in n_d caused by the removal of edges incident to the matching edge. If the matching edge meets a vertex of degree 1 in $\hat{V}$ then no such edges meet V other than at the matching vertex. Otherwise we need to remove $k - 1$ configuration points from the V side where k is the degree of the matched vertex in $\hat{V}$. Removing one configuration point increases n_d by an average of $\frac{1}{E}((d+1)n_{d+1} - dn_d)$, i.e., the probability of this configuration point corresponds to a degree $d+1$ vertex (which then becomes degree d), minus the probability that this configuration point corresponds to a degree d vertex (which then becomes degree $d-1$). If all $k-1$ removed configuration points corresponded to distinct vertices, then we could just multiply $\frac{1}{E}((d+1)n_{d+1} - dn_d)$ by $k-1$. There is however an error when $k > 2$ as it is possible for the degree of a vertex to drop by more than 1 when there are multiple edges in B. We bound the expected contribution of this by the number of pairs of configuration points in $\hat{V}$, namely $\binom{k-1}{2} = O(\Delta^2)$, multiplied by the probability that they match to the same vertex in V, which is $O(\Delta/E)$.

We can rewrite (3.7) in the form

$$\mathbb{E}[\Delta n_d(t) \mid \mathcal{F}_{t-1}] = h_d\left(\tfrac{n_1(t)}{n},\dots,\tfrac{n_\Delta(t)}{n},\tfrac{\hat{n}_1(t)}{n},\dots,\tfrac{\hat{n}_\Delta(t)}{n}\right) + O(1/n)$$

with

$$h_d(\zeta,\hat{\zeta}) = -\delta_{d1}\eta - (1-\eta)\phi^{-1}d\zeta_d$$
$$+ (1-\hat{\eta})\phi^{-2}((d+1)\zeta_{d+1} - d\zeta_d)\sum(k-1)k\hat{\zeta}_k, \qquad (3.8)$$

where $\zeta = (\zeta_0,\dots,\zeta_\Delta)$, $\hat{\zeta} = (\hat{\zeta}_0,\dots,\hat{\zeta}_\Delta)$, $\phi = \sum d\zeta_d = \sum d\hat{\zeta}_d$, $\eta = \zeta_1/\max\{\delta,\zeta_1+\hat{\zeta}_1\}$, and $\hat{\eta} = \hat{\zeta}_1/\max\{\delta,\zeta_1+\hat{\zeta}_1\}$. We define $\hat{h}_d$ similarly and note that h_d and $\hat{h}_d$ are Lipschitz on the domain

$$D = \{(\zeta,\hat{\zeta}) \in [0,\infty)^{2\Delta+2} : \phi \geq \sqrt{\delta}\}.$$

Hence we can apply Theorem 5.1 of [20] (with $\gamma = 0$, $\beta = \Delta$, $\lambda = n^{-1/4}$) to deduce that

$$n_d(t) = n\zeta_d(t) + O(n^{3/4}), \qquad \hat{n}_d(t) = n\hat{\zeta}_d(t) + O(n^{3/4}),$$

with probability $1 - e^{-\Omega(n^{1/4})}$ uniformly in the set of t for which $\sum d\zeta_d(t) \geq \sqrt{\delta}$, where $\zeta_d(t)$ and $\hat{\zeta}_d(t)$ are the (unique) solutions to the system of differential equations

$$\frac{d}{dt}\zeta_d(t) = h_d(\zeta(t),\hat{\zeta}(t)), \qquad \frac{d}{dt}\hat{\zeta}_d(t) = \hat{h}_d(\zeta(t),\hat{\zeta}(t)), \qquad d = 0,\dots,\Delta,$$

with initial conditions $\zeta_d(0) = z_d$, $\hat{\zeta}_d(0) = \hat{z}_d$.

Lemma 3.8 *Running the modified Karp-Sipser's algorithm yields with high probability at any stage a bipartite graph such that there exist α, β, γ, c*

and $\hat{\alpha}, \hat{\beta}, \hat{\gamma}, \hat{c}$ so that the degree generating functions of the bipartite graph are $f(\alpha + \beta x) - \gamma x - c + \varepsilon(x)$ and $\hat{f}(\hat{\alpha} + \hat{\beta}x) - \hat{\gamma}x - \hat{c} + \hat{\varepsilon}(x)$, where $\varepsilon(x) = \sum_{i=2}^{\Delta} \varepsilon_i x^i$, $\hat{\varepsilon}(x) = \sum_{i=2}^{\Delta} \hat{\varepsilon}_i x^i$ with $\varepsilon_i, \hat{\varepsilon}_i = o(1)$. Moreover, the coefficients $\alpha, \beta, \gamma, \hat{\alpha}, \hat{\beta}, \hat{\gamma}$ evolve according to the coupled differential equations

$$\left. \begin{array}{l} \frac{d}{dt}\alpha = \beta\hat{\beta}^2(1-\hat{\eta})\phi^{-2}\hat{f}''(\hat{\alpha}+\hat{\beta}), \\[2mm] \frac{d}{dt}(\alpha+\beta) = -(1-\eta)\phi^{-1}, \\[2mm] \frac{d}{dt}\gamma = \eta + \frac{\gamma}{\beta}\frac{d\beta}{dt}, \end{array} \right\} \qquad (3.9)$$

where

$$\eta = \zeta_1 / \max\{\delta, \zeta_1 + \hat{\zeta}_1\}, \quad \zeta_1 = \beta f'(\alpha) - \gamma,$$

$$\phi = \beta f'(\alpha + \beta) - \gamma = \hat{\beta}\hat{f}'(\hat{\alpha} + \hat{\beta}) - \hat{\gamma}$$

and similarly for $\hat{\alpha}, \hat{\beta}, \hat{\gamma}, \hat{\eta}, \hat{\zeta}_1$.

Proof Write $f_t(x) = \sum_{d=0}^{\Delta} \zeta_d(t)x^d$, $\hat{f}_t(x) = \sum_{d=0}^{\Delta} \hat{\zeta}_d(t)x^d$. Now from (3.8),

$$\frac{d}{dt}f_t(x) = -\eta x - (1-\eta)\frac{1}{f_t'(1)}xf_t'(x) + (1-\hat{\eta})\frac{\hat{f}_t''(1)}{\hat{f}_t'(1)^2}(1-x)f_t'(x).$$

as $\phi = f_t'(1) = \hat{f}_t'(1)$. Now this is of the form $\frac{d}{dt}f_t(x) = -A(t)x - B(t)xf_t' + C(t)(1-x)f_t'$, whose solution is easily seen to be of the form

$$f_t(x) = f(\alpha + \beta x) - \gamma x - c$$

for some functions $\alpha(t)$, $\beta(t)$, $\gamma(t)$, and $c(t)$ satisfying (3.9). $\qquad \square$

Proof of Theorem 3.1 We run the 1st phase of Karp–Sipser for t_0 rounds, $t_0 = o(\log n)$, $t_0 \to \infty$. After this phase the generating function for the degrees in V is given approximately by $f_{t_0}(x)$ which by Theorem 3.7(d) is of the form $f(\alpha + \beta x) - f'(\alpha)x - c$ for some constant c. Here $\alpha = \hat{w}_2$, $\beta = 1 - \hat{w}_1 - \hat{w}_2$. The number of vertices in V that become isolated up to time t is by Lemma 3.5 the difference between the number of lonely vertices in V and the number of popular vertices in $\hat{V}$, which by Theorem 3.7(b) is just

$$(f(\hat{w}_2) - \hat{f}(1) + \hat{f}(1 - w_1) + \hat{f}'(1 - w_1)w_1)n + o(n).$$

By Theorem 3.7(c) there are $(f(1-\hat{w}_1) - f(\hat{w}_2) - f'(\hat{w}_2)\hat{w}_2)n + o(n)$ vertices with degree at least 2, and if all, or almost all, of these are matched then the size of the matching will be

$$\{f(1) - (f(\hat{w}_2) - \hat{f}(1) + \hat{f}(1 - w_1) + \hat{f}'(1 - w_1)w_1)\}n + o(n) = (\xi - o(1))n.$$

Assume that $\xi \leq \hat{\xi}$, so that after phase 1 there are at least as many non-isolated vertices in $\hat{V}$ as in V. We then run phase 2 and bound the number of vertices of V that become isolated. We note that if throughout the evolution of the process $\zeta_1(t) \leq \delta$, then the number of vertices that become isolated is small. Indeed, by (3.8),

$$\frac{d}{dt}\zeta_0 = h_0(\zeta(t), \hat{\zeta}(t)) = (1 - \hat{\eta})\phi^{-2}\zeta_1 \sum (k-1)k\hat{\zeta}_k$$

and $\sum(k-1)k\hat{\zeta}_k \leq \Delta\phi$, so $\frac{d}{dt}\zeta_0 = O(\zeta_1/\phi) = O(\sqrt{\delta})$ when $\phi \geq \sqrt{\delta}$. Thus $\zeta_0(t) = O(n\sqrt{\delta})$ for all relevant t, and hence is $o(n)$ as $\delta \to 0$.

Thus it is enough to show that if $\zeta_1 \geq \delta$ then $\frac{d\zeta_1}{dt} \leq 0$, or equivalently $h_1(\zeta, \hat{\zeta}) \leq 0$. From (3.8), this is equivalent to

$$-\eta - (1 - \eta)\phi^{-1}\zeta_1 + (1 - \hat{\eta})\phi^{-2}(2\zeta_2 - \zeta_1)\sum(k-1)k\hat{\zeta}_k \leq 0.$$

However, $\zeta_1 \geq \delta$ implies that $1 - \hat{\eta} = \eta$, so it is enough if

$$\phi^{-2}2\zeta_2 \sum(k-1)k\hat{\zeta}_k \leq 1.$$

Now $\phi = f'_t(1) = \hat{f}'_t(1)$, $\sum(k-1)k\hat{\zeta}_k = f''_t(1)$, and $2\zeta_2 = f''_t(0)$. Thus it is enough that

$$\frac{f''_t(0)}{f'_t(1)} \cdot \frac{\hat{f}''_t(1)}{\hat{f}'_t(1)} \leq 1 \tag{3.10}$$

for all t. Now $f''_t(0) = \beta^2 f''(\alpha)$, $\hat{f}''(1) = \hat{\beta}^2 \hat{f}''(\hat{\alpha} + \hat{\beta})$, and $f'_t(1) \geq \beta(f'(\alpha + \beta) - f'(\alpha))$, $\hat{f}'_t(1) \geq \hat{\beta}(\hat{f}'(\hat{\alpha} + \hat{\beta}) - \hat{f}'(\hat{\alpha}))$. Thus (3.10) is implied by (3.4). Note that it is enough to require this only along the trajectory of $(\alpha, \beta, \hat{\alpha}, \hat{\beta})$ taken by the algorithm in phase 2. This trajectory is given by Lemma 3.8. $\qquad\square$

Proof of Theorem 3.2 In the symmetric case we have $\hat{f} = f$, $\hat{\alpha} = \alpha$, $\hat{\beta} = \beta$. Thus it is enough to check (3.4) for all $0 \leq \alpha < \alpha + \beta \leq 1$ as this covers all possible trajectories for any $\delta > 0$. In the symmetric case this is equivalent to

$$f''(\alpha)f''(\alpha + \beta) \leq ((f'(\alpha + \beta) - f'(\alpha))/\beta)^2. \tag{3.11}$$

The result then follows from the following lemma. $\qquad\square$

Lemma 3.9 *Equation (3.11) holds for all $0 \leq \alpha < \alpha + \beta \leq 1$ if and only if $(f''(x))^{-1/2}$ is convex in $[0,1]$.*

Proof First we note that, as $f(x)$ is a polynomial and hence infinitely differentiable, the convexity of $f''(x)^{-1/2}$ is equivalent to the condition

$$2f''(x)f^{(4)}(x) \le 3f^{(3)}(x)^2. \tag{3.12}$$

Indeed

$$\frac{d^2}{dx^2}f''(x)^{-1/2} = -\frac{1}{2}\frac{d}{dx}f''(x)^{-3/2}f^{(3)}(x)$$

$$= \frac{1}{4}f''(x)^{-5/2}(3f^{(3)}(x)^2 - 2f''(x)f^{(4)}(x)).$$

Now suppose (3.11) holds for all $0 \le \alpha < \alpha + \beta \le 1$. Expanding both sides of (3.11) as a Taylor series up to β^2 gives

$$f''(\alpha)(f''(\alpha) + \beta f^{(3)}(\alpha) + \tfrac{\beta^2}{2}f^{(4)}(\alpha) + O(\beta^3))$$

$$\le (f''(\alpha) + \tfrac{\beta}{2}f^{(3)}(\alpha) + \tfrac{\beta^2}{6}f^{(4)}(\alpha) + O(\beta^3))^2,$$

which after some simplification gives

$$2f''(\alpha)f^{(4)}(\alpha) \le 3f^{(3)}(\alpha)^2 + O(\beta). \tag{3.13}$$

Letting $\beta \to 0$ gives (3.12) for all $x = \alpha \in [0,1)$ (and hence also at $x = 1$ by continuity) as required.

Now assume $f''(x)^{-1/2}$ is convex. Then for $t \in [0,1]$,

$$f''(\alpha + t\beta)^{-1/2} \le c_0(1-t) + c_1 t,$$

where $c_0 = f''(\alpha)^{-1/2}$ and $c_1 = f''(\alpha + \beta)^{-1/2}$. Thus

$$f'(\alpha + \beta) - f'(\alpha) = \int_\alpha^{\alpha+\beta} f''(x)\, dx$$

$$\ge \beta \int_0^1 (c_0(1-t) + c_1 t)^{-2}\, dt$$

$$= \beta\Big[-(c_1 - c_0)^{-1}(c_0(1-t) + c_1 t)^{-1} \Big]_0^1$$

$$= \frac{\beta}{c_1 - c_0}\Big(\frac{1}{c_0} - \frac{1}{c_1}\Big)$$

$$= \frac{\beta}{c_0 c_1} = \beta(f''(\alpha)f''(\alpha + \beta))^{1/2}$$

which is just (3.11).   □

The proof of Theorem 3.3 is exactly analogous, so we omit it. Finally we show that the condition used in [4], namely the log-concavity of the sequence dz_d, is enough to deduce the convexity of $f''(x)^{-1/2}$ and hence apply Theorem 3.3.

Lemma 3.10 *If the sequence dz_d is log-concave then $f''(x)^{-1/2}$ is convex on $[0, 1]$.*

Proof Let $g(x) := f'(x) = \sum_{d=1}^{\Delta} dz_d x^{d-1}$. If we write $g(x) = \sum a_n x^n$, then log-concavity of (dz_d) is equivalent to the log-concavity of (a_n). Write $g(x + h) = \sum a_n(x) h^n$, so that $a_n(x) = \frac{1}{n!} \frac{d^n}{dx^n} g(x)$ and $a_n(0) = a_n$. We first show that for all $x \geq 0$, $(a_n(x))$ is log-concave. Log-concavity of $(a_n(x))$ at one value of x is equivalent to $\varepsilon_{n,k}(x) := a_{n+k-1}(x) a_{n+1}(x) - a_{n+k}(x) a_n(x) \geq 0$ for all $k \geq 2$, $n \geq 0$, see for example [13]. Now $\frac{d}{dx} a_n(x) = (n+1) a_{n+1}(x)$. Hence

$$
\begin{aligned}
\frac{d}{dx} \varepsilon_{n,k}(x) &= (n+k) a_{n+k}(x) a_{n+1}(x) + (n+2) a_{n+k-1}(x) a_{n+2}(x) \\
&\quad - (n+k+1) a_{n+k+1}(x) a_{n+1}(x) - (n+1) a_{n+k}(x) a_{n+1}(x) \\
&= (k-1) a_{n+k}(x) a_{n+1}(x) + (n+2) a_{n+k-1}(x) a_{n+2}(x) \\
&\quad - (n+k+1) a_{n+k+1}(x) a_{n+1}(x) \\
&= (k-1) \varepsilon_{n+1,k}(x) + (n+2) \epsilon_{n+2,k-1}(x)
\end{aligned}
$$

As $\varepsilon_{n,1}(x) = 0$, and $\epsilon_{n,k}(0) \geq 0$ for $k \geq 2$, we deduce that $\varepsilon_{n,k}(x) \geq 0$ for all $x \geq 0$ and $k \geq 2$.

Now as $(a_n(x))$ is log-concave, $a_1(x) a_3(x) \leq a_2(x)^2$. It follows that $\frac{1}{6} g'(x) g^{(3)}(x) \leq \frac{1}{4} g''(x)^2$, or equivalently $2 f''(x) f^{(4)}(x) \leq 3 f^{(3)}(x)^2$, as required. □

4   Matchings with an adversary

In this section we are interested in the problem of finding the largest matching in a random bipartite graph after an adversary has deleted some edges. More precisely, we consider a random bipartite graph $G = (A \cup B, E)$ with $|A| = n$ and $|B| = (1 + \varepsilon)n$. Each vertex in A is adjacent to d neighbours chosen uniformly at random from B. We allow repetition, so this is a multigraph. An adversary is then able to remove a single edge adjacent to each vertex of A, with the aim of minimising the size of the largest matching. We want find to the smallest ε such that a matching of size n exists. Note that this problem corresponds to finding a small control set in a directed network where each node has d-out-neighbours and an adversary can manipulate one out-arc at each node. Our result then implies that we can control roughly $(1 - 4 \log d/d^2)n$ nodes with a single node, which implies that there is a control set of size about $4 \log d/d^2$.

Theorem 4.1 ([2]) *Let $G = (A \cup B, E)$ with $|A| = n$ and $|B| = (1 + \varepsilon)n$ be a random bipartite (multi-)graph in which each vertex in A chooses d vertices uniformly at random with repetition from B. For each $\eta > 0$ there exists a d_0 such that for all $d \geq d_0$, if $\varepsilon > (4 + \eta) \log d/d^2$, then with high probability an adversary who deletes one edge incident to each vertex in A cannot destroy all matchings of size n. On the other hand if $\varepsilon < (4 - \eta) \log d/d^2$, then with high probability such an adversary can destroy all matchings of size n.*

The problem of finding such resilient matchings is closely related to finding a maximum strongly independent set in d-uniform hypergraphs $\mathcal{H}^d_{\tilde{n},n}$ on $\tilde{n} \geq n$ vertices and n edges. A strongly independent set in a d-uniform hypergraph is a set of vertices such that no hyperedge contains two or more vertices of this set, see for example [3]. Here, the partition class B of the bipartite graph G corresponds to the vertices of the hypergraph and the neighbourhoods of the vertices of A correspond to the edges (multi-edges are a small nuissance but one can show with Markov's inequality that whp there are not many, say less than $\sqrt{n}$) . Clearly, an adversary can isolate all vertices corresponding to a strongly independent set, and hence if the maximum size of a strongly independent set is β then the adversary can force the size of a maximum matching to be at most $\tilde{n} - \beta$. We will show that $\mathcal{H}^d_{\tilde{n},n}$ contains whp a strongly independent set of size $(4 - \eta)n \log d/d^2$ where η can be chosen arbitrarily small if d is sufficiently large.

As a note on notation, asymptotically, we are interested in the probabilistic results as $n \to \infty$, and so by $o(1)$ we mean a function that tends to 0 as $n \to \infty$, but since we are also considering $d \to \infty$ (and at the same time $\varepsilon \to 0$), we may also require the use of little o notation to denote the size of terms which do not depend on n, in which case we will label them $o_d(f(d))$ to indicate that the asymptotics depend on d rather than n. As an example $\varepsilon = o_d(1)$, but neither depend on n and hence are asymptotically constant in terms of n.

To prove that there exists a matching of size $n = |A|$ if $|B|$ is sufficiently large we use Hall's theorem [10]. Recall that by Hall's theorem for a bipartite graphs G with partition classes X and Y, a matching of size $|X|$ exists if and only if, for all $X' \subseteq X, |\Gamma(X')| \geq |X'|$. Here $\Gamma(X')$ is the set of neighbours of X' (in Y). With a careful (and lengthy) analysis we show in [2] that Hall's condition is satisfied and obtain the following theorem.

Theorem 4.2 *Let G be the random bipartite graph with partition sets A and B of size n and $\tilde{n} = (1+\varepsilon)n$ respectively, and each vertex of A chooses d vertices uniformly at random with repetition from B. An adversary deletes a single edge incident to each vertex of A to obtain G'. For each $\eta > 0$ there exists a d_0 such that for all $d \geq d_0$ and $\varepsilon > (4 + \eta)\log d/d^2$, a matching of size n still exists in G' with probability tending to 1 as $n \to \infty$.*

To prove that there is no mathcing of size $n = |A|$ if $|B|$ is too small we consider strongly indpendent sets in random d-uniform hypergraphs as discussed above.

Theorem 4.3 *For each $\eta > 0$ there exists a d_0 such that for all $d \geq d_0$, there exists a strongly independent set in the random d-uniform hypergraph $\mathcal{H}_{\tilde{n},n}^d$ consisting of $\tilde{n} \geq n$ vertices and n edges, which with high probability is at least of size*

$$(4 - \eta)\frac{\log d}{d^2}n.$$

The proof uses an approach by Frieze [8] suggested by Luczak. We use a partition P of the vertex set consisting of parts $P_1, \ldots, P_{n'}$ of size m where m grows asymptotically faster than $(\log d)^2$ but slower than $d^2/\log d$, that is, $m = o_d(d^2/\log d)$ and $m = \omega_d((\log d)^2)$. A set is a P-set if it is strongly independent and contains at most one vertex from each P_i for $i \in \{1, \ldots, n'\}$. Let β be the maximum size of a P-set. By Azuma's inequality we have that

$$\mathbb{P}\left(|\beta - \mathbb{E}(\beta)| \geq t\right) \leq 2e^{-t^2/2n'} \tag{4.1}$$

using the martingale that exposes the hyperedges incident to P_i at step i, $i = 1, \ldots, n'$, and noting that β can change by at most 1 if we change hyperedges incident to a single P_i.

Let X_k be the random variable that counts the number of P-sets of size k. Using the second moment method we show that for $\eta > 0$ and sufficiently large d,

$$\mathbb{P}(X_k > 0) > 2\exp\left(-2^9 \left(\tfrac{\log d}{d}\right)^4 n\right) \tag{4.2}$$

where

$$k = (4 - \psi)\frac{\log d}{d^2}n,$$

and $\eta = 2\psi$.

This implies the lower bound by setting $t = 2^5 \left(\frac{\log d}{d}\right)^2 n/\sqrt{m}$ as

$$t^2/2n' = \frac{2^{10}n^2}{2n'm}\left(\frac{\log d}{d}\right)^4 = \frac{2^{10}n^2}{2n}\left(\frac{\log d}{d}\right)^4 = 2^9\left(\frac{\log d}{d}\right)^4 n.$$

From this it follows that the probability that $X_k > 0$ is larger than the probability that β, (which is the largest k for which $X_k \neq 0$) lies further from the expectation than t. Thus k must be at most t greater than $\mathbb{E}(\beta)$. By Azuma's inequality we know that with high probability $|\beta - \mathbb{E}(\beta)| < t$ and hence with high probability $\beta > k - 2t$. Hence if $m/(\log d)^2 \to \infty$, we have that for sufficiently large d, whp

$$\beta > (4 - \psi - 2^6(\log d)/\sqrt{m})\frac{\log d}{d^2}n \geq (4 - 2\psi)\frac{\log d}{d^2}n = (4 - \eta)\frac{\log d}{d^2}n.$$

4.1 Conclusions/Open problems

Although we have proven a threshold for the lower and upper bounds which are asymptotically equal as $d \to \infty$, it seems likely, that a threshold should exist for each d. In other words, we conjecture that there exists constants c_d for $d \geq 3$ such that for all $\eta > 0$, if $\varepsilon > c_d + \eta$ then whp a matching of size n can be found, while for $\varepsilon < c_d - \eta$ there is whp a strategy for the adversary that reduces the size of the maximal matching below n. If this conjecture is true then we know $c_d = (4 + o_d(1))(\log d)/d^2$. Note that this conjecture fails for $d = 2$. Indeed, the adversary can simply delete a random choice of edge from each vertex in A and then whp in the resulting graph we have two vertices in A with the same remaining neighbour in B. On the other hand it is not hard to see that Theorem 4.2 can be strengthened so as to give a finite bound on ε even for $d = 3$. The proof required d to be large at several points, however, following the strategy of the proof, it is easy to show that $d \geq 3$ is enough to get a finite bound. For example, if $\tilde{n} \geq 10dn$ then, defining e_s to be the expected number of witnesses for Hall's theorem of size s and $p := \frac{s-1}{\tilde{n}}$, we have,

$$e_s \leq \binom{n}{s}\binom{\tilde{n}}{s-1}((qd+p)p^{d-1})^s$$
$$\leq (en/s)^s(e\tilde{n}/(s-1))^{s-1}(dp^2)^s$$
$$\leq (e/10dp)^s(e/p)^{s-1}(dp^2)^s$$
$$\leq (p/e)(e^2/10)^s$$
$$\leq (s/ne)(e^2/10)^s.$$

Noting that

$$\sum_{s=1}^{n}(s/ne)(e^2/10)^s = O\left(\frac{1}{n}\right),$$

it is clear that we have $e_s = o(1)$ as required.

Although we have identified the threshold for which an adversary can and cannot destroy the complete matching when subject to the restriction of removing a single edge incident to each vertex of A, there remains a number of interesting problems that would arise from allowing the adversary greater or differing powers in modifying G. The case where the adversary is able to delete n edges globally allows the adversary to easily isolate a linear proportion of the vertices, while equally, a matching still exists that covers a linear proportion of the vertices. In both cases a simple greedy algorithm provides fairly simple bounds, but finding the exact size of the largest remaining matching seems challenging and certainly would require further insight in tackling.

Another problem to consider would be the case of analysing the size of the maximum matching for values of d and ε for which we have shown that the adversary can eliminate a matching of size n. The use of our graph model was motivated by its use in [9] which analysed the size of the maximum matching in the same model but without an adversary removing edges, for all values of d, and it would be interesting to know what the behaviour of the size of the maximum matching becomes for small values of d once an adversary is introduced.

Acknowledgement

The first author was partially supported by NSF grant DMS 1301614. We would also like to thank the referee for the careful reading of the manuscript and the helpful comments.

References

[1] J. Aronson, A. Frieze and B. Pittel, *Maximum matchings in sparse random graphs: Karp–Sipser revisited*, Random Structures & Algorithms **12** (1998), 111–177.

[2] P.N. Balister, S. Gerke and A. McDowell, *Adversarial resilience of matchings in bipartite random graphs*, submitted.

[3] C. Berge, *Hypergraphs*, volume 45 of North-Holland Mathematical Library, North-Holland Publishing Co., Amsterdam, 1989.

[4] T. Bohman, A.M. Frieze, *Karp–Sipser on random graphs with a fixed degree sequence*, Combin. Probab. Comput. **20** (2011), 721–741.

[5] C. Bordenave, M. Lelarge, and J. Salez, *Matchings on infinite graphs*, Probab. Theory Related Fields **157** (2013), no. 1–2, 183–208.

[6] J.-M. Dion, C. Commault, and J. van der Woude, *Generic properties and control of linear structured systems*, Automatica J. IFAC, **39** (2003), no. 7, 1125–1144.

[7] J. Edmonds, *Paths, trees and flowers*, Canad. J. Math. **17** (1965), 449–467.

[8] A.M. Frieze, *On the independence number of random graphs*, Discrete Math. **81** (1990), no. 2, 171–175.

[9] A. Frieze and P. Melsted, *Maximum matchings in random bipartite graphs and the space utilization of Cuckoo Hash tables*, Random Structures & Algorithms **41** (2012), no. 3, 334–364.

[10] P. Hall, *On representatives of subsets*, J. Lond. Math. Soc. **s1-10** (1935), no. 1, 26–30.

[11] S. Janson, T. Łuczak, A. Ruciński, *Random graphs*, Wiley-Intersci. Ser. Discrete Math. Optim. Wiley-Interscience, New York (2000).

[12] R.E. Kalman, *Mathematical description of linear dynamical systems*, J. SIAM Ser. A Control **1** (1963) 152–192.

[13] B. Sagan, *Log-concave sequences of symmetric functions and analogs of the Jacobi-Trudi determinants*, Trans. Amer. Math. Soc. **329** (1992) 795–811.

[14] R. Karp, M. Sipser, *Maximum matchings in sparse random graphs*, Proceedings of the 22nd IEEE Symposium on the Foundations of Computer Science (1981) 364–375.

[15] C.-T. Lin, *Structural controllability*, IEEE Trans. Automatic Control **19** (1974) 201–208.

[16] Y.-Y. Liu, J.-J. Slotine, A.-L. Barabási, *Controllability of complex networks*, Nature **473** (2011) 167–173.

[17] S. Micali and V. Vazirani, *An $O(|V|^{\frac{1}{2}}|E|)$ algorithm for finding maximum matchings in general graphs*, Proceedings of the 21st IEEE Symposium on the Foundations of Computer Science (1980) 17–27.

[18] J. Salez, *Weighted enumeration of spanning subgraphs in locally tree-like graphs*, Random Structures & Algorithms **43** (2013), no. 3, 377–397.

[19] R.W. Shields and J.B. Pearson, *Structural controllability of multi-input linear systems*, IEEE Trans. Automatic Control **21** (1976), 203–212.

[20] N.C. Wormald, *The differential equation method for random graph processes and greedy algorithms*, In M. Karoński and H.J. Prömel, editors, Lectures on approximation and randomized algorithms, Warsaw (1999) 75–152.

University of Memphis
Department of Math Sciences
Memphis, TN 38152, USA
pbalistr@memphis.edu

Royal Holloway University of London
Mathematics Department
Egham TW20 0EX, UK
stefanie.gerke@rhul.ac.uk

Some old and new problems in combinatorial geometry I: around Borsuk's problem

Gil Kalai[1]

Abstract

Borsuk [16] asked in 1933 if every set of diameter 1 in $\mathbf{R}^d$ can be covered by $d + 1$ sets of smaller diameter. In 1993, a negative solution, based on a theorem by Frankl and Wilson [42], was given by Kahn and Kalai [65]. In this paper I will present questions related to Borsuk's problem.

1 Introduction

The title of this paper is borrowed from Paul Erdős who used it (or a similar title) in many lectures and papers, e.g., [39]. I will describe several open problems in the interface between combinatorics and geometry, mainly convex geometry. In this part, I describe and pose questions related to the Borsuk conjecture. The selection of problems is based on my own idiosyncratic tastes. For a fuller picture, the reader is advised to read the review papers on Borsuk's problem and related questions by Raigorodskii [105, 106, 107, 110, 112]. Among other excellent sources are [13, 14, 18, 89, 99].

Karol Borsuk [16] asked in 1933 if every set of diameter 1 in $\mathbf{R}^d$ can be covered by $d + 1$ sets of smaller diameter. That the answer is positive was widely believed and referred to as the *Borsuk conjecture*. However, some people, including Rogers, Danzer, and Erdős, suggested that a counterexample might be obtained from some clever combinatorial configuration. In hindsight, the problem is related to several questions that Erdős asked and its solution was a great triumph for Erdősian mathematics.

2 Better lower bounds to Borsuk's problem

2.1 The asymptotics

Let $f(d)$ be the smallest integer such that every set of diameter one in $\mathbf{R}^d$ can be covered by $f(d)$ sets of smaller diameter. The set of vertices of a regular simplex of diameter one demonstrates that $f(d) \geq d + 1$. The

[1]Work supported by an ERC advanced grant, an ISF grant, and an NSF grant.

famous Borsuk–Ulam theorem [16] asserts that the d-dimensional ball of diameter 1 cannot be covered by d sets of smaller diameter. The Borsuk–Ulam theorem has many important applications in many areas of mathematics. See Matousek's book [90] for applications and connections to combinatorics. In the same paper [16] Borsuk asked if $f(d) \leq d + 1$. This was proved for $d = 2, 3$. It was shown by Kahn and Kalai [65] that $f(d) \geq 1.2^{\sqrt{d}}$, by Lassak [84] that $f(d) \leq 2^{d-1} + 1$ and by Schramm [116] that $f(d) \leq (\sqrt{3/2} + o(1))^d$.

Problem 2.1 *Is $f(d)$ exponential in d?*

The best shot (in my opinion) at an example leading to a positive answer is:

(a) Start with binary linear codes of length n (based on algebraic-geometry codes) with the property that the number of maximal-weight codewords is exponential in n.

(b) Show that the code cannot be covered by less than an exponential number of sets that do not realize the maximum distance.

Part (a) should not be difficult, given that it is known that for certain AG-codes the number of minimal-weight codewords is exponential in n [5]. Part (b) can be difficult, but the algebraic techniques used for the Frankl and Wilson theorem may apply.

The Kahn–Kalai counterexample and many of the subsequent results depend on the Frankl–Wilson [42] theorem or on some related algebraically-based combinatorial results. (One can rely also on the Frankl–Rödl theorem [41], which allows much greater generality but not as good quantitative estimates.) We will come back to these results later on.

Let $g(d)$ be the smallest integer such that every *finite* set of diameter one in $\mathbf{R}^d$ can be covered by $g(d)$ sets of smaller diameter.

Problem 2.2 *Is $f(d) = g(d)$?*

I am not aware of any reduction from infinite sets to finite sets, and indeed the proof of Borsuk's conjecture for $d = 2, 3$ is easier if one considers only finite sets. On the other hand, the counterexamples are based on finite configurations. Perhaps one can demonstrate a gap between the finite and infinite behavior for some extension or variation of the problem, e.g., for arbitrary metric spaces. (Our knowledge of $f(d)$ does not seem accurate enough to hope to prove that such a gap exists for the original problem.)

2.2   Larman's conjecture

The counterexample to Borsuk's conjecture is based on the special case where the set consists of 0-1 vectors of fixed weight. Here, the conjecture has an appealing combinatorial formulation.

Problem 2.3 (Larman's conjecture) *Let $\mathcal{F}$ be a t-intersecting family of k sets from $[n]$. Then $\mathcal{F}$ can be covered by n $(t+1)$-intersecting subfamilies.*

We now know that Larman's conjecture does not hold in general. However:

Problem 2.4 *Is Larman's conjecture true for $t = 1$?*

For more discussion of the combinatorics of Larman's conjecture and related combinatorial questions on the packing and coloring of graphs and hypergraphs, see [63]. We can sort of "dualize" the $t = 1$ case of Larman's conjecture by replacing "intersecting" (i.e., "every pair of sets has at least one common element") by "nearly disjoint" (namely, "every pair of distinct sets has at most one common elements") and thus recover the famous:

Conjecture 2.5 (Erdős–Faber–Lovász) *Let $\mathcal{F}$ be a family of nearly disjoint k-sets from $[n]$. Then $\mathcal{F}$ is the union of n matchings.*

While the Erdős–Faber–Lovász conjecture is still open it is known that n is the right number for the fractional version of the problem [66], and that $(1 + o(1))n$ matchings suffice [62]. Such results are not available for the $t = 1$ case of Larman's conjecture but a counterexample to a certain strong form of the conjecture is known [64].

2.3   Embedding the elliptic metric into an Euclidean one

Let $\mathcal{E}_n$ be the *elliptic* metric space of lines through the origin in $\mathbf{R}^n$ where the distance between two lines is the (smallest) angle between them. So the diameter of $\mathcal{E}_n$ is $\pi/2$ and the famous Frankl–Wilson theorem implies that $\mathcal{E}_n$ cannot be covered by less than an exponential number (in d) of sets of diameter smaller than $\pi/2$. The proof by Kahn and Kalai can be seen as adding a single simple fact: $\mathcal{E}$ can be embedded into an Euclidean space $\mathbf{R}^{n(n+1)/2}$ by the map

$$x \to x \otimes x.$$

The distance between $x \otimes x$ and $y \otimes y$ is a simple monotonic function of the distance between x and y. (Here $\| x \|_2 = 1$ and we note that x and $-x$ are mapped onto the same point.)

The original counterexample, C_1, is the image under this map for (normalized) ± 1 vectors of length n with $n/2$ '1's (n divisible by 4). Another example, C_2, is the image of all ± 1 vectors, and we can also look at C_3 the image of all unit vectors. All these geometric objects are familiar: C_3 is the unit vectors in the cone of rank-one positive semi-definite matrices, C_2 is called the *cut polytope,* and C_1 is the polytope of balanced cuts [33].

We can ask if there are more economic embeddings of the elliptic space into a Euclidean space. Namely, is there an embedding $\varphi : \mathcal{E}_n \to \mathbf{R}^m$, $m = o(n^2)$, such that $\| \varphi(x) - \varphi(y) \|_2 = \varphi(d(x,y))$ for some (strictly) monotone function φ?

The answer to this question is negative by an important theorem of de Caen from 2000 [24].

Theorem 2.6 (de Caen) *There are quadratically many equiangular lines in $\mathcal{E}_n$.*

Weaker forms of embeddings of $\mathcal{E}_n$ into Euclidean spaces possibly with some symmetry-breaking may still lead to improved lower bounds for $f(d)$, and are of independent interest.

Problem 2.7 *Is there a continuous map $\varphi : \mathcal{E}_n \to \mathbf{R}^m$, $m = o(n^2)$ so that φ preserves the set of diameters of $\mathcal{E}_n$?*

2.4   Spherical sets without pairwise orthogonal vectors

Regarding $\mathcal{E}_n$ itself, Witsenhausen asked in 1974 [129] what is the maximum volume $\mu(A)$ of an n-dimensional spherical set A without a pair of orthogonal vectors. Witsenhausen proved that:

$$\mu(A) \leq 1/(n+1).$$

The following natural conjecture is very interesting:

Conjecture 2.8 *Let A be a measurable subset of S^n and suppose that A does not contain two orthogonal vectors. Then the volume of A is at most twice the volume of two spherical caps of radius $\pi/4$.*

Asymptotically this conjecture asserts that a subset of the n sphere of measure $(1/\sqrt{2} + o(1))^n$ must contain a pair of orthogonal vectors. If true, this can replace the Frankl–Wilson bound and will show that C_3 defined above is a counterexamples to Borsuk's conjecture for $d > 70$ or so. The Frankl–Wilson theorem gives that if $\mu(A) > 1.203...^{-n}$ then A contains two orthogonal vectors. It seems that the main challenge is to extend the linear algebra/polynomial method from 0-1 vectors to general vectors. One important step was taken by Raigorodskii [108] who improved the bound to 1.225^{-n}.

Remarkably, the upper bound of $1/3$ for the two-dimensional case stood unimproved for 40 years until very recently DeCorte and Pikhurko improved it to $0.31.. -[26]$! The proof uses Delsartes' linear programming method [31] combined with a combinatorial argument.

2.5 Borsuk's problem for spherical sets

Borsuk's problem itself has an important extension to spherical sets. Consider a set of Euclidean diameter 1 on a d-dimensional sphere S_r^{d-1} of radius r.

Problem 2.9 *What is the maximum number $f_r(d)$ of parts one needs to partition any set of diameter 1 on S_r^d?*

Obviously, one has $f_r(d) \le f(d)$ for any r, and we as well have $f_{1/2}(d) = d + 1$ due to Borsuk–Ulam theorem.

Kupavskii and Raigorodskii [76] proved the following theorem:

Theorem 2.10 *Given $k \in \mathbf{N}$, if $r > \frac{1}{2}\sqrt{\frac{2k+1}{2k}}$, then there exists $c > 1$ such that $f_r(d) \ge (c + o(1))^{2k\sqrt{d}}$. Moreover, there exist a $c > 0$ such that if $r > 1/2 + c\frac{\log\log d}{\log d}$, then for all sufficiently large d we have $f_r(d) > d + 1$.*

The proof is based on mappings involving multiple tensor products. We note that embeddings of similar nature via multiple tensor-products play a role also in the disproof of Khot and Vishnoi [69] of the Goemans–Linial conjecture.

Problem 2.11 *Is it the case that for every $r < 1/2$ there is a constant $C_r > 1$ such that*
$$f_r(d) \ge C_r^d?$$

2.6 Low dimensions and two-distance sets

The initial counterexample showed that the Borsuk conjecture is false for $n = 1325$ and all $n > 2014$ and there were gradual improvements over the years down to 946 (Nilli [96]), 903 (Weissbach), 561 (Raigorodski

[109]), 560 (Weissbach [126]), 323 (Hinrichs [57]), 321 (Pikhurko [102]), 298 (Hinrichs and Richter [58]). The construction of Hinrichs and the subsequent ones remarkably rely on the Leech lattice.

A two-distance set is a set of vectors in $\mathbf{R}^d$ that attain only two distances. Larman asked early on and asked again recently:

Problem 2.12 *Is the Borsuk conjecture correct for two-distance sets?*

This has proven to be a very fruitful question. In 2013 Bondarenko [15] found a two-distance set with 416 points in 65 dimensions that cannot be partitioned into less than 83 parts of smaller diameter. Remarkable! Jenrich [61] pushed the dimension down to 64. These constructions beatifully relies on known strongly-regular graphs.

Problem 2.13 *What is the smallest dimension for which Borsuk's conjecture fails? Is Borsuk's conjecture correct in dimension 4?*

In dimensions 2 and 3 Borsuk's conjecture is correct. Eggleston gave the first proof for dimension 3 [35], which was followed by simpler proofs by Grünbaum [47] and Heppes [54]. A simple proof for finite sets of points in 3-space was found by Heppes and Revesz [56]. For dimension 2 it follows from an earlier 1906 result that every set of diameter one can be embedded into a regular hexagon whose opposite edges are distance one apart. For a simpler argument see Pak's book [99]. Here too, for finite configurations the proof is very simple.

3   Upper bounds for Borsuk's problem and sets of constant width

3.1   Improving the upper bound

Lassak [84] proved that for every d, $f(d) \leq 2^{d-1} + 1$ (and this still gives the best-known bound when the dimension is not too large). Schramm [116] proved that every convex body of constant width 1 can be covered by $(\sqrt{3/2} + o(1))^d$ smaller homothets. It is a well-known fact [36] that every set of diameter one is contained in a set of constant width 1, and, therefore, for proving an upper bound on $f(d)$ it is enough to consider sets of constant width. Bourgain and Lindenstrauss [17] showed that every convex body in $\mathbf{R}^d$ of diameter 1 can be covered by $(\sqrt{3/2} + o(1))^d$ balls of diameter 1. Both these results show that $f(d) \leq (\sqrt{3/2} + o(1))^d$.

Problem 3.1 *Prove that $f(d) \leq C^d$ for some $C < \sqrt{3/2}$.*

We note that Danzer constructed a set of diameter 1 that requires exponentially many balls to cover. Danzer constructed a set for which $(1.003)^d$ balls needed, and Bourgain and Lindenstrauss in 1991 [17] found much better bound, $(1.064)^n$.

As for covering by smaller homothets we recall the famous:

Conjecture 3.2 (Hadwiger [52, 53]) *Every convex body K in $\mathbf{R}^d$ can be covered by 2^d smaller homothets of K.*

The case of sets of constant width is of particular interest:

Problem 3.3 *Are there $\epsilon > 0$ and sets of constant width in $\mathbf{R}^n$ that require at least $(1 + \epsilon)^n$ smaller homothets to cover?*

Note that a positive answer neither implies nor follows from a $(1 + \epsilon)^d$ lower bound for the Borsuk number $f(d)$.

3.2 Volumes of sets of constant width

Let us denote the volume of the n-ball of radius $1/2$ by V_n.

Problem 3.4 (Schramm [117]) *Is there some $\epsilon > 0$ such that for every $d > 1$ there exists a set K_d of constant width 1 in dimension d whose volume satisfies $VOL(K_d) \leq (1 - \epsilon)^d V_d$?*

Schramm raised a similar question for spherical sets of constant width and pointed out that a negative answer for spherical sets will push the $(3/2)^{d/2}$ upper bound for $f(d)$ to $(4/3)^{d/2}$.

4 Saving the Borsuk conjecture

4.1 Borsuk's conjecture under transversality

I would like to examine the possibility that Borsuk's conjecture is correct except for some "coincidental" sets. The question is how to properly define "coincidental," and we will now give it a try!

Let K be a set of points in $\mathbf{R}^d$ and let A be a set of pairs of points in K. We say that the pair (K, A) is *general* if for every continuous deformation of the distances on A there is a deformation K of K which realizes the deformed distances.

Remark This condition is related to the "strong Arnold property" (a.k.a. "transversality") in Colin de Verdiére's theory of invariants of graphs [21].

Conjecture 4.1 *If D is the set of diameters in K and (K, D) is general then K can be partitioned into $d + 1$ sets of smaller diameter.*

We further propose (somewhat more strongly) that this conjecture holds even when "continuous deformation" is replaced with "infinitesimal deformation."

The finite case is of special interest. A graph embedded in $\mathbf{R}^d$ is *stress-free* if we cannot assign not-all-zero weights to the edges such that the weighted sum of the edges containing any vertex v (regarded as vectors from v) is zero for every vertex v. Here we embed the vertices and regard the edges as straight line segments. (Edges may intersect.) Such a graph is called a "geometric graph." When we restrict the conjecture to finite configurations of points we get:

Conjecture 4.2 *If G is a stress-free geometric graph of diameters in $\mathbf{R}^d$ then G is $(d + 1)$-colorable.*

Remark A stress-free graph for embeddings into $\mathbf{R}^d$ has at most $dn - \binom{d+1}{2}$ edges and therefore its chromatic number is at most $2d - 1$.

4.2 A weak form of Borsuk's conjecture

Conjecture 4.3 *Every polytope P with m facets can be covered by m sets of smaller diameter.*

This conjecture was motivated by recent important works on projections of polytopes [40]. A positive answer will give an alternative path for showing that the cut polytope cannot be described as a projection of a polytope with only polynomially many facets.

4.3 Classes of bodies for which Borsuk's conjecture holds

Perhaps the most natural way to "save" Borsuk's conjecture is given by:

Problem 4.4 *Find large and interesting classes of convex bodies for which Borsuk's conjecture holds!*

Borsuk's conjecture is known to be true for centrally symmetric bodies, Hadwiger proved it for smooth convex bodies [53], and Boris Dekster proved the conjecture both for bodies of revolution [29] and for convex bodies with a belt of regular points [30].

4.4   Partitioning the unit ball and diametric codes

The unit ball in $\mathbf{R}^d$ can be covered by $d+1$ convex sets of smaller diameter. But how much smaller? We do not know the answer. Let $u(d)$ be the minimum value of t such that the unit ball in $\mathbf{R}^d$ can be covered by $d+1$ sets of diameter at most t.

Problem 4.5 *Determine the behavior of $u(d)$!*

The motivation for this question comes from an even stronger form of Borsuk's conjecture asserting that every set of diameter 1 can be covered by $d+1$ sets of diameter $u(d)$. It was also conjectured that the optimal covering for the sphere is described by a partition based on the Voronoi regions of a regular simplex that gives $u(n) \leq 1 - \Omega(1)/n$. This is known to be optimal in dimensions two and three and is open in higher dimensions. Larman and Tamvakis [81] showed by a volume argument that $u(n) \geq 1 - 3/2 \log n/n + O(1/n)$. See also [28].

It will be interesting to close the logarithmic gap for $u(d)$. I don't know what one should expect for the answer, and it will be quite exciting if the standard example is *not* optimal.

We can more pose general questions:

Problem 4.6   *(i) What is the smallest number of sets of diameter t that are needed to cover the unit sphere?*

(ii) What is the largest number of convex sets of width $\geq t$ that can be packed into the unit sphere? (The width of a convex set is the minimum distance between opposite supporting hyperplanes.)

Let $\Omega_n = \{0, 1\}^n$. We can ask the analogous questions about the binary cube.

Problem 4.7   *(i) What is the smallest number of sets of diameter t that are needed to cover Ω_n?*

(ii) What is the largest number of sets of width $\geq t$ that can be packed into Ω_n?

Here by "diameter" and "width" we refer to the Euclidean notions (for which, for Problem 4.7, "diameter" essentially coincides with the Hamming diameter).

5   Unit-distance graphs and complexes

For a subset A of $\mathbf{R}^n$ the *unit-distance graph* is a graph whose set of edges consists of pairs of points of A of distance 1. If all pairwise distances are at most 1, we call the unit-distance graph a *diameter graph*. If all the pairwise distances are at least 1 we call it a *kissing graph*. Borsuk's question is a question about coloring diameter graphs.

Problem 5.1 *What is the maximum number of edges, the maximum chromatic number, and the maximum minimal degree for the diameter graph, kissing graph, and unit-distance graph for a set of n points in $\mathbf{R}^d$?*

Finding the maximum number of edges in a planar unit distance graphs is a famous problem by Erdős [37]. Another famous problem by Hadwiger and Nelson is about the chromatic number of the planar unit distance graph and yet another famous question is if the minimal kissing number of a set of n points in $\mathbf{R}^d$ is exponential in d, see [22, 4].

We can define also the *unit-distance complex* to be the simplicial complex of cliques in the unit-distance graphs or, alternatively, the simplicial complex whose faces are sets of points in A that form regular simplices of diameter 1. And again when the diameter of A is 1 we call it the *diameter complex* and when the minimum distance is 1 we call it the *kissing complex*.

Problem 5.2 *What is the maximum number of r-faces for the diameter complex, kissing complex, and the unit-distance complex for a set of n points in $\mathbf{R}^d$?*

For the chromatic number of the unit-distance graph it makes a difference if we demand further that each color class be measurable. (This is referred to as the *measurable chromatic number*.) For progress on the chromatic number of unit-distance graphs, see [44, 72, 73, 74, 75]. For progress on the measurable chromatic number and related questions, see [8, 9, 10, 27].

Rosenfeld asked (see [115]):

Problem 5.3 *Does the graph whose vertex set is the set of points in the plane and whose edges represent points whose distance is an odd integer have a bounded chromatic number?*

For measurable chromatic numbers the answer is negative as follows from a theorem of Furstenberg, Katznelson and Weiss that asserts that every planar set of positive measure realizes all sufficiently large distances. See also [121] for a simple direct proof.

5.1 Schur's conjecture

A conjecture by Schur deals with an interesting special case:

Conjecture 5.4 (Schur) *The number of $(d-1)$-faces of every diameter complex for a set of n points in $\mathbf{R}^d$ is at most n.*

The planar case is an old result and it implies a positive answer to Borsuk's problem for finite planar sets. The proof is based on an observation that sets the metric aside: the edges of the diameter graph are pairwise intersecting and therefore we need to show that every geometric graph with n vertices and $n+1$ edges must have two disjoint edges. This result by Hopf and Pannwitz [59] from 1934 can be seen as the starting point of "geometric graph theory" [98]. Zvi Schur was a high school teacher who did research in his spare time. He managed to prove his conjecture in dimension 3 (see [118]) but in his writing he mentioned that "the power of my methods diminishes as the dimension goes up." The paper [118] includes also a proof that in any dimension the number of d-faces of the diameter complex is at most one. Schur's conjecture has recently been proven by Kupavskii and Polyanski [77]! (For $d=4$ it was proved by Bulankina, Kupavskii, and Polyanskii [19].) A key step in Kupavskii and Polyanskii's work is proving the $k=m=d-1$ case of the following additional conjecture by Schur, still open in the general case.

Conjecture 5.5 (Schur) *Let S_1 and S_2 be two regular simplices of dimensions k and m in $\mathbf{R}^d$ such that their union has diameter 1. Then S_1 and S_2 share at least $\min(0, k+2m-2d+1)$ vertices for $k \geq m$.*

Heppes and Révész proved that the number of edges in the diameter graph of n points in space is $2n-2$. This gives an easy proof of Borsuk's conjecture for finite sets of points in $\mathbf{R}^3$.

A natural weakening of Borsuk's conjecture is:

Problem 5.6 *What is the smallest $r=r(d)$ such that every set of diameter 1 in $\mathbf{R}^d$ can be covered by $d+1$ sets, none of which contains an r-dimensional simplex of diameter 1?*

Unit-distance graphs and especially diameter graphs and complexes are closely related to the study of *ball polytopes*. Those are convex bodies that can be described as the intersection of unit balls. A systematic study of ball polytopes was initiated by Cároly Bezdek around 2004 and they were also studied by Kupitz, Martini, and Perles, see [11, 12, 79]. Ball polytopes are also related to sets of constant width.

5.2 Tangent graphs and complexes for collections of balls (of different radii)

We can try further to adjust the problems discussed in this section to the case where we have a collection A of points in $\mathbf{R}^d$ and a close ball centered around each point. Two balls can be in three mutual positions (that we care about): They can be disjoint, they can have intersecting interiors, or they can be tangential.

The *tangent graph* is a graph whose set of vertices is A and a pair of vertices are adjacent if the corresponding balls are tangential. Note that if all balls have the same radius $1/2$, then the tangent graph is the unit-distance graph. As before we can consider also the *tangent complex* - the simplicial complex described by cliques in the tangent graph.

Problem 5.7   (i) *What is the maximum number of edges in a tangent graph (especially in the plane)? What is its maximum chromatic number (especially in the plane)?*

   (ii) *If every two balls intersect, then the tangent graph is a generalization of the diameter graph. Again we can ask for the maximum number of edges, cliques of size r, and the chromatic number. Again we can ask if when the graph is stress-free the chromatic number is at most $d+1$.*

   (iii) *If every two balls have disjoint interiors then the tangent graph is a generalization of the kissing graph. Again we can ask for the maximum number of edges, cliques of size r, the maximum minimal degree, and the chromatic number.*

In the plane we can find n points and n lines with $n^{C4/3}$ incidences and the famous Szemeredi–Trotter theorem (see, e.g., [124]) asserts that this is best possible. Now, we can replace each point by a small circle, arrange for the lines incident to the points to be tangential to them, and regard the lines as circles as well. This shows that tangent graphs with n vertices in the plane can have as much as $n^{C4/3}$ edges. It is conjectured by Pinchasi, Sharir, and others that

Conjecture 5.8   (i) *Planar tangent graphs with n vertices can have at most $n^{4/3}\operatorname{polylog}(n)$ edges.*

   (ii) *More generally, m red discs and n blue discs (special case: n blue points), can touch at most $((mn)^{2/3}+m+n)\operatorname{polylog}(m,n)$ times.*

This conjecture proposes a profound extension of the Szemeredi–Trotter theorem. The best known upper bound $n^{3/2} \log n$ is by Markus and Tardos [88], following an earlier argument by Pinchási and Radoicić [104]. This particular approach based on a certain "forbidden configurations" – a self crossing 4-cycle – cannot lead to better exponents. Sharir found a beautiful connection with Erdos's distinct distances problem [51] which also shows that the polylog(n, m) term cannot be eliminated. Indeed, assume you have n points with just x distances. Then draw around each point x circles whose radii are the x possible distances and then you get a collection of $m = nx$ circles and n points with n^2 incidences (because every point lies on n circles exactly). Therefore: $n^2 \leq \text{polylog}(xn)(n(xn))^{2/3} + n + xn$ which implies $x \geq n \text{polylog}(n)$.

For circles that pairwise intersect, Pinchasi [103] proved a Gallai–Sylvester conjecture by Bezdek asserting that (for more than 5 circles) there is always a circle tangential to at most two other circles. This was the starting point of important studies [6, 1] concerning arrangement of circles and pseudo-circles in the plane. Alon, Last, Pinchasi, and Sharir [6] showed an upper bound of $2n - 2$ for the number of edges in the tangent graph for pairwise intersecting circles.

The problem considered in this section can be asked under greater generality in at least two ways: one important generalization is to consider two circles adjacent if their intersection is an *empty lens*, that is, not intersected by the boundary of another disc. Another generalization is for pseudocircles (where both notions of adjacency essentially coincide).

Let me end with the following problem:

Conjecture 5.9 (Ringle circle problem) *Tangent graphs for finite collections of circles in the plane such that no more than two circles pass through a point have bounded chromatic numbers.*

Recently, Pinchasi proved that without the assumption that no more than two circles pass through a point, the chromatic number is $O(\log^2 n)$, where n is the number of vertices of the graphs. Pinchasi also gave an example where (again, dropping the extra assumption) you need $\log n$ colors.

6   Other metric spaces

6.1   Very symmetric spaces

We already discussed Borsuk's problem for spherical sets. We can also ask

Problem 6.1 *Study the Borsuk problem, and other questions considered above, for very symmetric spaces like the hyperbolic space, the Grassmanian, and $GL(n)$.*

The Grassmanian, the space of k-dimensional linear spaces of $\mathbf{R}^n$ is of special interest. The "distance" between two vector spaces can be seen as a vector of k angles, and there may be several interesting ways to extend the questions considered here. (The case $k = 1$ brings us back to the Elliptic space).

6.2 Normed spaces

Given a metric space X and a real number t we can consider the *Borsuk number* $b(X, t)$ defined as the smallest integer such that every subset of diameter t in X can be covered by $b(X, t)$ sets of smaller diameter. There are interesting results and questions regarding Borsuk's numbers of various metric spaces. Let $a(X, t)$ be the maximum cardinality of an *equilateral* subset $Y \subset X$ of diameter t (namely, a set so that every pairwise distance between distinct points in Y is t). Of course, $a(X, t) \leq b(X, t)$. Understanding $a(X, t)$ for various metric spaces is of great interest. Kusner conjectured that an equilateral set in ℓ_1^n has at most $2n$ elements and an equilateral set in ℓ_p^n has size at most $n + 1$ for $p, 1 < p < \infty$. Smyth found the first polynomial upper bound for the size of an equilateral set in ℓ_1 which followed by an important result by Alon and Pudlak [7]:

Theorem 6.2 (Alon and Pudlak) *For an odd integer p, an equilateral set in ℓ_p^n has at most $c_p n \log n$ points.*

When we move to general normed spaces there are very basic things we do not know. It is widely conjectured that:

Conjecture 6.3 *Every normed n-dimensional space has an equilateral set of $n + 1$ points.*

For more on this conjecture see Swanepoel [122].

Petty [101] proved the $n = 3$ case and his proof is based on the topological fact that a Jordan curve in the plane enclosing the origin cannot be contracted without passing through the origin at some stage. Makeév proved the four-dimensional case using more topology. Brass and Dekster proved independently a $(\log n)^{1/3}$ lower bound and a major improvement by Swanepoel and Villa [123] improved the lower bound to $\exp(c \log n)^{1/2}$. I would not be surprised if Conjecture 6.3 is false. It is known that 2^n is an upper bound for the size of an equilateral set for a normed n-dimensional space.

Let me end this section with a beautiful result of Matoušek about unit distances in normed space. One of the most famous problems in geometry is Erdős' unit distance problem of finding the maximum number of unit distances among n points in the plane. This question can be asked with respect to every planar norm with unit ball K. It is known that for every norm the number of edges can be as large as $\theta(n \log n)$ and here we state a breakthrough theorem by Matoušek [91]:

Theorem 6.4 *There are norms (in fact, for most norms in a Baire category sense) for which the maximum number of unit distances on n points is $O(n \log n \log \log n)$.*

7 Around Frankl–Wilson and Frankl–Rödl

7.1 The combinatorics of cocycles and Turán numbers

The original counterexamples to Larman's conjecture (and Borsuk's conjecture) were based on cuts: we consider the family of edges of complete bipartite graphs with $4n$ vertices. (In one variant we consider balanced bipartite graphs, and in another, arbitrary bipartite graphs.) We now consider high-dimensional generalization of cuts in graphs.

A $((k-1)$-dimensional) *cocycle* is a k-uniform hypergraph G such that every $k+1$ vertices contains an even number of edges. Equivalently, you can start with an arbitrary $(k-1)$ uniform hypergraph H and consider the k-uniform hypergraph G of all k-sets that contain an odd number of edges from H. Cocyles are familiar objects from simplicial cohomology and they have also been studied by combinatorialists and mainly by Seidel [119].

For even k, let $f(n,k)$ be the largest number of edges in a $(k-1)$-dimensional cocycle with n vertices. (Note that when k is odd, the complete k-uniform hypergraph is a cocycle.) Let $T(n,k,k+1)$ be the maximum number of edges in a k-uniform hypergraph without having a complete sub-hypergaph with $(k+1)$ vertices.

Conjecture 7.1 ([68]) *When k is even, $T(n,k,k+1) = f(n,k)$.*

The best constructions for Turán numbers $T(n,2k,2k+1)$ are obtained by cocycles. Let me just consider the case where $k = 2$. For a while the best example was based on a planar drawing of K_n with the minimum number of crossings. For every such drawing the set of 4-sets of points without a crossing is an example for Turán's (5,4) problem because K_5

is non-planar. It is easy to see that this non-crossing hypergraph is also
a cocycle. In 1988 de Caen, Kreher, and Wiseman [25] found a better,
beautiful example: consider a $n/2$ by $n/2$ matrix M with ± 1 entries.
Your hypergraph vertices will correspond to rows and columns of M. It
will include all 4-tuples with 3 rows or with 3 columns and also all sets
with 2 rows and 2 columns such that the product of the four matrix entries
is -1. The expected number of edges in the hypergraph for a random ± 1
matrix is $(11/16 + o(1))\binom{n}{4}$.

As for upper bounds, the best-known upper bounds are stronger for
cocycles. Peled [100] used a flag-algebras technique to show that $f(n,4) \leq
\binom{n}{4}(0.6916 + o(1))$.

7.2   High-dimensional versions of the cut cone and the cone of rank-one PSD matrices

The counterexamples for Borsuk's conjecture were very familiar geometric objects [33]. The example based on bipartite graphs (where the
number of edges is arbitrary) is the *cut-polytope*. The image of the elliptic
space under the map $x \to x \otimes x$ is simply the set of unit vectors in the
cone of rank-one positive semidefinite matrices. The unit vectors in the
cone of cocycles is an interesting generalization of the cut polytope since
for graphs (1-dimensional complexes) it gives us the cut-polytope.

Problem 7.2 *Find and study a "high-dimensional" extension of the cone
of rank one PSD matrices (analogous to the cone of cocycles).*

One possibility is the following: start with an arbitrary real-valued
function g on $\binom{[n]}{k-1}$ and derive a real-valued function on $\binom{[n]}{k}$ by:

$$f(T) = \prod \{g(S) : S \subset T, |S| = k - 1\}.$$

Let $U_{k,n}$ be the cone of all such g's.
Speculative application to Borsuk's problem is given by:

Conjecture 7.3   *(i) The set of unit vectors in the cone of 3-cocycles
with n vertices demonstrates a Euclidean set in $\mathbf{R}^d$ that cannot be
covered by less than $exp(d^{4/5})$ sets of smaller diameter.*

*(ii) The set of norm-1 vectors in $U_{4,n}$ demonstrates a Euclidean set in
$\mathbf{R}^d$ that cannot be covered by less than $exp(d^{4/5})$ sets of smaller
diameter.*

7.3 The Frankl–Wilson and Frankl–Rödl theorems

We conclude this paper with the major technical tool needed for the disproof of Borsuk's conjecture, which is the Frankl–Wilson (or Frankl–Rödl) forbidden intersection theorem. Most of the counterexamples to Borsuk's conjecture in low dimensions are based on algebraic techniques "the polynomial method" (or some variant) which seem related to the technique used for the proof of Frankl–Wilson's theorem. (The only exception are the new examples based on strongly regular graphs.) The Frankl–Wilson theorem [42] is wonderful and miraculous and the Frankl–Rödl theorem [41] is great - it allows many extensions (but not with sharp constants). The proof of Frankl–Wilson is a terrific demonstration of the linear-algebra method. The proof of Frankl–Rödl is an ingenious application (bootstrapping of a kind) of isoperimetric results. Recently Keevash and Long [67] found a new proof of Frankl–Rödl's theorem based on the Frankl–Wilson theorem.

Problem 7.4 *Is there a proof of Frankl–Rödl's theorem based on Delsarte's linear-programming method [31]?*

The work of Evan and Pikhurko [26] mentioned above suggests that applying the linear-programming method with input coming from other combinatorial methods can lead to improved result.

It is time to state the Frankl–Rödl theorem.

Theorem 7.5 (Frankl–Rödl) *For every $\alpha, \beta, \gamma, \epsilon > 0$, there is $\delta > 0$ with the following property. Let $\mathcal{U}_1$ be the family of $[\alpha n]$-subsets of $[n]$, let $\mathcal{U}_2$ be the family of $[\beta n]$-subsets of $[n]$, and let X be the number of pairs of sets $A \in \mathcal{U}_1$, $B \in \mathcal{U}_2$ whose intersection is of size $[\gamma n]$.*
Let $\mathcal{F}, \mathcal{G}$ be two subfamilies of $\mathcal{U}_1$ and $\mathcal{U}_2$, respectively, with $|\mathcal{F}||\mathcal{G}| \geq (1 - \delta)^n |\mathcal{U}_1| \cdot |\mathcal{U}_2|$. Then the number of pairs (A, B), $A \in \mathcal{F}$ and $B \in \mathcal{G}$ whose intersection has $[\gamma n]$ elements is at least $(1 - \epsilon)^n X$.

An important special case is where $\alpha = \beta = 1/2$ and $\gamma = 1/4$. The Frankl–Rödl paper contains generalizations in various directions. We could have assumed that, e.g., $\mathcal{F}$ and $\mathcal{G}$ are families of partitions of $[n]$ into r parts instead of families of sets. It also contains interesting geometric applications. We will propose here two extensions of the Frankl–Rödl theorem.

7.4 Frankl–Rödl/Frankl–Wilson with sum restrictions

For $S \subset [n]$, we write $\| S \| = \sum \{ s : s \in S \}$. Let $\alpha_1, \alpha_2, \beta_1, \beta_2$ be reals such that $0 < \alpha_1, \alpha_2 < 1$, $0 < \beta_1, \beta_2 < 1$. . Consider the family

$\mathcal{G}$ of subsets of $[n]$ such that for every $S \in \mathcal{G}$ we have $|S| = [\alpha_1 n]$, and $\| S \| = [\alpha_2(\binom{n}{2})]$.

Let X be the number of pairs A and B in $\mathcal{G}$ with the properties:
(*) The intersection C of A and B has precisely $[\beta_1 n]$ elements.
(**) The sum of elements in C is precisely $[\beta_2(\binom{n}{2})]$.

Conjecture 7.6 (Frankl–Rödl/Frankl–Wilson with sum restrictions)
For every $\epsilon > 0$, there is $\delta > 0$ such that if you have a subfamily F of G of size $> (1 - \delta)^n |G|$, then the number of pairs of sets in F satisfying (*) and (**) is at least $(1 - \epsilon)^n X$.

Remark (February 2015): Eoin Long has recently reduced many cases of this conjecture to the original Frankl–Rödl theorem.

7.5 Frankl–Rödl/Frankl–Wilson for cocycles

Conjecture 7.7 (Frankl–Rödl/Frankl–Wilson theorem for cocycles)
For every $\epsilon, \gamma > 0$, there is $\delta > 0$ with the following property. Let $\mathcal{F}$ be the family of 3-cocycles. Let X be the number of pairs of elements in $\mathcal{F}$ whose symmetric difference has precisely $m = [\gamma\binom{n}{4}]$ sets. Then for every $\mathcal{G} \subset \mathcal{F}$ if $|\mathcal{G}| \geq (1 - \delta)^{\binom{n}{4}}|\mathcal{F}|$, the number of pairs of elements in $\mathcal{G}$ whose symmetric difference has precisely m sets is at least $(1 - \epsilon)^{\binom{n}{4}} X$.

The case of 1-cocycles is precisely the conclusion of Frankl–Wilson/Frankl–Rödl needed for Borsuk's conjecture, and a Frankl–Rödl theorem for 4-cycles may also be a way to push up the asymptotic lower bounds for Borsuk's problem via Conjecture 7.3.

8 Paul Erdős' way with people and with mathematical problems

There is a saying in the ancient Hebrew scriptures:

> Do not scorn any person and do not dismiss any thing, for there is no person who has not his hour, and there is no thing that has not its place.

Paul Erdős had an amazing way of practicing this saying, when it came to people, and likewise when it came to his beloved "things," - mathematical problems. And his way accounts for some of our finest hours.

Acknowledgment

I am very thankful to Evan DeCorte, Jeff Kahn, Andrey Kupavskii, Rom Pinchasi, Oleg Pikhurko, Andrei Raigorodskii, Micha Sharir, and Konrad Swanepoel for useful discussions and comments.

References

[1] P. Agarwal, E. Nevo, J. Pach, R. Pinchasi, M. Sharir, and S. Smorodinsky, Lenses in arrangements of pseudocircles and their applications, *J. ACM* 51 (2004), 139–186.

[2] P. K. Agarwal and J. Pach, *Combinatorial Geometry* John Wiley and Sons, New York, 1995.

[3] N. Alon, L. Babai, and H. Suzuki, Multilinear polynomials and Frankl–Ray-Chaudhuri–Wilson type intersection theorems, *J. Combin. Theory A* 58 (1991), 165–180.

[4] N. Alon, Packings with large minimum kissing numbers, *Discrete Math.* 175 (1997), 249.

[5] A. Ashikhmin, A. Barg, and S. Vladut, Linear codes with exponentialy many light vectors, *J. Combin. Theory A* 96 (2001) 396–399.

[6] N. Alon, H. Last, R. Pinchasi, and M. Sharir, On the complexity of arrangements of circles in the plane, *Discrete and Comput. Geometry* 26 (2001), 465-492.

[7] N. Alon and P. Pudlak, Equilateral sets in l_p^n, *Geom. Funct. Anal.* 13 (2003), 467-482.

[8] C. Bachoc, G. Nebe, F. M. de Oliveira Filho, and F. Vallentin, Lower bounds for measurable chromatic numbers, *Geom. Funct. Anal.* 19 (2009), 645-661.

[9] C. Bachoc, E. DeCorte, F. M. de Oliveira Filho, and F. Vallentin, Spectral bounds for the independence ratio and the chromatic number of an operator (2013), http://arxiv.org/abs/1301.1054.

[10] C. Bachoc, A. Passuello, and A. Thiery, The density of sets avoiding distance 1 in Euclidean space (2014), arXiv:1401.6140

[11] K. Bezdek, Z. Langi, M. Naszódi, and P. Papez, Ball-polyhedra. *Discrete Comput. Geom.* 38 (2007), 201–230.

[12] K. Bezdek and M. Naszódi, Rigidity of ball-polyhedra in Euclidean 3-space, *European J. Combin.* 27 (2006), 255–268.

[13] V. G. Boltyanski and I. Gohberg, *Results and Problems in Combinatorial Geometry*, Cambridge University Press, Cambridge, 1985.

[14] V.G. Boltyanski, H. Martini, and P.S. Soltan, *Excursions into combinatorial geometry*, Universitext, Springer, Berlin, 1997.

[15] A. V. Bondarenko, On Borsuk's conjecture for two-distance sets, arXiv:1305.2584. *Disc. Comp. Geom.*, to appear.

[16] K. Borsuk and Drei Sätze über die n-dimensionale euklidische Sphäre, *Fund. Math.* 20 (1933), 177–190.

[17] J. Bourgain and J. Lindenstrauss, On covering a set in $\mathbf{R}^N$ by balls of the same diameter, in *Geometric Aspects of Functional Analysis* (J. Lindenstrauss and V. Milman, eds.), Lecture Notes in Mathematics 1469, Springer, Berlin, 1991, pp. 138–144.

[18] P. Brass, W. Moser, and J. Pach, *Research problems in discrete geometry*, Springer, Berlin, 2005.

[19] V. V. Bulankina, A. B. Kupavskii, and A. A. Polyanskii, On Schur's conjecture in $\mathbf{R}^4$, *Dokl. Math.* 89 (2014), N1, 88–92.

[20] H. Busemann, Intrinsic area, *Ann. Math.* 48 (1947), 234–267.

[21] Y. Colin de Verdiére, Sur un nouvel invariant des graphes et un critére de planarité, *J. Combin. Th. B* 50 (1990): 11–21.

[22] J. H. Conway and N. J. A. Sloane, *Sphere Packings, Lattices and Groups*, Grundlehren Math. Wiss., vol. 290, Springer, New York, third ed., 1993.

[23] H. Croft, K. Falconer, and R. Guy, *Unsolved Problems in Geometry*, Springer, New York, 1991.

[24] D. de Caen, Large equiangular sets of lines in Euclidean space, *Electronic Journal of Combinatorics* 7 (2000), Paper R55, 3 pages.

[25] D. de Caen, D. L. Kreher, and J. Wiseman, On constructive upper bounds for the Turán numbers T(n,2r+1,2r), *Congressus Numerantium* 65 (1988), 277–280.

[26] E. De Corte and O. Pikhurko, Spherical sets avoiding a prescribed set of angles, preprint 2014.

[27] D. de Laat and F. Vallentin, A semidefinite programming hierarchy for packing problems in discrete geometry, arXiv:1311.3789 (2013).

[28] B. V. Dekster, Diameters of the pieces in Borsuk's covering *Geometriae Dedicata* 30 (1989), 35–41.

[29] B. V. Dekster, The Borsuk conjecture holds for convex bodies with a belt of regular points, *Geometriae Dedicata* 45 (1993), 301–306.

[30] B. V. Dekster, The Borsuk conjecture holds for bodies of revolution *Journal of Geometry* 52 (1995), 64–73.

[31] P. Delsarte, *An algebraic approach to the association schemes of coding theory*, Diss. Universite Catholique de Louvain (1973).

[32] P. Delsarte, J. M. Goethals, and J. J. Seidel. Spherical codes and designs, *Geometriae Dedicata* 6 (1977), 363–388.

[33] M. Deza and M. Laurent, *Geometry of Cuts and Metrics,* Algorithms and Combinatorics, Springer, Berlin, 1997.

[34] V. L. Dolnikov, Some properties of graphs of diameters, *Discrete Comput. Geom.* 24 (2000), 293–299.

[35] H. G. Eggleston, Covering a three-dimensional set with sets of smaller diameter, *J. London Math. Soc.* 30 (1955), 11–24.

[36] H. G. Eggleston, *Convexity,* Cambridge University Press, 1958.

[37] P. Erdős, On sets of distances of n points, *Amer. Math. Monthly* 53 (1946), 248–250.

[38] P. Erdős, On the combinatorial problems I would like to see solved, *Combinatorica* 1 (1981), 25–42.

[39] P. Erdős, Some old and new problems in combinatorial geometry, *Ann. of Disc. Math.* 57 (1984), 129–136.

[40] S. Fiorini, S. Massar, S. Pokutta, H. Tiwary, and R. de Wolf, Linear vs. semidefinite extended formulations: exponential separation and strong lower bounds, preprint, 2011.

[41] P. Frankl and V. Rödl, Forbidden intersections, *Trans. Amer. Math. Soc.* 300 (1987), 259–286.

[42] P. Frankl and R. Wilson, Intersection theorems with geometric consequences, *Combinatorica* 1 (1981), 259–286.

168 G. Kalai

[43] Z. Füredi, J. C. Lagarias, and F. Morgan, Singularities of minimal surfaces and networks and related extremal problems in Minkowski space, *Discrete and Computational Geometry* Amer. Math. Soc., Providence, 1991, pp. 95–109.

[44] E. S. Gorskaya, I. M. Mitricheva, V. Yu. Protasov, and A. M. Raigorodskii, Estimating the chromatic numbers of Euclidean spaces by methods of convex minimization, *Sb. Math.* 200 (2009), N6, 783–801.

[45] P. M. Gruber and C. G. Lekkerkerker, *Geometry of Numbers*, North-Holland, Amsterdam, 1987.

[46] B. Grünbaum, A proof of Vászonyi's conjecture, *Bull. Res. Council Israel, Sect. A* 6 (1956), 77–78.

[47] B. Grünbaum, A simple proof of Borsuk's conjecture in three dimensions, *Mathematical Proceedings of the Cambridge Philosophical Society* 53 (1957), 776–778.

[48] B. Grünbaum, Borsuk's partition conjecture in Minkowski planes, *Bull. Res. Council Israel* (1957/1958), pp. 25–30.

[49] B. Grünbaum, Borsuk's problem and related questions, *Convexity, Proc. Sympos. Pure Math.*, vol. 7, Amer. Math. Soc, Providence, RI, 1963.

[50] A. E. Guterman, V. K. Lyubimov, A. M. Raigorodskii, and A. S. Usachev, On the independence numbers of distance graphs with vertices at $\{-1,0,1\}^n$: estimates, conjectures, and applications to the Borsuk and Nelson – Erdős – Hadwiger problems, *J. of Math. Sci.* 165 (2010), N6, 689–709.

[51] L. Guth and N. H. Katz, On the Erdős distinct distances problem in the plane, *Ann. Math.* 181 (2015), 155-190.

[52] H. Hadwiger, Ein Überdeckungssatz für den Euklidischen Raum, *Portugaliae Math.* 4 (1944), 140–144.

[53] H. Hadwiger, Überdeckung einer Menge durch Mengen kleineren Durchmessers, *Comm. Math. Helv.*, 18 (1945/46), 73–75; Mitteilung betreffend meine Note: Überdeckung einer Menge durch Mengen kleineren Durchmessers, *Comm. Math. Helv.* 19 (1946/47), 72–73.

[54] A. Heppes, Térbeli ponthalmazok felosztása kisebbátmérőjű részhalmazok összegére, *A magyar tudományos akadémia* 7 (1957), 413–416.

[55] A. Heppes, Beweis einer Vermutung von A. Vázsonyi, *Acta Math. Acad. Sci. Hungar.* 7 (1957), 463–466.

[56] A. Heppes, P. Révész, Zum Borsukschen Zerteilungsproblem, *Acta Math. Acad. Sci. Hung.* 7 (1956), 159–162.

[57] A. Hinrichs, Spherical codes and Borsuk's conjecture, *Discrete Math.* 243 (2002), 253–256.

[58] A. Hinrichs and C. Richter, New sets with large Borsuk numbers, *Discrete Math.* 270 (2003), 137-147

[59] H. Hopf and E. Pannwitz: Aufgabe Nr. 167, *Jahresbericht d. Deutsch. Math.-Verein.* 43 (1934), 114.

[60] M. Hujter and Z. Lángi, On the multiple Borsuk numbers of sets, *Israel J. Math.* 199 (2014), 219–239.

[61] T. Jenrich, A 64-dimensional two-distance counterexample to Borsuk's conjecture, arxiv:1308.0206.

[62] J. Kahn, Coloring nearly-disjoint hypergraphs with $n + o(n)$ colors, *J. Combin. Th. A* 59 (1992), 31–39,

[63] J. Kahn, Asymptotics of Hypergraph Matching, Covering and Coloring Problems, *Proceedings of the International Congress of Mathematicians* 1995, pp. 1353–1362.

[64] J. Kahn and G. Kalai, On a problem of Füredi and Seymour on covering intersecting families by pairs, *Jour. Comb. Th. Ser A.* 68 (1994), 317–339.

[65] J. Kahn and G. Kalai, A counterexample to Borsuk's conjecture, *Bull. Amer. Math. Soc.* 29 (1993), 60–62.

[66] J. Kahn and P. D. Seymour, A fractional version of the Erdős-Faber-Lovász conjecture *Combin.* 12 (1992), 155–160.

[67] P. Keevash and E. Long, Frankl–Rödl type theorems for codes and permutations, preprint.

[68] G. Kalai, A new approach to Turan's Problem (research problem) *Graphs and Comb.* 1 (1985), 107–109.

[69] S. Khot and N. Vishnoi, The unique games conjecture, integrality gap for cut problems and embeddability of negative type metrics into l1. In *The 46th Annual Symposium on Foundations of Computer Science* 2005.

[70] V. Klee and S. Wagon, *Old and new unsolved problems in plane geometry and number theory*, Math. Association of America, 1991.

[71] R. Knast, An approximative theorem for Borsuk's conjecture, *Proc. Cambridge Phil. Soc.* (1974), N1, 75–76.

[72] A. B. Kupavskii, On coloring spheres embedded into $\mathbb{R}^n$, *Sb. Math.* 202 (2011), N6, 83–110.

[73] A. B. Kupavskii, On lifting of estimation of chromatic number of $\mathbb{R}^n$ in higher dimension, *Doklady Math.* 429 (2009), N3, 305–308.

[74] A. B. Kupavskii, On the chromatic number of $\mathbb{R}^n$ with an arbitrary norm, *Discrete Math.* 311 (2011), 437–440.

[75] A. B. Kupavskii and A. M. Raigorodskii, On the chromatic number of $\mathbb{R}^9$, *J. of Math. Sci.* 163 (2009), N6, 720–731.

[76] A.B.Kupavskii and A.M.Raigorodskii, Counterexamples to Borsuk's conjecture on spheres of small radii, *Moscow Journal of Combinatorics and Number Theory 2* N4 (2012), 27–48.

[77] A. B. Kupavskii and A. A. Polyanskii, Proof of Schur's conjecture in $\mathbf{R}^d$, arXiv:1402.3694.

[78] Y. S. Kupitz, H. Martini, and M. A. Perles: Finite sets in $\mathbf{R}^d$ with many diameters a survey. In: *Proceedings of the International Conference on Mathematics and Applications* (ICMA-MU 2005, Bangkok), Mahidol University Press, Bangkok, 2005, 91–112.

[79] Y. S. Kupitz, H. Martini, and M. A. Perles, Ball polytopes and the Vázsonyi problem. *Acta Mathematica Hungarica* 126 (2010), 99–163.

[80] Y. S. Kupitz, H. Martini, and B. Wegner, Diameter graphs and full equiintersectors in classical geometries. *Rendiconti del Circolo Matematico di Palermo* (2), Suppl. 70, Part II (2002), 65–74.

[81] D. Larman and N. Tamvakis, The decomposition of the n-sphere and the boundaries of plane convex domains, *Ann. Discrete Math.* 20 (1984), 209–214.

[82] D. G. Larman, A note on the realization of distances within sets in Euclidean space, *Comment. Math. Helvet.* 53 (1978), 529–535.

[83] D. G. Larman and C. A. Rogers, The realization of distances within sets in Euclidean space, *Mathematika* 19 (1972), 1–24.

[84] M. Lassak, An estimate concerning Borsuk's partition problem, *Bull. Acad. Polon. Sci. Ser. Math.* 30 (1982), 449–451.

[85] L. Lovaśz, Self-dual polytopes and the chromatic number of distance graphs on the sphere, *Acta Sci. Math.* 45 (1983), 317–323.

[86] M. Mann, Hunting unit-distance graphs in rational n-spaces, *Geombinatorics* 13 (2003), N2, 49–53.

[87] M. S. Melnikov, Dependence of volume and diameter of sets in an n-dimensional Banach space (Russian), *Uspehi Mat. Nauk* 18 (1963), 165-170.

[88] A. Marcus and G. Tardos, Intersection reverse sequences and geometric applications *Jour. Combin. Th. A* 113 (2006), 675–691.

[89] J. Matoušek, *Lectures on Discrete Geometry,* Springer, May 2002.

[90] J. Matoušek, *Using the Borsuk – Ulam theorem,* Universitext, Springer, Berlin, 2003.

[91] J. Matoušek, The number of unit distances is almost linear for most norms, *Advances in Mathematics* 226 (2011), 2618–2628.

[92] F. Morić and J. Pach, Remarks on Schurs Conjecture, in: *Computational Geometry and Graphs* Lecture Notes in Computer Science Volume 8296, 2013, pp. 120–131

[93] N. G. Moshchevitin and A. M. Raigorodskii, On colouring the space $\mathbb{R}^n$ with several forbidden distances, *Math. Notes* 81 (2007), N5, 656–664.

[94] V. F. Moskva and A. M. Raigorodskii, New lower bounds for the independence numbers of distance graphs with vertices at $\{-1, 0, 1\}^n$, *Math. Notes* 89 (2011), N2, 307–308.

[95] O. Nechushtan, Note on the space chromatic number, *Discrete Math.* 256 (2002), 499–507.

[96] A. Nilli, On Borsuk problem, in Jerusalem Combinatorics 1993 (H. Barcelo et G. Kalai, eds) 209–210, Contemporary Math. 178, AMS, Providence, 1994.

[97] F. M. de Oliveira Filho and F. Vallentin. Fourier analysis, linear programming, and densities of distance avoiding sets in $\mathbf{R}^n$, *J. Eur. Math. Soc.* 12 (2010) 1417–1428.

[98] J. Pach, The Beginnings of Geometric Graph Theory, *Erdős centennial*, Bolyai Soc. Math. Studies 25, 2013.

[99] I. Pak, *Lectures on Discrete and Polyhedral Geometry*, forthcoming.

[100] Y. Peled, M.Sc thesis, Hebrew University of Jerusalem, 2012.

[101] C. M. Petty, Equilateral sets in Minkowski spaces, *Proc. Amer. Math. Soc.* 29 (1971), 369–374.

[102] O. Pikhurko, Borsuk's conjecture fails in dimensions 321 and 322, arXiv: CO/0202112, 2002.

[103] R. Pinchasi, Gallai–Sylvester Theorem for Pairwise Intersecting Unit Circles, *Discrete and Computational Geometry* 28 (2002), 607–624.

[104] R. Pinchasi and R. Radoičić, On the Number of Edges in a Topological Graph with no Self-intersecting Cycle of Length 4, in *19th ACM Symposium on Computational Geometry*, San Diego, USA, 2003, pp 98–103.

[105] A. M. Raigorodskii, Around Borsuk's conjecture, *Itogi Nauki i Tekhniki, Ser. "Contemp. Math."* 23 (2007), 147–164; English transl. in J. of Math. Sci., 154 (2008), N4, 604–623.

[106] A. M. Raigorodskii, Cliques and cycles in distance graphs and graphs of diameters, *Discrete Geometry and Algebraic Combinatorics*, AMS, Contemporary Mathematics, 625 (2014), 93–109.

[107] A. M. Raigorodskii, Coloring Distance Graphs and Graphs of Diameters, *Thirty Essays on Geometric Graph Theory*, J. Pach ed., Springer, 2013, 429–460.

[108] A. M. Raigorodskii, *On a bound in Borsuk's problem*, Uspekhi Mat. Nauk, 54 (1999), N2, 185–186; English transl. in Russian Math. Surveys, 54 (1999), N2, 453–454.

[109] A. M. Raigorodskii, *On the dimension in Borsuk's problem*, Russian Math. Surveys, 52 (1997), N6, 1324–1325.

[110] A. M. Raigorodskii, The Borsuk partition problem: the seventieth anniversary, *Mathematical Intelligencer* 26 (2004), N3, 4–12.

[111] A. M. Raigorodskii, *The Borsuk problem and the chromatic numbers of some metric spaces*, Uspekhi Mat. Nauk, 56 (2001), N1, 107–146; English transl. in Russian Math. Surveys, 56 (2001), N1, 103–139.

[112] A. M. Raigorodskii, Three lectures on the Borsuk partition problem, *London Mathematical Society Lecture Note Series,* 347 (2007), 202–248.

[113] C. A. Rogers, Covering a sphere with spheres, *Mathematika* 10 (1963), 157–164.

[114] C. A. Rogers, Symmetrical sets of constant width and their partitions, *Mathematika* 18 (1971), 105–111.

[115] M. Rosenfeld, Odd integral distances among points in the plane, *Geombinatorics* 5 (1996), 156–159.

[116] O. Schramm, Illuminating sets of constant width, *Mathematica* 35 (1988), 180–199.

[117] O. Schramm, On the volume of sets having constant width *Israel J. of Math.* Volume 63 (1988), 178–182.

[118] Z. Schur, M. A. Perles, H. Martini, and Y. S. Kupitz, On the number of maximal regular simplices determined by n points in $\mathbf{R}^d$. In: *Discrete and Computational Geometry The Goodman-Pollack Festschrift,* Eds. B. Aronov, S. Basu, J. Pach, and M. Sharir, Springer, New York et al., 2003, 767–787.

[119] J. J. Seidel, A survey of two-graphs, in: Teorie Combinatorie (Proc. Intern. Coll., Roma 1973), Accad. Nac. Lincei, Roma, 1976, pp. 481–511

[120] P. S. Soltan, Analogues of regular simplexes in normed spaces (Russian), *Dokl. Akad. Nauk SSSR* 222 (1975), 1303-1305. English translation: Soviet Math. Dokl. 16 (1975), 787–789.

[121] J. Steinhardt, On Coloring the Odd-Distance Graph *Electronic Journal of Combinatorics* 16:N12 (2009).

[122] K. J. Swanepoel, Equilateral sets in finite-dimensional normed spaces. In: *Seminar of Mathematical Analysis,* eds. Daniel Girela lvarez, Genaro Lpez Acedo, Rafael Villa Caro, Secretariado de Publicationes, Universidad de Sevilla, Seville, 2004, pp. 195-237.

[123] K. J. Swanepoel and R. Villa, A lower bound for the equilateral number of normed spaces, *Proc. of the Amer. Math. Soc.* 136 (2008), 127–131.

[124] L. A. Székely, Erdős on unit distances and the Szemerédi – Trotter theorems, *Paul Erdős and his Mathematics,* Bolyai Series Budapest, J. Bolyai Math. Soc., Springer, 11 (2002), 649–666.

[125] L. A. Székely, N.C. Wormald, Bounds on the measurable chromatic number of $\mathbb{R}^n$, *Discrete Math.* 75 (1989), 343–372.

[126] B. Weissbach, Sets with large Borsuk number, *Beitrage Algebra Geom.* 41 (2000), 417-423.

[127] D. R. Woodall, Distances realized by sets covering the plane, *J. Combin. Theory A* 14 (1973), 187–200.

[128] N. Wormald, A 4-Chromatic Graph With a Special Plane Drawing, *Australian Mathematics Society (Series A)* 28 (1979), 1–8.

[129] H. S. Witsenhausen. Spherical sets without orthogonal point pairs, *American Mathematical Monthly* (1974): 1101–1102.

[130] G. M. Ziegler, Coloring Hamming graphs, optimal binary codes, and the 0/1 - Borsuk problem in low dimensions, *Lect. Notes Comput. Sci.* 2122 (2001), 159–171.

Einstein Institute of Mathematics, Hebrew University of Jerusalem, and Department of Mathematics, Yale University.
kalai@math.huji.ac.il

Randomly generated groups

Tomasz Łuczak

Abstract

We discuss some older and a few recent results related to randomly generated groups. Although most of them are of topological and geometric flavour the main aim of this work is to present them in combinatorial settings.

1 Introduction

For the last half of the century the theory of randomly generated discrete structures has established itself as a vital part of combinatorics. Random graphs and hypergraphs and, more generally, combinatorial, algebraic, and geometric structures generated randomly have been used widely not only to provide numerous examples of objects of exotic properties but also as the way of studying and understanding large non-random systems which often can be decomposed into a small number of pseudorandom parts (see, for instance, Tao [37]). However, until recently, in the theory of random structures as known to combinatorialists random groups have not appeared very frequently (one is tempted to say, sporadically) although Gromov's model of the random group has already been introduced in the early eighties. The main reason was, undoubtedly, the fact that the world of combinatorialists seemed to be quite distant from the land of geometers and topologists and, despite many efforts of a few distinguished mathematicians familiar with both territories, combinatorialists did not believe that one can get basic understanding of the subject without much effort. This landscape has dramatically changed over the last few years. Topological combinatorics (or combinatorial topology) has been developing rapidly; many new projects have been started and a substantial number of articles have been published; combinatorialists have started to use topological terminology and more and more topological works are using advanced combinatorial tools. The aim of this article is just to spread the news. So it is not exactly a survey or even an introduction to this quickly evolving area – the reader who looks for this type of work is referred to a somewhat old but still excellent survey of Ollivier ([34], see also [35]) and the recent paper of Kahle [22]. This article should be considered rather as an encouragement, or a guide to the subject, written by a combinatorialist and addressed to the combinatorial community.

175

2 How to generate a random finite group

2.1 Random graphs

The best understood and most widely studied models in the theory of random discrete structures are the binomial and the uniform models of random graphs. Let us recall briefly the definition and basic properties of these two 'almost equivalent' probabilistic models. The vertex set for both of them is the set $[n] = \{1, 2, \ldots, n\}$. In order to construct the *binomial random graph* $G(n, p)$ one should include each of $\binom{n}{2}$ pairs of vertices in the set of edges of $G(n, p)$ independently with probability p. The *uniform random graph* $G(n, M)$ is a graph chosen uniformly at random from the family of $\binom{\binom{n}{2}}{M}$ graphs with the vertex set $[n]$ and exactly M edges. Let us also point out that, formally, both these graphs should be treated as a probability space, i.e. the family of graphs with the probabilistic measure defined on them, but in this article we stick to an informal style – the meticulous reader may look at [21] to ensure that each notion is correctly defined (which is not always obvious, especially when dealing with infinite structures). We only remark that each graph on n vertices appears with the same probability as $G(n, 1/2)$.

Typically we allow the parameters p and M to depend on n and say that for a given function $M(n)$ $[p(n)]$, the random graph $G(n, M)$ $[G(n, p)]$ has some property *aas* (*asymptotically almost surely*) if the probability that it has this property tends to 1 as $n \to \infty$. We adopt this notion throughout the paper, i.e. we always say that some random structure, which depends on parameter ℓ, has some property aas if it has this property with probability $1 - o_\ell(1)$, where $o_\ell(1)$, often abbreviated as $o(1)$, is a quantity which tends to 0 as $\ell \to \infty$. As we have already mentioned, for many properties the asymptotic behaviour of $G(n, p)$ and $G(n, M)$ are equivalent provided $M \sim p\binom{n}{2}$ (see [21] for details).

Very often to study properties of random graphs we use the idea of the random graph process, and the notion of threshold functions. The random graph process $\mathcal{G}_n = \{G(n, M) : 0 \leq M \leq \binom{n}{2}\}$ is the Markov chain whose states are graphs. It starts with the empty graph $G(n, 0)$ with vertex set $[n]$ and at each step $G(n, M+1)$ is obtained from $G(n, M)$ by adding to the previous graph an edge which is chosen uniformly at random among all pairs of vertices which are not edges of $G(n, M)$. It is easy to see that each graph with vertex set $[n]$ and M edges is equally likely to appear as $G(n, M)$ and so our notation is consistent since the M-th state of the process $\mathcal{G}_n$ is just the uniform random graph $G(n, M)$. Now let us take a graph property $\mathcal{A}$ and assume for a moment that it is *increasing*, i.e. that if a graph G has this property, then each supergraph

of G on the same set of vertices has this property as well. The property
that a graph is connected, or that it contains a cycle of length three are
examples of increasing properties. Typically our aim is to characterize
the moment $\mathcal{M}(\mathcal{G}, \mathcal{A}; n)$ of the random process $\mathcal{G}_n$ when the property $\mathcal{A}$
appears; note that, since the process is random, $\mathcal{M}(\mathcal{G}, \mathcal{A}, n)$ is a random
variable. The median $M_{\mathcal{A}}(n)$ of $\mathcal{M}(\mathcal{G}, \mathcal{A}, n)$ is called the *threshold function* of $\mathcal{A}$. As observed by Bollobás and Thomason (see [11] or [21]), for
increasing properties $\mathcal{A}$, the following holds.

Theorem 2.1 *If a property $\mathcal{A}$ is increasing, then*

$$\lim_{n \to \infty} \Pr(G(n, M(n)) \ has \ \mathcal{A}) = \begin{cases} 0 & if \quad M(n)/M_{\mathcal{A}}(n) \to 0 \\ 1 & if \quad M(n)/M_{\mathcal{A}}(n) \to \infty. \end{cases}$$

However very often the random variable $\mathcal{M}(\mathcal{G}, \mathcal{A}, n)$ is more sharply con-
centrated around its median $M_{\mathcal{A}}(n)$ and for some function $g(n)$ such that
$g(n)/M_{\mathcal{A}}(n) \to 0$ we have

$$\lim_{n \to \infty} \Pr(G(n, M(n)) \ has \ \mathcal{A}) = \begin{cases} 0 & if \quad M(n) \le M_{\mathcal{A}}(n) - g(n) \\ 1 & if \quad M(n) \ge M_{\mathcal{A}}(n) + g(n). \end{cases}$$

We remark that in such cases, typically, $M_{\mathcal{A}}(n)$ is very close to the expectation of $\mathcal{M}(\mathcal{G}, \mathcal{A}, n)$. A threshold of the above type we call *sharp*, while if
such a function $g(n)$ does not exist we call the threshold *coarse*. A power-
ful result of Friedgut [16] states that each property with coarse threshold
must be 'local' in some precise but somewhat technical way (see [16] for
details). An example of a sharp threshold is the classical result of Erdős
and Rényi which states that if $M = \frac{n}{2}(\log n + a)$, where a is a constant,
then the probability that $G(n, M)$ is connected tends to $\exp(-e^{-a})$ as
$n \to \infty$. On the other hand the property that a graph contains a cycle
of length three has a coarse threshold since whenever $M = cn$, where c
is a positive constant, the probability that $G(n, M)$ contains no triangles
tends to $\exp(-4c^3/3)$. Finally, let us also mention that the graph process
and the threshold functions of both types can be defined in a similar way
also for $G(n, p)$.

Finally, let us introduce yet another type of properties we shall refer to
later on. We say that an increasing property $\mathcal{A}$ is *explosive* if there exists a
constant $\delta > 0$ such that aas in the random graph process $\mathcal{G}_n = \{G(n, M) :
0 \le M \le \binom{n}{2}\}$ there exists $\hat{M}(\mathcal{G}, \mathcal{A})$ such that $G(n, \hat{M}(\mathcal{G}, \mathcal{A}))$ contains no
subgraphs with property $\mathcal{A}$, while $G(n, \hat{M}(\mathcal{G}, \mathcal{A}) + 1)$ contains a subgraph
with property $\mathcal{A}$ which has at least δn vertices. Thus, aas adding one edge
to a graph in which no subgraph has property $\mathcal{A}$ creates a giant subgraph

which has property $\mathcal{A}$. Although such a sudden appearance of a large subgraph with property $\mathcal{A}$ is a quite spectacular phenomenon, typically the fact that a property is explosive is easy to prove. To this end it is enough to verify that for some function $M = M(n)$ and some constant $\epsilon > 0$:

(i) aas $G(n, \epsilon M)$ contains no subgraph with property $\mathcal{A}$;

(ii) aas $G(n, M)$ contains a subgraph with property $\mathcal{A}$;

(iii) aas $G(n, M)$ contains no subgraphs with property $\mathcal{A}$ on fewer than δn vertices.

An example of an explosive property is the property that the minimum degree of a graph is at least k, where $k \geq 3$ (see Łuczak [29]).

2.2   A few approaches which do not work too well

Can we mimic the definition of either $G(n, M)$ or $G(n, p)$ to introduce an interesting model of a finite random group? It is not at all obvious how to do it – below we discuss a few ways which do not quite seem to work.

Let us make first one technical remark. Both $G(n, p)$ and $G(n, M)$ are models of random *labelled* graphs, while when studying groups we often do not distinguish between isomorphic groups (such an approach is analogous to considering graphs which are unlabelled). There exists an unlabelled analogue $G^u(n, M)$ of $G(n, M)$, when we classify all graphs with M edges and the vertex set $[n]$ up to isomorphism and select uniformly at random one of the isomorphic classes, but it is considered much less natural and is not so frequently studied (see Łuczak [30]). Note however that formally both $G^u(n, M)$ and $G(n, M)$ are probability distributions on the set of graphs with vertex set $[n]$, only in this case of $G^u(n, M)$ the distribution is defined in a slightly more complicated way. Thus, in order to construct a random group of at most n elements, we can always restrict ourselves to subgroups of S_n, which contains isomorphic copies of all groups on n elements, and define a probability distribution on this family.

Thus, suppose for instance that we aim for the definition analogous to $G(n, 1/2)$ and choose a group uniformly at random from all groups with n elements (i.e. from all isomorphic types of subgroups of S_n with n elements). Unfortunately, the number and properties of such groups would strongly depend on arithmetic properties of n (e.g. if n is prime, then all groups of n elements are isomorphic to $\mathbf{Z}/n\mathbf{Z}$) so we cannot expect to get interesting asymptotic results for such a model. To select a group uniformly at random among all groups on at most n elements seems to be a better choice, but we have no idea how to study properties of such groups – in fact even counting them is a hard and challenging problem (cf. Blackburn *et al.* [9]).

Since it is hard to control the number of elements in 'random group' let us try another natural approach. Let us choose uniformly at random a few elements in S_n, and consider the subgroup generated by them. However, by well-known theorem of Dixon [14], just two randomly generated permutations aas generate either the whole group S_n or the alternating group A_n. This and related results (see, for instance, Łuczak and Pyber [31]), show that we can hardly hope that any method of similar flavour would lead to interesting models of random groups.

Yet another idea is to go back to a random graph and study a group related to this random object. The most obvious choice to get a group from a graph is to consider its automorphism group. Unfortunately, in this way we shall not construct interesting examples of groups either. Once $G(n,p)$ becomes connected its automorphism group becomes aas trivial and before this point the structure of its automorphism group is not too involved as well, since it is basically related to the presence of small isolated components and trees attached to the invariant part of the unique largest component (cf. Łuczak [26]). A random tree has a large automorphism group but its structure is not too exciting either and neither are the automorphism groups of other random graph models.

2.3   The cyclic space of a graph

However, there is another natural structure which is related to a graph: its cycle space. As we shall soon see, it can be used as a starting point to define several groups which are definitely worth studying. Let us recall that the *cycle space* consists of all subgraphs with all degrees even, viewed as 0-1 vectors in the $e(G)$-dimensional vector space over the two-element field $\mathbb{F}_2$, where $e(G)$ stands for the number of edges of G. The dimension of the space is $e(G) - v(G) + c(G)$, where $v(G)$ and $c(G)$ are the number of vertices and the number of components of G respectively, and there is a natural basis of the space which consists of cycles (i.e. of the indicator functions of cycles) – one needs to take a spanning forest F of G, and for each edge $e \in E(G) \setminus E(F)$ take the unique cycle in the graph $E(F) \cup \{e\}$.

The additive group of this vector space is rather easy to study. For instance, in order to determine if it is trivial, one usually uses the following obvious fact.

Observation 2.2 *The graph contains no cycles if and only if each subgraph of G contains a vertex of degree one.*

Thus, one can quickly decide if the cycle space of a graph G contains just one zero vector – it is enough to peel out pendant vertices from G and check if at the end of the process we get a graph which consists of isolated points.

Another natural question related to the cycle space is whether some class $\mathbb{J}$ of cycles in G, such as, for instance, the family of cycles of a prescribed length (say, 3 or possibly n) span the whole cycle space.

Let us look at the behaviour of the cycle space of the random graph $G(n,p)$ (the same applies for $G(n,M)$ when $M = \lfloor p\binom{n}{2}\rfloor$). The existence of cycles in $G(n,p)$ has been studied already in one of the first articles devoted to these models, the seminal paper of Erdős and Rényi [15], who proved the following result.

Theorem 2.3 *If $0 \leq c \leq 1$ is a constant, then the probability that $G(n, c/n)$ contains no cycles tends to*

$$0 < \beta(c) = (\sqrt{1-c})e^{-c/2-c^2/4}$$

as $n \to \infty$. In particular, if $np \geq 1$, then aas $G(n,p)$ contains a cycle.

Furthermore, if $np = c$ for some constant $c < 1$, then aas no two cycles belong to the same component, while for $c > 1$ the dimension of the cycle space is of order $\Omega(n)$.

Let us also mention that, due to the ground-breaking work of Bollobás [10], later supplemented by Łuczak [27] and [28], we also know quite precisely the structure of $G(n,p)$ 'near the phase transition point' when $np \to 1$ (see [21]).

The other problem, when we are to decide if the family $\mathbb{J}$ spans the cycle space of $G(n,p)$, is more challenging. Clearly, if this is the case, then each edge which belongs to a cycle must belong to at least one cycle from the family $\mathbb{J}$. This necessary condition is also sufficient for triangles, as proved recently by De Marco, Hamm and Kahn [13].

Theorem 2.4 *Let $\epsilon > 0$ be a positive constant. Then for*

$$p \leq (1 - \epsilon)\sqrt{3\ln n/(2n)}$$

aas $G(n,p)$ contains an edge which belongs to no triangle while for

$$p \geq (1 + \epsilon)\sqrt{3\ln n/(2n)}$$

aas the triangles spans the cycle space in $G(n,p)$.

As for the other extreme case, when the family $\mathbb{J}$ consists of Hamiltonian cycles, the necessary condition is not quite sufficient because of parity issues. First of all, if n is even, then the cycle space spanned by Hamiltonian cycles contains no odd cycles, so one needs to consider only the case when n is odd. Moreover, it is easy to prove that at the moment when the minimum degree of a random graph reaches two, aas each of its edges belongs to a Hamiltonian cycle but also to another cycle of even length. Now let us suppose that a graph G with odd number of vertices has a vertex of degree two, and let f be an edge incident to this vertex which belongs to an even cycle C. Then C cannot belong to the cycle space generated by the Hamiltonian cycles of G. Indeed, suppose that this is the case. Each Hamiltonian cycle contains f so C must be a sum of odd number of Hamiltonian cycles but it is impossible since a sum of odd number of odd sets has an odd number of edges. It turns out however that the cycle space of $G(n,p)$ is aas generated by its Hamiltonian cycles as soon as the minimum degree of $G(n,p)$ is at least three (see Heinig and Luczak [19]).

We conclude this subsection with a short side remark that, instead of considering the cycle space of some random graph, one may also study the structure of the random element of the cycle space of a given graph G. Surprisingly, as shown by Grimmett and Janson [17], such element can be effectively generated and analysed both for finite as well as for 'locally finite' graphs G.

2.4   From graphs to hypergraphs

The notion of a cycle space can be naturally generalized to the case of hypergraphs. Let us take a k-uniform hypergraph with vertex set $[n]$, i.e. a family of k-element subsets of $[n]$, and consider all its subhypergraphs F which have the property that each $(k-1)$-element subset of vertices is contained in even number of hyperedges of F. It is easy to see that the symmetric difference of any pair of such subhypergraphs also enjoys the same property, so the family of all such subhypergraphs can be viewed as a vector space over $\mathbb{F}_2$.

It turns out that once $k \geq 3$, such a group for a k-uniform hypergraph is much harder to study. Let us restrict to the case $k = 3$, and for a 3-uniform hypergraph G denote this group by $H_2(G) = H_2(G; \mathbb{F}_2)$. Equivalently, $H_2(G)$ can be described in the following algebraic way. Consider the incidence matrix $\mathbf{M}(G)$ of G, where the rows are indexed by pairs of vertices, and columns correspond to edges of G, so that, since each triple contains three pairs of vertices, each column contains precisely three ones. Then $H_2(G)$ can be identified with the set of vectors $\mathbf{v}$ such that $\mathbf{M}(G)\mathbf{v} = 0$, i.e. with the right kernel of $\mathbf{M}(G)$.

Although in this article we try to avoid topological notions as much as possible we remark here that the notation $H_2(G)$ comes from the fact that it is just the homology group over $\mathbb{F}_2$ of the simplicial complex corresponding to G. The subscript 2 means that we are interested in the top homology of a 2-dimensional complex, i.e. we study the dependence of columns of the matrix which correspond to two dimensional simplices, i.e. 3-element hyperedges of G.

Let us start the description of $H_2(G)$ with a remark that if G is *collapsible*, i.e. if in each subhypergraph F of G one can find a pair of vertices which is contained in precisely one edge of F, then, clearly, $H_2(G)$ is trivial. However, there are simple examples showing that for $k \geq 3$, the converse does not hold and so the result analogous to Observation 2.2 is not true. In fact, as we shall see shortly, examples of non-collapsible G with trivial $H_2(G)$ are quite common.

In order to justify this claim we consider the behaviour of $H_2(G_3(n, M))$, where $G_3(n, M)$ is the 3-uniform hypergraph chosen uniformly at random from the family of all 3-uniform hypergraphs with vertex set $[n]$ and exactly M edges. If $M > \binom{n}{2}$, then the incidence matrix $\mathbf{M}(G_3(n, M))$ has more columns than rows and so some rows must be dependent; consequently, $H_2(G_3(n, M))$ contains non-trivial elements. On the other hand, one can use the upper bound for the number of collapsible hypergraphs with $\bar{n}$ vertices and $\bar{M}$ edges (see Aronshtam *et al.* [7]) to show that:

(i)   there exists a small positive constant $c > 0$ such that aas $G_3(n, cn^2)$, after possibly removing small number of tetrahedrons, is collapsible;

(ii) for each (large) constant C, there exists a (small) constant $\delta > 0$ such that aas each non-collapsible subgraph of $G_3(n, Cn^2)$ with fewer than δn vertices is a union of vertex disjoint tetrahedrons.

The above statements imply that the property which states that 'G, after removing all tetrahedrons, is collapsible', as well as the property 'G, after removing all tetrahedrons, contains a subgraph such that each pair of vertices is contained in an even number of its edges', are both explosive. Let us comment briefly on the 'tetrahedron issue' which emerged in the statements above. If $M = \Theta(n^2)$ then, with probability bounded away from zero, $G_3(n, M)$ contains a very few vertex-disjoint tetrahedrons. However these structures are small, rare, and lie far from each other, so they do not affect very much the global properties of $G_3(n, M)$ we are interested in.

Once we know that both above properties appear in the graph process for $M = \Theta(n^2)$ we may want to find the exact threshold for it. It has been done for collapsibility: a lower bound for the threshold was given by Aronshtam *et al.* [7] and the matching upper bound was found by Aronshtam and Linial [5].

Theorem 2.5 *There exists an explicitly computable constant $c_1 = 0.40\ldots$ such that for each positive constant $\epsilon > 0$:*

(i) *for $M \leq (1 - \epsilon)c_1 n^2$ aas $G(n, M)$ collapses to either an empty graph, or a union of a few tetrahedrons;*

(ii) *for $M \geq (1 + \epsilon)c_1 n^2$ aas $G(n, M)$ contains a non-collapsible subhypergraph on $\Omega(n)$ vertices.*

In order to study the behaviour of $H_2(G_3(n, M))$ note first that below the collapsibility threshold the size of $H_2(G_3(n, M))$ is determined by small number of tetrahedrons contained in $G_3(n, M)$, but at some point adding a single edge will increase it from $O(1)$ to $\exp(\Omega(n))$. Aronshtam and Linial [6] found an upper bound for this moment of the evolution of $G_3(n, M)$ and the matching lower bound was given by Linial and Peled [25].

Theorem 2.6 *There exists an explicitly computable constant $c_2 = 0.45\ldots$ such that for each positive constant $\epsilon > 0$:*

(i) *for $M \leq (1 - \epsilon)c_2 n^2$ aas $H_2(G_3(n, M))$ is determined by tetrahedrons contained in $G_3(n, M)$ and so $|H_2(G_3(n, M))| = O(1)$;*

(ii) *for $M \geq (1 + \epsilon)c_2 n^2$ there exists a constant $d > 0$ such that aas $|H_2(G_3(n, M))| = e^{(1 + o(1))dn}$.*

Note that the above two results imply that for any c, $c_1 < c < c_2$, aas $G_3(n, cn)$, after removing tetrahedrons, is an example of a hypergraph G for which $H_2(G) = 0$ yet it is not collapsible.

Let us also mention that the above results are valid also for k-uniform random hypergraphs $G_k(n, M)$. In this case Aronshtam and Linial [5] found the collapsibility threshold (which, in this case, is given by a formula $M = (c_{1,k} + o(1))n^{k-1}$ for some constant $c_{1,k} > 0$), and Linial and Peled [25] determined the moment $M = (c_{2,k} + o(1))n^{k-1}$ starting from which $H_{k-1}(G_k(n, M))$ grows exponentially with n. It is worthwhile to mention that the relative length of the period when the random hypergraph is not collapsible but its top homology is basically trivial grows with k; in fact, for large k we have $c_{1,k} = \Theta_k(\log k)$ while $c_{2,k} = \Theta_k(k)$.

2.5   From the right kernel to the left one

Let us recall that, for a 3-uniform hypergraph G, $H_2(G)$ stands for the set of vectors $\mathbf{v}$ such that $\mathbf{M}(G)\mathbf{v} = 0$, where G is the incidence matrix of G whose rows correspond to pairs of vertices, and columns are related to the edges of G. A natural question emerges about the vector space which consists of vectors $\mathbf{w}$ which are in the left kernel of the matrix, i.e. such that $\mathbf{w}\mathbf{M}(G) = 0$. This additive group of the vector space we denote by $H_1(G) = H_1(G; \mathbb{F}_2)$ (note that now we add rows, which correspond to pairs of vertices i.e. 1-dimensional simplices). Clearly, since we are looking for the dependencies between rows, $H_1(G)$ decreases when we add edges to G. Hence, in the random hypergraph setting, we ask for the property that $H_1(G_3(n, M)) = 0$. Note that if one row of $\mathbf{M}(G)$ consists of zeros, i.e. if there exists a pair of vertices of G which is not contained in any of the edges then, clearly, a vector which consists of one 1 and $\binom{n}{2} - 1$ zeros belongs to $H_1(G)$, and so it is non-trivial. Linial and Meshulam [24] proved that, in fact, aas in the random graph process $H_1(G_3(n, M)) = 0$ as soon as each pair of vertices belongs to at least one of the edges of $G_3(n, M)$. Thus, the threshold function for the property that $H_1(G_3) = 0$ is the same as the threshold function for the property that each pair of vertices is contained in some edge of a 3-uniform hypergraph G_3. This result was later generalized for k-uniform hypergraphs by Meshulam and Wallach [32]. We state it for the binomial model $G_k(n, p)$ which denotes the k-uniform hypergraph with vertex set $[n]$ in which each k-tuple is chosen as an edge independently with probability p.

Theorem 2.7 *Let $k \geq 2$ and let $G_k(n, p)$ denote the random k-uniform hypergraph. Consider its incidence matrix $\mathbf{M}_k(n, p)$ where rows are labelled by $(k-1)$-tuples of vertices and columns correspond to edges of $G_k(n, p)$. Then for every function $\omega(n) \to \infty$ the following holds.*

If $np \leq k \log n - \omega(n)$, then aas $\mathbf{M}_k(n, p)$ has a row consisting of zeros, while for $np \geq k \log n + \omega(n)$, aas the left kernel of $\mathbf{M}_k(n, p)$ is trivial.

2.6   A remark on flag complexes

Although random hypergraphs $G_k(n, M)$ or $G_k(n, p)$ are very natural models of random families of k-element subsets of $[n]$ there are other simple models of such families as well. For instance, one can take the random graph $G(n, \rho)$ and consider all k-subsets which induce k-cliques in $G(n, \rho)$. Let us denote this family by $G_{(k)}(n, \rho)$. It is easy to see that any k-element subset A of $[n]$ belongs to $G_{(k)}(n, \rho)$ with probability $\rho^{\binom{k}{2}}$, but for sets which share at least two vertices these events are not independent but positively correlated. Nonetheless one expects that at least for small values of ρ, the limit behaviours of both $G_{(k)}(n,\rho)$ and $G_k(n,\rho^{\binom{k}{2}})$ are similar and this is usually the case, although very often one needs to use different arguments for each of the models. Random flag complexes are hypergraphs (or simplicial complexes) obtained from random graphs or hypergraphs of lower dimension, where for each clique we add to it a new, larger hyperedge (a simplex of larger dimension). Note that since simplexes in random flag complexes are of different sizes, they can and usually do have a richer structure than, say, $G_k(n,p)$, where we choose randomly only k-subsets of $[n]$. The theory of random flag complexes is rapidly developing and offers a lot of interesting questions (for more details see Kahle's survey [22]).

3   Infinite groups

Unlike in graph theory, which is built around finite structures and where the theory of infinite graphs and hypergraphs is considered a separate and somewhat elusive area, in group theory infinite groups appear naturally and are considered at least as interesting and important as finite ones. In fact the random group, as introduced by Gromov, is typically an infinite one, although, as we shall see shortly, it is defined by 'finite means'. In this part of the article we present some results concerning such groups, which are basically the fundamental groups of randomly generated cell complexes. However we start first with two remarks on the groups we considered in the previous part of the paper.

3.1   Infinite graphs

One way to get infinite groups similar to the finite ones described above is to replace finite random graphs by infinite ones. It turns out however that if p is a constant, then almost surely there exists only a unique, up to isomorphism, random graph with countable vertex set analogous to $G(n,p)$ (see Cameron [12] for more information on the properties of this fascinating object). The same is true for countable random k-uniform hypergraphs. Consequently, they have a uniquely defined automorphism group (which, in this case, is far from trivial), and homology groups.

3.2   Homologies over other groups/fields

Let us recall that in the previous chapter our starting point was the cycle space of a graph G, i.e. the vector space which consisted of indicator functions of subgraphs of G with all degrees even. In particular, we observed that this space is the same as the right kernel of the incidence matrix $\mathbf{M}(G)$ over the two-element field $\mathbb{F}_2$. But possibly we can replace $\mathbb{F}_2$ by some other fields or rings and get other groups which, when the group we used instead of $\mathbb{F}_2$ is infinite, will result in an infinite group as well.

Needless to say, as we learn in a basic course in topology, this indeed can be done if we are careful enough. Let us look first at the case where we use the three-element field $\mathbb{F}_3$ and let us speculate what should be the 'correct' analogous combinatorial question to the one we had for $\mathbb{F}_2$, where we ask if there exists a non-trivial subgraph of G with all degrees even, i.e. if the group which consists of such subgraphs is trivial. Thus, for $\mathbb{F}_3$ one can ask about finding non-trivial subgraphs with all degrees divisible by three. It looks like an interesting problem and in $G(n,p)$ the exact thresholds for this explosive property are not known. However, the family of subgraphs with all degrees divisible by three does not have a group structure, so it does not fit to the topological framework. We can also check if the right kernel of the incidence matrix $\mathbf{M}(G)$ over $\mathbb{F}_3$ is trivial. In this case the kernel has a group structure but the related combinatorial question is somewhat artificial because we allow edges to be replaced by double ones. It turns out that the 'right' topological question should be stated in a similar yet somewhat different way. Namely, we should ask whether G admits a non-trivial $\mathbb{F}_3$-flow. Let us recall that for an abelian group S an S-flow in a graph $G = (V, E)$ is a function $f : V \times V \to S$ which has the following properties:

- $f(v,w) = 0$ if $\{v,w\} \notin E$,

- $f(v,w) + f(w,v) = 0$ if $\{v,w\} \in E$,

- for every $v \in V$ we have $\sum_{w \in V} f(v,w) = 0$.

It is easy to see that the family of all flows with addition forms a group (a vector space if S is a field) and for $S = \mathbb{F}_2$ it is just the group of all subgraphs of even degrees. Note also that for $\mathbb{F}_3$ this group is related to the right kernel of the matrix $\mathbf{M}(G, \mathbb{F}_3)$, where in each column, instead of two 1's, we put one 1 and one 2.

Thus, we may go through all the types of questions we considered in the previous chapter for $\mathbb{F}_2$ and ask them in a more general setting. The answer for many of them is the same if we replace $\mathbb{F}_2$ by any other field, finite or infinite; in fact proofs of some of the theorems mentioned above work also for homology groups over other fields. On the other hand, there are also questions which become much more involved if we replace $\mathbb{F}_2$ by, say, $\mathbb{Z}$, because of the 'torsion factor' which often emerges when one studies homologies over groups which are not fields. We will not elaborate on this fact (more details can be found, say, in Kahle [22]) but just give a simple example of this phenomenon. Consider $\mathbb{Z}$-flows over the wheel W_4 which consists of a cycle C_4 of length four and a centre adjacent to all elements of the cycle. Then a (constant) $\mathbb{Z}$-flow on C_4 can be represented as a sum of unit flows on Hamiltonian cycles of W_4 if and only if its value is divisible by three.

3.3   The fundamental group of $G_3(n, p)$

Let us recall that $G_k(n, p)$ can be viewed as a (random) simplicial complex with vertex set $[n]$ which contains all subsets of $[n]$ with at most $k-1$ elements and some (random) family of k-sets (i.e. some random family of $(k-1)$-dimensional simplices). The threshold for the property that the fundamental group $\pi_1(G_3(n, p))$ of this complex is non-trivial was given by Babson *et al.* [8].

Theorem 3.1 *Let $\epsilon > 0$ be a constant and $\omega(n)$ be any function which tends to infinity as $n \to \infty$.*

(i) If $p(n) \leq n^{-1/2-\epsilon}$, then aas the fundamental group $\pi_1(G_3(n, p))$ is an infinite hyperbolic group,

(ii) If $p(n) \geq n^{-1/2}\sqrt{3(\log n + \omega(n))}$, then aas $\pi_1(G_3(n, p)) = 0$.

In order to understand the first part of the theorem we need to explain what it means for a group to be hyperbolic. Fortunately, hyperbolicity can be defined using combinatorial means. Let us take a finitely generated group Γ, the symmetric set of its generators S, and consider the Cayley graph $C(\Gamma, S)$. A group is *hyperbolic* (in the sense of Gromov) if the metric generated in $C(\Gamma, S)$ looks like a metric of a hyperbolic space. More precisely, for two vertices x, y of $C(\Gamma, S)$, we denote by $P(x, y)$ the shortest path between them in $C(\Gamma, S)$. Then Γ is *hyperbolic* if there exists a constant δ such that for any three vertices x, y, z, all vertices of $P(x, y)$ lie within distance δ from $P(x, z) \cup P(y, z)$. It is not hard to see that hyperbolicity is well defined, i.e. this definition does not depend on the choice of generators S (although the constant δ does). Clearly each finite group is hyperbolic (with $\delta = |\Gamma|$) and free groups are hyperbolic as well (with $\delta = 0$).

Let us also mention that the proof of Theorem 3.1 is rather hard and some of its ingredients work only for 2-dimensional simplicial complexes. For larger k the behaviour of $\pi_1(G_k(n, p))$ is, at this moment, not very well understood.

3.4 Żuk's model of triangular random groups

In this section we define another model of the random group, $\Gamma(n,p)$, which basically was introduced in [38] by Żuk who was inspired by Gromov's model we describe below. Let us recall that the random group $\pi_1(G_3(n,p))$ is the fundamental group of the simplicial complex in which we add random simplices of dimension two (i.e. triangles) to the 1-dimensional simplicial complex which corresponds to the complete graph on n vertices. In Żuk's model $\Gamma(n,p)$ we randomly add 2-dimensional cells to the 1-dimensional wedge whose fundamental group is the free group with n generators.

$\Gamma(n,p)$ can be equivalently defined in terms of group presentations. Let S be the set of n generators and $R = R(n,p)$ denote the random set of cyclically reduced words of length three over the alphabet $S \cup S^{-1}$, where each such word is chosen to be included in $R(n,p)$ independently with probability p. Then $\Gamma(n,p)$ is defined as the group which has presentation $\langle S|R(n,p)\rangle$. Thus, for instance, if $p = 0$, then $\Gamma(n,p)$ is the free group with n generators. Note however that if p is so small that, say, $R(n,p)$ consists of edge disjoint triangles, then $\Gamma(n,p)$ is a free group as well. Indeed, for each $r \in R$ one can choose one of the generators s contained in it and remove s from S and r from R. Since s belonged only to r we have $\langle S \setminus \{s\}|R \setminus \{r\}\rangle \cong \langle S|R\rangle$. Consequently, if the set $R(n,p)$ is sparse enough, $\Gamma(n,p)$ is the free group with $n - |R(n,p)|$ generators.

The threshold for the property that $\Gamma(n,p)$ is not free, as well as the threshold for collapsing of $\Gamma(n,p)$, was found by Żuk [38]. In fact his result is much stronger.

Theorem 3.2 *Let $\epsilon > 0$ be a positive constant.*

(i) If $p \leq n^{-2-\epsilon}$, then aas $\Gamma(n,p)$ is free.

(ii) If $n^{-2+\epsilon} \leq p \leq n^{-3/2-\epsilon}$, then aas $\Gamma(n,p)$ is an infinite hyperbolic group with property (T).

(iii) If $p \geq n^{-3/2+\epsilon}$, then aas $\Gamma(n,p)$ is trivial.

In the statement of the result the notion of Kazhdan's property (T) appeared, which, unfortunately, has no nice compact combinatorial interpretation. Here we only point out that groups with property (T) are somewhat analogous to graph expanders (in fact these two notions are strongly related) and for the precise definition we refer the reader to any textbook of group theory. We also remark that a free group does not have property (T), but if a group Γ has (T) then, for each normal subgroup H of Γ, the group Γ/H has this property as well. Consequently, if for some

p aas $\Gamma(n,p)$ has (T), then for each $p' \geq p$ the random group $\Gamma(n,p')$ aas has this property as well.

Looking at Theorem 3.2 one wonders what happens in the 'critical regions' when $p \sim n^{-3/2}$ and $p \sim n^{-2}$. The case $p \sim n^{-2}$ was first studied by Antoniuk et al. [4] who discovered a 'new period' in the evolution of the random group.

Theorem 3.3 *There are constants c, C, c', C' for which the following holds.*

(i) *If $p \leq c/n^2$, then aas $\Gamma(n,p)$ is free.*

(ii) *If $C/n^2 \leq p \leq c' \log n/n^2$, then aas $\Gamma(n,p)$ is neither free, nor has property (T).*

(iii) *If $p \geq C' \log n/n^2$, then aas $\Gamma(n,p)$ has property (T).*

It turns out that after some work one can find the values of the constants in the above result. Antoniuk et al. [2] computed the value of c.

Theorem 3.4 *There exists an explicitly computable constant c such that for every positive constant $\epsilon > 0$ the following holds.*

(i) *If $p \leq (c - \epsilon)/n^2$, then aas $\Gamma(n,p)$ is free.*

(ii) *If $p \geq (c + \epsilon)/n^2$, then aas $\Gamma(n,p)$ is not free.*

The constant $C' = c'$ was determined by Hoffman et al. [20].

Theorem 3.5 *For every positive constant $\epsilon > 0$ the following holds.*

(i) *If $p \leq (\frac{1}{24} - \epsilon) \log n/n^2$, then aas $\Gamma(n,p)$ does not have property (T).*

(ii) *If $p \geq (\frac{1}{24} + \epsilon) \log n/n^2$, then aas $\Gamma(n,p)$ has property (T).*

The critical period $p \sim n^{-3/2}$ when the group is collapsing is far more intriguing. Antoniuk et al. [1] showed that this property has a sharp threshold and the following statement holds.

Theorem 3.6 *There exists a function $\gamma = \gamma(n)$ such that for every positive constant $\epsilon > 0$ the following holds.*

(i) *If $p \leq (\gamma(n) - \epsilon)n^{-3/2}$, then aas $\Gamma(n,p)$ is not trivial.*

(ii) *If $p \geq (\gamma(n) + \epsilon)n^{-3/2}$, then aas $\Gamma(n,p)$ is trivial.*

It is also known (see Antoniuk et al. [4]) that the function $\gamma(n)$ in Theorem 3.6 is bounded, but the natural conjecture that $\gamma(n) \to c'' > 0$ as $n \to \infty$ remains wide open.

3.5 Gromov's general model of random groups

Żuk's model $\Gamma(n,p)$ is a special case of a more general model of a random group. Thus, let $\Gamma(k,\ell;p)$ denote the (random) group given by presentation $\langle S|R(k,\ell;p)\rangle$, where the set S contains k generators and $R(k,\ell;p)$ is a random family of cyclically reduced words of length ℓ, where each such word is chosen independently with probability p. Then, $\Gamma(n,p)$ becomes in our new notation $\Gamma(n,3;p)$. The first model of a random group was introduced by Gromov [18] who studied $\Gamma(k,\ell;p)$, where $k \geq 2$ was fixed but the parameter ℓ, the length of words in $R(k,\ell;p)$, tended to infinity. We should also mention that both Gromov's and Żuk's models were defined in a slightly different way. Both these authors instead of p used the *logarithmic density* δ and took as the set of words the randomly chosen set $\bar{R}(k,l;\delta)$ of $(2k-1)^{\delta\ell}$ cyclically reduced words of length ℓ. Let us denote this model by $\bar{\Gamma}(k,\ell;\delta)$. It turns out that the models with different k and ℓ but the same density are closely related due to the following observation of Żuk [38].

Rule of Thumb *If r is a natural number, then for some normal subgroup H of $\bar{\Gamma}(rk,\ell;\delta)$ we should have*

$$\bar{\Gamma}(rk,\ell;\delta)/H \cong \bar{\Gamma}(k,r\ell;\delta')$$

where δ is very close to δ'.

The rule is an immediate consequence of the fact that each word from $R(k,r\ell;\delta)$ in the presentation of $\bar{\Gamma}(k,r\ell;\delta) = \langle S|R(k,r\ell;\delta)\rangle$ of length $r\ell$ one can cut into ℓ subwords and consider each such subword as a new generator. However, one should be a bit careful when applying it. Thus, for instance, since Theorem 3.2(iii) restated in our language states that for $\delta > 1/2$, Żuk's group $\bar{\Gamma}(n,3;\delta)$ is aas trivial, then Gromov's group $\bar{\Gamma}(3,n;\delta)$ should be aas trivial for $\delta > 1/2$. Because of the parity issue it is only the case for odd n; for even n the group collapses to $\mathbb{F}_2$, as stated in the seminal result of Gromov's [18] (see also [23]).

Theorem 3.7

(i) *If $\delta < 1/2$, then $\bar{\Gamma}(3,n;\delta)$ is aas infinite and hyperbolic.*

(ii) *If $\delta > 1/2$, then $\bar{\Gamma}(3,n;\delta)$ has aas at most two elements.*

Gromov's model seems to be considerably harder to study than Żuk's model, at least by combinatorial means. For instance, we still do not know what is the threshold density δ_T above which $\bar{\Gamma}(3,n;\delta)$ has property (T).

Theorem 3.2(ii) together with the Rule implies that $\delta_T \leq 1/3$. On the other hand it is known (see Ollivier and Wise [36]) that $\delta_T \geq 1/5$. Thus, because of the Rule it would be worthwhile to study the analogous threshold $\delta_{T,\ell}$ for the model $\bar{\Gamma}(n, \ell; \delta)$ for large but fixed ℓ and $n \to \infty$ hoping that $\delta_{T,\ell} \to \delta_T$ as $\ell \to \infty$. However, the asymptotic behaviour of $\bar{\Gamma}(n, \ell; \delta)$ when $\ell \geq 5$ is fixed is not very well understood (for the work on $\bar{\Gamma}(n, 4; \delta)$ see Odrzygóźdź [33]). Some of the arguments used for $\bar{\Gamma}(n, 3; p)$ can be generalized also for this model; for instance, one can show that the intermediate phase described in Theorem 3.3(ii) lasts much longer for $\bar{\Gamma}(n, \ell; \delta)$ when ℓ is large. However, at this moment finding $\delta_{T,\ell}$ is still far out of our reach.

4 Final remarks

Because of the character of this work only a small fraction of the results concerning randomly generated groups have been presented here; furthermore, this choice was very strongly influenced by the personal preferences of the author of the article. Thus, we conclude the paper with a few words on results and problems we decided to omit.

As we have already pointed out we almost exclusively dealt with the field $\mathbb{F}_2$ which is very special; many results in the literature concern the other fields and groups (especially $\mathbb{Q}$ and $\mathbb{Z}$).

For random graphs and hypergraphs we used the standard random graph models $G_k(n, p)$ and $G_k(n, M)$. It is a natural choice but by no means the only possible one. One can instead consider for instance random regular graphs, random inhomogeneous graphs, random geometric graphs, random lifts as well as the product of random graphs – the topological properties of some of these models have already been explored but much more is still to be done.

Finally, there are plenty of group properties (such as, for instance, the asphericity, the cohomological dimension, or Serre's property) we have not mentioned here but which have been studied for some models of random groups.

One can find more information on many of these topics in Kahle's survey [22], although the reader should be aware that the literature on the subject is growing very fast.

Acknowledgements

The author wishes to thank all combinatorialists, topologists, and geometers, who patiently tried to teach him the topological and geometric language. In particular, I would like to thank Nati Linial, Roy Meshulam, Peter Heining, and Wojciech Politarczyk for illuminating discussions. I am especially grateful to Sylwia Antoniuk and an anonymous reviewer for their numerous remarks and comments on earlier versions of the manuscript.

References

[1] S. Antoniuk, E. Friedgut, T. Łuczak, *A sharp threshold for collapse of the random triangular group*, arXiv:1403.3516.

[2] S. Antoniuk, T.Łuczak, T. Prytula, P. Przytycki, B. Zalewski, *When a random triangular group is free?*, in preparation.

[3] S. Antoniuk, T. Łuczak, J. Świątkowski, *Collapse of random triangular groups: a closer look*, Bull. Lond. Math. Soc. **46** (2014), 761–764.

[4] S. Antoniuk, T. Łuczak, J. Świątkowski, *Random triangular groups at density 1/3*, Compositio Mathematica **151** (2015), 167–178.

[5] L. Aronshtam, N. Linial, *The threshold for collapsibility in random complexes*, Random Structures & Algorithms, to appear.

[6] L. Aronshtam, N. Linial, *When does the top homology of a random simplicial complex vanish?*, Random Structures & Algorithms **46** (2015), 26–35.

[7] L. Aronshtam, N. Linial, T. Łuczak, R. Meshulam, *Collapsibility and vanishing of top homology in random simplicial complexes*, Discrete Comput. Geom. **49** (2013), 317–334.

[8] E. Babson, C. Hoffman, M. Kahle, *The fundamental group of random 2-complexes*, J. Amer. Math. Soc. **24** (2011), 1–28.

[9] S.R. Blackburn, P.M. Neumann, G. Venkataraman, "Enumeration of finite groups". Cambridge Tracts in Mathematics, 173. Cambridge University Press, Cambridge, 2007.

[10] B. Bollobás, *The evolution of random graphs*, Trans. Amer. Math. Soc. **286** (1984), 257–274.

[11] B. Bollobás, A. Thomason, *Threshold functions*, Combinatorica **7** (1987), 35–38.

[12] P.J. Cameron, *The random graph revisited*, European Congress of Mathematics, Vol. I (Barcelona, 2000), 267–274, Progr. Math., 201, Birkhäuser, Basel, 2001.

[13] B. DeMarco, A. Hamm, J. Kahn, *On the triangle space of a random graph*, J. Combin. **4** (2013), 229–249.

[14] J.D. Dixon, *The probability of generating the symmetric group*, Math. Z. **110** (1969), 199–205.

[15] P. Erdős, A. Rényi, *On the evolution of random graphs*, Magyar Tud. Akad. Mat. Kutató Int. Közl. **5** (1960), 17–61.

[16] E. Friedgut, *Sharp thresholds of graph properties, and the k-sat problem. With an appendix by Jean Bourgain.* J. Amer. Math. Soc. **12** (1999), 1017–1054.

[17] G. Grimmett, S. Janson; Random even graphs. Electron. J. Combin. 16 (2009), no. 1, Research Paper, 46, 19 pp.

[18] M. Gromov, *Asymptotic invariants of infinite groups. Geometric Group Theory*, London Math. Soc. Lecture Note Ser. 182 (1993), 1–295.

[19] P. Heinig, T. Łuczak, *Hamiltonian space of random graphs*, in preparation.

[20] C. Hoffman, M. Kahle, E. Paquette, *Spectral gaps of random graphs and applications to random topology*, arXiv:1201.0425.

[21] S. Janson, T. Łuczak, A. Ruciński, "*Random Graphs*", Wiley, New York, 2000.

[22] M. Kahle, *Topology of random simplicial complexes: a survey.* To appear in AMS Contemporary Volumes in Mathematics, Nov 2014, arXiv:1301.7165.

[23] M. Kotowski, M. Kotowski, *Random groups and property (T): Żuk's theorem revisited*, J. London Math. Soc. **88** (2013), 396–416.

[24] N. Linial, R. Meshulam, *Homological connectivity of random 2-complexes*, Combinatorica **26** (2006), 475–487.

[25] N. Linial, Y. Peled, *On the phase transition in random simplicial complexes*, arXiv:1410.1281.

[26] T. Łuczak, *The automorphisms group of random graphs with given number of edges*, Math. Proc. Camb. Phil. Soc. **104** (1988), 441–449.

[27] T. Łuczak, *Components behavior near the critical point of the random graph process*, Random Structures & Algorithms **1** (1990), 287–310.

194   Tomasz Łuczak

[28] T. Łuczak, *Cycles in a random graph near the critical point*, Random Structures & Algorithms **2** (1991), 421–440.

[29] T. Łuczak, *Size and connectivity of the k-core of a random graph*, Discrete Math. **91** (1991), 61–68.

[30] T. Łuczak, *How to deal with unlabelled random graphs*, J. Graph Theory **15** (1991), 303–316.

[31] T. Łuczak, L. Pyber, *On random generation of the symmetric group*, Combinatorics, Probability & Computing **2** (1993), 505–512.

[32] R. Meshulam, N. Wallach, *Homological connectivity of random k-dimensional complexes*, Random Structures & Algorithms **34** (2009), 408–417.

[33] T. Odrzygóźdź *The square model for random groups*, arXiv:1405.2773.

[34] Y. Ollivier, *A January 2005 invitation to random groups*, Ensaios Matematicos [Mathematical Surveys] 10, Sociedade Brasileira de Matematica, Rio de Janeiro, 2005.

[35] Y. Ollivier, *Random group update*, http://www.yann-ollivier.org/rech/publs/rgupdates.pdf.

[36] Y. Ollivier, D.T. Wise, *Cubulating random groups at density less than 1/6*, Trans. Amer. Math. Soc. **363** (2011), no. 9, 4701–4733.

[37] T. Tao, *The dichotomy between structure and randomness, arithmetic progressions, and the primes*, International Congress of Mathematicians. Vol. I, 581–608, Eur. Math. Soc., Zürich, 2007.

[38] A. Żuk, *Property (T) and Kazhdan constants for discrete groups*, Geom. Funct. Anal. **13** (2003), 643–670.

Adam Mickiewicz University
Faculty of Mathematics & Computer Science
ul. Umultowska 87
61-614 Poznań, Poland
tomasz@amu.edu.pl

Curves over finite fields and linear recurring sequences

Omran Ahmadi and Gary McGuire

Abstract

We investigate what happens when we apply the theory of linear recurring sequences to certain sequences that arise from curves over finite fields. The sequences we will study are $a_n := \#C(\mathbb{F}_{q^n}) - (q^n + 1)$ where $\#C(\mathbb{F}_{q^n})$ is the number of $\mathbb{F}_{q^n}$-rational points on a curve C defined over $\mathbb{F}_q$.

1 Introduction

Let $\mathbb{K}$ be a field. A *linear recurring sequence of order d over* $\mathbb{K}$ is a sequence $\mathcal{A} = (a_n)_{n \geq 1}$ with elements in $\mathbb{K}$ which satisfy the homogenous recurrence relation

$$a_{n+d} = c_1 a_{n+d-1} + c_2 a_{n+d-2} + \cdots + c_d a_n \text{ for } n \geq 1, \quad (1.1)$$

where $c_1, \ldots, c_d$ are in $\mathbb{K}$ and $c_d \neq 0$. The recurrence (1.1) and the initial values $a_1, \ldots, a_d$ determine the complete infinite sequence.

The *characteristic polynomial* of the sequence (1.1) is defined to be the polynomial

$$\chi_{\mathcal{A}}(t) = t^d - c_1 t^{d-1} - \cdots - c_{d-1} t - c_d \in \mathbb{K}[t].$$

This does not depend on the initial values.

A linear recurring sequence is called *ultimately periodic of period T* if there exists an integer number N such that $a_{n+T} = a_n$ for any $n \geq N$. If $N = 0$, then the sequence is called *periodic*. Periodic sequences have been subject to intensive study in the past because of their many applications in communications, cryptography, radar and sonar (see [6]).

When $\mathbb{K}$ is a finite field, then any linear recurring sequence of order d is periodic. If $|\mathbb{K}| = q$, then the period of any linear recurring sequence of order d is at most $q^d - 1$. A linear recurring sequence $\mathcal{A}$ of order d over the finite field $\mathbb{F}_q$ with q elements is of period $q^d - 1$ if and only if $\chi_{\mathcal{A}}$ is a primitive polynomial over $\mathbb{F}_q$, i.e, any root of $\chi_{\mathcal{A}}(t)$ is a generator of the cyclic group $\mathbb{F}_{q^d}^*$ which is the set of the nonzero elements of the degree d

extension of $\mathbb{F}_q$. The degree d extension of $\mathbb{F}_q$ has q^d elements and usually is denoted by $\mathbb{F}_{q^d}$.

Sequences of great importance in practice are the *binary* linear recurring sequences which are linear recurring sequences defined over the binary field $\mathbb{F}_2$ with two elements. Among the binary sequences, the most interesting sequences are the so called *maximum length sequences* (MLS) which are the binary sequences with maximal period, i.e, they are binary sequences of order d and period $2^d - 1$. Sometimes these are called m-sequences.

Binary sequences which are used in practice should possess some good correlation properties. If $\mathcal{A} = (a_n)_{n \geq 1}$ and $\mathcal{B} = (b_n)_{n \geq 1}$ are two periodic binary sequences of period T, then their cross-correlation is defined as

$$C(\mathcal{A}, \mathcal{B}) = \sum_{i=1}^{T} (-1)^{a_i + b_i}.$$

Roughly speaking, the cross-correlation of two sequences measures the similarity of the sequences. For example, we have $C(\mathcal{A}, \mathcal{A}) = T$ for any binary sequence of period T. One of the important notions related to cross-correlation is the notion of auto-correlation. Auto-correlation measure the similarity of a periodic sequence $\mathcal{A}$ and periodic sequences obtained from shifting the terms of $\mathcal{A}$. If $\mathcal{A} = (a_n)_{n \geq 1}$ is a periodic binary sequence of period T, then its shift by τ positions denoted $\mathcal{A}_\tau = (a_n^\tau)_{n \geq 1}$ where $a_i^\tau = a_{i+\tau}$ is also a periodic sequence of period T. The *auto-correlation* of the sequence $\mathcal{A}$ with respect to its shift by τ positions is defined as

$$C_{\mathcal{A}}(\tau) = \sum_{i=1}^{T} (-1)^{a_i + a_{i+\tau}}.$$

A periodic binary sequence of order d and period T is good for applications if besides other properties its correlation function is two valued, i.e.,

$$C_{\mathcal{A}}(\tau) = \begin{cases} T, & \text{if } \tau \equiv 0 \pmod{T} \\ L, & \text{if } \tau \not\equiv 0 \pmod{T} \end{cases}$$

for some integer L.

From a given periodic sequence defined over the finite field $\mathbb{F}_q$ one can construct many periodic sequences. One obvious method is by shifting the sequence. Another less obvious method is by decimating a given sequence. If $\mathcal{A} = (a_n)_{n \geq 1}$ is a periodic sequence over the finite field $\mathbb{F}_q$, then the s-decimation of $\mathcal{A}$ which is the linear recurring sequence $\mathcal{B} = (a_{ns})_{n \geq 1}$ will be a periodic sequence over $\mathbb{F}_q$.

In many scenarios one has to compute the auto-correlation function of a sequence, or its cross-correlation with a decimation of it. Some linear recurring sequences defined over finite fields can be represented using trace function of finite fields which is very helpful for computing the auto-correlation of a sequence or cross-correlation of two sequences. More specifically, suppose that $\mathcal{A} = (a_n)_{n \geq 1}$ is a linear recurring sequence of order d over the finite field $\mathbb{F}_q$ with an irreducible characteristic polynomial $\chi_{\mathcal{A}}(t)$ over $\mathbb{F}_q$ and let α be a root of $\chi_{\mathcal{A}}(t)$ in $\mathbb{F}_{q^d}$. Then for some $\beta \in \mathbb{F}_{q^d}$ we have

$$a_n = Tr(\beta \alpha^n); n \geq 1 \tag{1.2}$$

where $Tr(\cdot)$ is the absolute trace function from $\mathbb{F}_{q^d}$ to $\mathbb{F}_q$ and for any $x \in \mathbb{F}_{q^d}$ is defined by

$$Tr(x) = x + x^q + x^{q^2} + \ldots + x^{q^{d-1}}.$$

Notice that if the trace representation of the sequence $\mathcal{A}$ is given by (1.2), then the trace representation of s-decimation $\mathcal{B} = (b_n)_{n \geq 1}$ of $\mathcal{A}$ is easily obtained and

$$b_n = a_{ns} = Tr(\beta \alpha^{ns}); n \geq 1. \tag{1.3}$$

As it mentioned earlier in many scenarios one wants to compute the cross-correlation of a binary MLS $\mathcal{A}$ of order d and its s-decimation $\mathcal{A}^s$. If s and $2^d - 1$ are coprime, it turns out that $\mathcal{A}^s$ is also an MLS with a period the same as that of $\mathcal{A}$ equal to $2^d - 1$. Thus using the trace representation of the two sequences the cross-correlation of $\mathcal{A}$ and $\mathcal{A}^s$ is

$$C(\mathcal{A}, \mathcal{A}^s) = \sum_{n=1}^{2^d-1} (-1)^{a_n + b_n} = \sum_{n=1}^{2^d-1} (-1)^{Tr(\beta \alpha^n) + Tr(\beta \alpha^{ns})}.$$

From the facts that $Tr(\cdot)$ is an additive function meaning that $Tr(x+y) = Tr(x) + Tr(y)$, β is non-zero and α is a primitive element of $\mathbb{F}_{2^d}$, it follows that when n runs through 1 to $2^d - 1$, then α^n runs through all the nonzero elements of $\mathbb{F}_{2^d}$. Thus denoting the set of nonzero elements of $\mathbb{F}_{2^d}$ by $\mathbb{F}_{2^d}^*$ we have

$$C(\mathcal{A}, \mathcal{A}^s) = \sum_{x \in \mathbb{F}_{2^d}^*} (-1)^{Tr(x) + Tr(bx^s)} = \sum_{x \in \mathbb{F}_{2^d}^*} (-1)^{Tr(bx + x^s)},$$

where $b = \beta^{\frac{s-1}{s}}$. Now if $Tr(bu + u^s) = 0$ for $u \in \mathbb{F}_{2^d}$, then there are two points on the following curve considered over $\mathbb{F}_{2^d}$

$$C : y^2 + y = x^s + bx$$

with x-coordinate equal to u, and if $Tr(bu + u^s) = 1$ then there is no point on C with x-coordinate equal to u. Thus we have

$$\#C(\mathbb{F}_{2^d}) = 2^d + 1 - C(\mathcal{A}, \mathcal{A}^s),$$

where $\#C(\mathbb{F}_{2^d})$ denotes the number of the points on C either having both coordinates in $\mathbb{F}_{2^d}$ or being at infinity. Hence computing the cross-correlation of two sequences $\mathcal{A}$ and $\mathcal{A}^s$ is the same as computing the number of points on a curve defined over a finite field. Actually, there are situations where one is forced to consider the sequences corresponding to two different curves.

This paper is partly motivated by a family of curves from a previous paper [9] by the second author. The curves are over $\mathbb{F}_2$ and are defined for any integer $k \geq 1$ by

$$C_k : y^2 + y = x^{2^k+1} + x. \tag{1.4}$$

It is proved in [9] that $\#C_1(\mathbb{F}_{2^d}) = \#C_k(\mathbb{F}_{2^d})$ for every d that is relatively prime to k. In particular, the sequences corresponding to the two different decimations 3 and $2^k + 1$ have the same crosscorrelation infinitely many times.

Motivated by these examples, in this paper we will define a certain linear recurring sequence of integers that arises from any algebraic curve defined over a finite field (see Section 3). The sequence itself has been well studied over the years. We will study this sequence from the point of view of a linear recurring sequence. The study of linear recurring sequences over finite fields has led us to curves, which has led us to linear recurring sequences of integers.

In Section 2 we recall some background information, and in Section 3 we give the definition. Sections 4, 5 and 6 consider the characteristic and minimal polynomial. In Section 7 we consider the case of maximal curves. Section 8 considers subsequences of the sequence, and Section 9 applies the Skolem–Mahler–Lech theorem and relates it to subsequences. Section 10 considers recovery of the sequence of a curve from a subsequence.

2   Background

We recall basic information here about linear recurring sequences over the integers, and also about curves over finite fields.

2.1   Background on linear recurring sequences

A *linear recurring sequence of order d* over the integers is a sequence of integers $(a_n)_{n \geq 1}$ satisfying the homogeneous recurring relation

$$a_{n+d} = c_1 a_{n+d-1} + c_2 a_{n+d-2} + \cdots + c_d a_n \text{ for } n \geq 1, \qquad (2.1)$$

where $c_1, \ldots, c_d$ are integers, $c_d \neq 0$. Linear recurring sequences may be defined in a similar manner over other fields and rings. In the rest of this paper we will be concerned only with integer sequences, so all sequences are assumed to be integer sequences from now on.

The recurrence (2.1) and the initial values $a_1, \ldots, a_d$ determine the complete infinite sequence. The best known example of a linear recurring sequence is the recursion

$$a_{n+2} = a_{n+1} + a_n$$

for $n \geq 1$. With initial values $a_1 = 1, a_2 = 1$ the numbers are called the *Fibonacci numbers*. With initial values $a_1 = 1, a_2 = 3$ the numbers are called the *Lucas numbers*. The correct notation for a particular linear recurring sequence $\mathcal{A} = (a_n)_{n \geq 1}$ defined by (2.1) is really

$$\mathcal{A} = \mathcal{A}(a_1, \ldots, a_d; c_1, \ldots, c_d).$$

This notation emphasizes the dependence on the recurring and the initial values.

The *characteristic polynomial* of the sequence (2.1) is defined to be the polynomial

$$\chi(t) = t^d - c_1 t^{d-1} - \cdots - c_{d-1} t - c_d \in \mathbb{Z}[t].$$

This does not depend on the initial values.

For a given linear recurring sequence $\mathcal{A}$, there is a unique (monic) characteristic polynomial of minimal degree, which is called the *minimal polynomial* of the sequence. The minimal polynomial of the sequence $\mathcal{A}$ divides the characteristic polynomial of every recurring relation satisfied by the sequence.

We fix an algebraic closure $\overline{\mathbb{Q}}$ of $\mathbb{Q}$. The following theorem is well known. See, e.g, Chapter 1 of [4].

Theorem 2.1 *Let* $\chi(t) = t^d - c_1 t^{d-1} - \cdots - c_{d-1} t - c_d$ *be a polynomial with* $c_d \neq 0$. *Factor* $\chi(t)$ *as* $\chi(t) = \prod_{i=1}^{r} (t - \alpha_i)^{m_i}$ *over* $\overline{\mathbb{Q}}$, *where the* α_i

*are distinct, and the m_i are positive integers. Then a sequence $(a_n)_{n \geq 1}$
satisfies the linear recurring with characteristic polynomial $\chi(t)$ if and only
if there exist polynomials $P_1(n), P_2(n), \ldots, P_r(n)$, where $P_i(n)$ has degree
$\leq m_i - 1$, such that*

$$a_n = P_1(n)\alpha_1^n + \cdots + P_r(n)\alpha_r^n \quad \text{for every } n \geq 1. \qquad (2.2)$$

If $\chi(t)$ has distinct roots then all the $P_i(n)$ are constant polynomials. A
choice of the coefficients in the $P_i(n)$ corresponds to a choice of the initial d
values of the sequence. The expression in (2.2) is often called a *generalized
power sum*. In general, the polynomials $P_i(x)$ may have coefficients in the
splitting field $\mathbb{Q}(\alpha_1, \ldots, \alpha_r)$.

There is another way to express the dependence on $a_1, \ldots, a_d, c_1, \ldots, c_d$
using rational functions. Given any linear recurring sequence

$$\mathcal{A} = \mathcal{A}(a_1, \ldots, a_d; c_1, \ldots, c_d)$$

we define two polynomials with integer coefficients

$$i(t) = a_1 t + (a_2 - a_1 c_1)t^2 + (a_3 - a_2 c_1 - a_1 c_2)t^3 \cdots + (a_d - a_{d-1}c_1 \cdots - a_1 c_{d-1})t^d$$

and

$$\chi^*(t) = 1 - c_1 t - c_2 t^2 - \cdots - c_d t^d,$$

(the reciprocal or reverse polynomial of the characteristic polynomial $\chi(t)$).
Then the formal Taylor series expansion of $i(t)/\chi^*(t)$ is

$$\frac{i(t)}{\chi^*(t)} = \sum_{i=1}^{\infty} a_i t^i \qquad (2.3)$$

as is easily verified by formal multiplication. Conversely, any rational function with numerator of degree at most d defines a linear recurring sequence
in this way. The following summary of this is Theorem 1.5 from [4].

Theorem 2.2 *A sequence $\mathcal{A}$ satisfies the recurring relation (2.1) with
characteristic polynomial $\chi(t) \in \mathbb{Z}[t]$ if and only if it is the sequence of
Taylor coefficients of a (formal) power series representing a rational function $\frac{i(t)}{\chi^*(t)}$, where $i(t)$ is a polynomial of degree less than d determined by
the initial values of $\mathcal{A}$.*

Example 2.3 The sequence of Fibonnacci numbers corresponds to the
rational function $\frac{x}{1-x-x^2}$. The minimal polynomial is $x^2 - x - 1$.

2.2   Background on curves

Let $q = p^a$ where p is a prime, and let $\mathbb{F}_q$ denote the finite field with q elements. Let $C = C(\mathbb{F}_q)$ be a projective smooth absolutely irreducible curve of genus g defined over $\mathbb{F}_q$ (see [12, 13] for further explanation of these terms). For any $n \geq 1$ let $C(\mathbb{F}_{q^n})$ be the set of $\mathbb{F}_{q^n}$-rational points of C, and let $\#C(\mathbb{F}_{q^n})$ be the cardinality of this set.

The *zeta function* of C is defined by

$$Z_C(t) = exp\left(\sum_{n\geq 1} \#C(\mathbb{F}_{q^n})\frac{t^n}{n}\right).$$

It is well known that $Z_C(t)$ can be written in the form

$$\frac{L_C(t)}{(1-t)(1-qt)} \tag{2.4}$$

where $L_C(t) \in \mathbb{Z}[t]$ (called the *L-polynomial* of C) is of degree $2g$. It is traditional to write

$$L_C(t) = \prod_{i=1}^{2g}(1 - \alpha_i t),$$

and the α_i are called the *Frobenius eigenvalues* of C. We briefly recall some further well-known facts about L-polynomials (see [12, 13] for example). If $L^{(n)}(t)$ denotes the L-polynomial of $C(\mathbb{F}_{q^n})$, then

$$L^{(n)}(t) = \prod_{i=1}^{2g}(1 - \alpha_i^n t). \tag{2.5}$$

The number of rational points for all $n \geq 1$ is given by

$$\#C(\mathbb{F}_{q^n}) = q^n + 1 - \sum_{i=1}^{2g}\alpha_i^n.$$

The algebraic integers α_i can be labelled so that $\overline{\alpha_i} = \alpha_{i+g}$ and $\alpha_i\alpha_{i+g} = q$. The Riemann Hypothesis for curves over finite fields, first proved by Weil, states that $|\alpha_i| = \sqrt{q}$ (see [13] for a proof).

We always assume the genus is the geometric genus, so it is unchanged when we extend the base field to $\mathbb{F}_{q^n}$.

We also remark that smoothness (non-singularity) is not essential for the results; however the singular points must be 'resolved'. Any absolutely irreducible singular curve has a nonsingular model obtained by the resolution of singularities, and the zeta function will always refer to the zeta function of the nonsingular model. See [5] for a discussion of these ideas. For a reducible curve, one should consider the irreducible components separately.

3   The sequence of a curve

We define the linear recurring sequence that is the subject of this paper.

Definition 3.1 Given a curve C defined over $\mathbb{F}_q$, the sequence

$$a_n := \#C(\mathbb{F}_{q^n}) - (q^n + 1) = -\sum_{i=1}^{2g} \alpha_i^n$$

will be called the *sequence* of C.

By Theorem 2.1 $(a_n)_{n \geq 1}$ is a linear recurring sequence of integers.

Example 3.2 This example will be revisited throughout the paper. Let $C_1 : y^2 + y = x^3 + x$ be defined over $\mathbb{F}_2$, which is a curve of genus 1.

The L-polynomial of C_1 is $2t^2 + 2t + 1$.

The characteristic polynomial $\chi(t) = t^2 + 2t + 2$ with the two initial values $a_1 = -2, a_2 = 0$ defines the sequence of C_1:

$$-2, 0, 4, -8, 8, 0, -16, 32, -32, 0, 64, -128, 128, 0, -256, 512...$$

Theorem 3.3 *A characteristic polynomial of the sequence of C is the reciprocal polynomial of the L-polynomial of C.*

Proof This follows from Theorem 2.1, or from the rational function expression of a linear recurring sequence (2.3), and the expression of the zeta function of a curve as a rational function (2.4).   □

Let us denote the reciprocal polynomial of the L-polynomial of C as $\chi_C(t)$, i.e., $\chi_C(t) = t^{2g}L(1/t)$. Write $L(t) = \sum_{j=0}^{2g} c_j t^j$ and $\chi_C(t) = \sum_{j=0}^{2g} c_j t^{2g-j}$, where $c_0 = 1$.

4   Degree g versus $2g$

The characteristic polynomial of the sequence of a curve has degree $2g$, which would appear to imply that we need $2g$ initial values to determine the sequence. However, it is known from the theory of curves that the first g values are enough to determine the zeta function and the L-polynomial. More precisely, using the Riemann Hypothesis for curves over finite fields, it can be shown (see [13]) that

$$\chi_C(t) = t^{2g} + c_1 t^{2g-1} + \cdots + c_{g-1} t^{g+1} + c_g t^g + q c_{g-1} t^{g-1} + \cdots + c_1 q^{g-1} t + q^g.$$

In other words, if $\chi_C(t) = \sum_{j=0}^{2g} c_j x^{2g-j}$ then $c_0 = 1$ and for $j = 0, \ldots, g-1$ we have $c_{2g-j} = q^{g-j} c_j$.

The L-polynomial is $L(t) = \sum_{j=0}^{2g} c_j t^j = \chi^*(t)$ (the reciprocal of the characteristic polynomial).

As in the previous section, we let

$$a_n = \#C(\mathbb{F}_{q^n}) - (q^n + 1) = -\sum_{i=1}^{2g} \alpha_i^n$$

be the sequence of a curve C. The characteristic polynomial has degree $2g$ but has only g undetermined coefficients. Knowing those coefficients is equivalent to knowing the first g values $a_1, \ldots, a_g$ of the sequence. The relation between the first g values and the coefficients c_i is convolutional:

$$i c_i = \sum_{k=0}^{i-1} c_k a_{i-k}, \quad i = 2, 3, \ldots, g \tag{4.1}$$

where $c_0 = 1$. The first three coefficients (for example) are given by

$$c_1 = a_1$$

$$2c_2 = a_2 + c_1 a_1 = a_2 + a_1^2$$

$$3c_3 = a_3 + c_1 a_2 + c_2 a_1.$$

So the entire sequence a_n of C depends only on the first g terms $a_1, \ldots, a_g$ of the sequence, but the dependence is nonlinear. The first g terms determine the characteristic polynomial (equivalently, the L-polynomial), by (4.1). Equation (2.4) gives

$$\log L(t) = \sum_{n \geq 1} a_n \frac{t^n}{n}$$

which shows that each a_n is then determined. In particular, $a_{g+1}, \ldots, a_{2g}$ are determined by $a_1, \ldots, a_g$ (in a nonlinear way). The characteristic polynomial has degree $2g$, which means that each term a_n depends on the previous $2g$ values $a_{n-1}, \ldots, a_{n-2g}$ in a linear way. With $2g$ initial values, the complete linear recurring sequence is determined.

5   Elliptic curves

For a curve of genus 1, an elliptic curve, if we let $N_k = a_k + (q^k + 1) = \#C(\mathbb{F}_{q^k})$, then each N_k can be determined in terms of N_1. Musiker [11] writes this relationship as

$$N_k = \sum_{i=1}^{k} (-1)^i P_{k,i}(q) N_1^i.$$

One corollary of this equation is that N_1 divides N_k. If we write $N_k = N_1 \cdot N_k'$ then Musiker shows that we can understand the cofactor N_k' as the cardinality of the trace-zero subgroup (the points P such that $\phi_k(P) = O$ where $\phi_k(P) = \sum_{i=0}^{k-1} \pi^i(P)$). Musiker also gives a combinatorial interpretation of the $P_{k,i}(q)$.

6   Minimal polynomial, linear complexity

In light of the issue of g versus $2g$ in Section 4, it is natural to ask whether the minimal polynomial of the linear recurring sequence of a curve is the same as the characteristic polynomial of degree $2g$. Alternatively, we ask for the smallest recursion that generates the sequence of a curve, the depth of which is known as the linear complexity of the sequence. The answer is that sometimes the characteristic polynomial of degree $2g$ (the reciprocal polynomial of the L-polynomial) is the minimal polynomial, and sometimes not. For example, if the characteristic polynomial is irreducible over $\mathbb{Z}$ then it must be the minimal polynomial. On the other hand, the sequence of the elliptic curve $y^2 + y = x^3 + x$ considered as being defined over $\mathbb{F}_{16}$ is $-8, 32, 128, -512, \ldots$ and has minimal polynomial $t + 4$.

6.1   General method for finding minimal polynomial

If one knows the characteristic polynomial $\chi(t) = \sum_{i=0}^{d} c_i t^{d-i}$ (with $c_0 = 1$) of any linear recurring sequence, and the initial values $a_1, a_2, \ldots, a_d$

then it is possible to determine the minimal polynomial by defining

$$h(t) = \sum_{j=0}^{d-1} \left(\sum_{i=1}^{d-j} c_{i+j} a_i \right) t^j.$$

If $d(t) = \gcd(\chi(t), h(t))$ then the minimal polynomial is $\chi(t)/d(t)$ (see [10]).

Example 6.1 Consider the sequence defined by $a_1 = 1$, $a_2 = 2$, and $a_n = 2a_{n-1} - a_{n-2}$ for $n \geq 3$. The characteristic polynomial is $\chi(t) = t^2 - 2t + 1$ and so $c_1 = -2$, $c_2 = 1$. Therefore

$$h(t) = (c_1 a_1 + c_2 a_2) + c_2 a_1 t = t$$

and $\gcd(\chi(t), h(t)) = 1$. This shows that $(t-1)^2$ is the minimal polynomial of the sequence (which is easy to see anyway).

6.2   Genus 2

For most curves of genus 2, the minimal polynomial is equal to the characteristic polynomial, and has degree 4. When this is not the case, the characteristic polynomial is reducible and one might expect that the minimal polynomial would be a factor that is the characteristic polynomial of a genus 1 curve. However, we have the following theorem.

Theorem 6.2 *For a curve of genus 2, the minimal polynomial cannot equal the characteristic polynomial of a curve of genus 1.*

Proof Suppose that there is a curve of genus 2, C, with characteristic polynomial $x^4 + c_1 x^3 + c_2 x^2 + qc_1 x + q^2$, whose sequence actually has minimal polynomial $t^2 + At + q$. Suppose also that there is an elliptic curve E with characteristic polynomial $x^2 + Ax + q$. This is the characteristic polynomial of a linear recurring sequence ($e_n = -Ae_{n-1} - qe_{n-2}$) with initial values which we will denote e_1, e_2. As we stated earlier, $e_1 = A$, and e_2 can be determined from e_1 in a nonlinear fashion, by the equation $2q = e_2 + e_1^2$.

We would then have a situation where the first two values of the sequence of E are the same as the first two values of the sequence of C. The first two values of the sequence of C are $a_1 = c_1$ and $a_2 = 2c_2 - a_1^2$. Therefore we would have $a_1 = c_1 = A = e_1$ and $2c_2 - a_1^2 = 2q - e_1^2$, which implies $c_2 = q$. However, the factorization

$$x^4 + c_1 x^3 + c_2 x^2 + qc_1 x + q^2 = (x^2 + Ax + q)(x^2 + A'x + A'')$$

with these restrictions yields a contradiction ($c_1 = A$ implies $A' = 0$ and the constant terms imply $A'' = q$, but this gives $c_2 = 2q$).   □

See Example 8.6 for a genus 2 example.

6.3   A general question

Suppose $g_1 < g_2$. It is natural to wonder how much the sequence of a curve of genus g_1 can agree with the sequence of a curve of genus g_2. In particular, one wonders whether they can agree infinitely often (see Section 9).

Here we consider a more specific question: can the sequence for a curve of genus g_1 agree with the sequence for a curve of genus g_2 for the first $2g_2$ terms? Theorem 6.2 shows that the answer is no (where $g_1 = 1$ and $g_2 = 2$).

However let us look at an example where the sequences 'almost' agree for $2g_2$ terms.

Example 6.3 Over $\mathbb{F}_4 = \{0, 1, \omega, \omega^2\}$, the curve $y^3 + wy = x^4 + x^2 + wx$ has genus 3 and characteristic polynomial

$$x^6 + 2x^5 + 4x^4 + 16x^2 + 32x + 64 = (x^4 + 16)(x^2 + 2x + 4)$$

The first 6 terms of the sequence of this curve are $-2, -4, 16, -80, -32, 128$.

Example 6.4 The curve $y^2 + y = x^5 + wx$ has genus 2 and characteristic polynomial

$$x^4 + 16.$$

The first 6 terms of the sequence of this curve are $0, 0, 0, -64, 0, 0$.

Example 6.5 The curve $y^2 + wy = x^3 + 1$ has genus 1 and characteristic polynomial

$$x^2 + 2x + 4.$$

The first 6 terms of the sequence of this curve are $-2, -4, 16, -16, -32, 128$.

The sequences in Examples 6.3 and 6.5 agree everywhere in the first $2g_2 = 6$ terms except for one term.

6.4 Sums of sequences

The sequence in Example 6.3 is the sum of the two sequences in Examples 6.4 and 6.5. This corresponds to the characteristic polynomial being the product of the two smaller characteristic polynomials. In general, the following theorem is well known.

Theorem 6.6 *If $(a_n)_{n\geq 1}$ and $(b_n)_{n\geq 1}$ are any linear recurring sequences with relatively prime characteristic polynomials $\chi_a(t)$ and $\chi_b(t)$, then the sequence $(a_n + b_n)_{n\geq 1}$ is a linear recurring sequence with characteristic polynomial $\chi_a(t) \cdot \chi_b(t)$.*

7 Maximal curves

Recall that

$$\#C(\mathbb{F}_q) - (q+1) = -\sum_{i=1}^{2g} \alpha_i$$

as we stated in Section 2. The Riemann Hypothesis ($|\alpha_i| = \sqrt{q}$) and triangle inequality immediately give the Hasse–Weil bound

$$|\#C(\mathbb{F}_q) - (q+1)| \leq 2gq^{1/2}.$$

It follows that

$$(q+1) - 2gq^{1/2} \leq \#C(\mathbb{F}_q) \leq (q+1) + 2gq^{1/2}$$

and C is called *maximal* (*minimal*) when equality holds in the right (left) inequality.

The following theorem is well known (see [13] for a proof).

Theorem 7.1 *For a curve C of genus g defined over $\mathbb{F}_q$ the following are equivalent.*

1. *C is maximal over $\mathbb{F}_q$*
2. *$L(t) = (1 + \sqrt{q}t)^{2g}$*
3. *$\alpha_i = -\sqrt{q}$ for $i = 1, \ldots, 2g$*
4. *$\#C(\mathbb{F}_q) - (q+1) = 2gq^{1/2}$*
5. *$\#C(\mathbb{F}_{q^s}) - (q^s + 1) = (-1)^{s-1}2gq^{s/2}$*

It follows that for maximal curves, the minimal polynomial is $t + \sqrt{q}$. The converse is true also, in the sense that if the minimal polynomial has degree 1 then it must be $t \pm \sqrt{q}$ and the curve is maximal or minimal.

Example 7.2 Let $C_k : y^2 + y = x^{2^k+1} + x$ be defined over $\mathbb{F}_2$. The L-polynomial of C_1 is $2t^2 + 2t + 1$. Because C_1 is a supersingular elliptic curve, it becomes maximal over an extension of $\mathbb{F}_2$. The roots of the characteristic polynomial over $\mathbb{F}_2$ (which is $t^2 + 2t + 2$) are of the form $\sqrt{2}\,\zeta_8$ where ζ_8 is an eighth root of unity, By property (2.5) the characteristic polynomial over $\mathbb{F}_{2^r}$ has roots $(\sqrt{2}\,\zeta_8)^r$, and it follows that C_1 is maximal over $\mathbb{F}_{2^r}$ if and only if r is 4 modulo 8. This means that the characteristic polynomial of C_1 over $\mathbb{F}_{2^r}$ when r is 4 modulo 8 is $(t + 2^{r/2})^2$. The minimal polynomial in this case is $t + 2^{r/2}$.

8 Subsequences

The study of subsequences is very much related to the question of whether two sequences can agree infinitely often. We consider subsequences of the form $(a_{sn+j})_{n\geq 1}$, which are called lacunary sequences by Young [14]. It is known that all linear recurring subsequences of a linear recurring sequence must have this form (c.f. [4] ch. 1).

We now present some results relating the characteristic and minimal polynomials of a subsequence to the corresponding polynomials of the original sequence.

Theorem 8.1 *Let $(a_n)_{n\geq 1}$ be a linear recurring sequence with characteristic polynomial $\chi(t)$. Let $s > 1$ be a positive integer, and let j be an integer with $0 \leq j < s$. Then the characteristic polynomial of the subsequence $(a_{sn+j})_{n\geq 1}$ is the polynomial whose roots are the s-th powers of the roots of $\chi(t)$.*

Proof To see this, the representation (2.2) for $(a_n)_{n\geq 1}$ is

$$a_n = P_1(n)\alpha_1^n + \cdots + P_r(n)\alpha_r^n,$$

and so for the subsequence we may write

$$a_{ns+j} = Q_1(n)\alpha_1^{ns} + \cdots + Q_r(n)\alpha_r^{ns},$$

where $Q_i(n) = \alpha_i^j P_i(n)$. By Theorem 2.1 again, the characteristic polynomial of this sequence has roots $\alpha_1^s, \ldots, \alpha_r^s$. $\square$

Example 8.2 The characteristic polynomial $\chi(t) = t^2 - t - 1$ with initial conditions $a_1 = 1, a_2 = 1$ defines a well-known sequence

$$1, 1, 2, 3, 5, 8, 13, 21, 34, 55, \ldots$$

To get the polynomial whose roots are the squares we calculate

$$\chi(t)\chi(-t) = (t^2 - t - 1)(t^2 + t - 1) = t^4 - 3t^2 + 1$$

and replacing t^2 by t gives $t^2 - 3t + 1$. With initial values $1, 3$ this gives a sequence $1, 3, 8, 21, 55, \ldots$ which is the subsequence $(a_{2n})_{n \geq 1}$. With initial values $1, 2$ this gives a sequence $1, 2, 5, 13, 34, \ldots$ which is the subsequence $(a_{2n+1})_{n \geq 0}$.

Remark For the sequence $(a_n)_{n \geq 1}$ of a curve C defined over $\mathbb{F}_q$, the subsequence $(a_{sn})_{n \geq 1}$ gives us the numbers of rational points on C over $\mathbb{F}_{q^s}$ and all its extensions. Therefore, if $L(t) = \prod_{i=1}^{2g}(1 - \alpha_i t)$ is the L-polynomial of $C(\mathbb{F}_q)$, and if $L^{(s)}(t)$ denotes the L-polynomial of $C(\mathbb{F}_{q^s})$, it follows from Theorem 8.1 that $L^{(s)}(t) = \prod_{i=1}^{2g}(1 - \alpha_i^s t)$. This provides a proof of (2.5).

Definition 8.3 Let $\chi(t)$ be the characteristic polynomial of a linear recurring sequence $(a_n)_{n \geq 1}$ defined by the recurring (2.1). Let s be a positive integer. We say that $(a_n)_{n \geq 1}$ is *s-nondegenerate* if the s-th powers of the roots of $\chi(t)$ are distinct. A linear recurring sequence is said to be *nondegenerate* if it is s-nondegenerate for all $s \geq 1$ (c.f. [4] ch. 1).

Thus, a sequence is 1-nondegenerate if $\chi(t)$ has distinct roots.

Let $f^{(s)}(t)$ denote the polynomial whose roots are the s-th powers of the roots of a polynomial $f(t)$.

Corollary 8.4 *Let $(a_n)_{n \geq 1}$ be a linear recurring sequence with minimal polynomial $m(t)$ of degree d. Let $s > 1$ be a positive integer, and let j be an integer with $0 \leq j < s$. Then the minimal polynomial of the subsequence $(a_{sn+j})_{n \geq 1}$ has degree at most d.*

If $(a_n)_{n \geq 1}$ is s-nondegenerate then the degree of the minimal polynomial of the subsequence $(a_{sn+j})_{n \geq 1}$ is equal to d.

Proof The first statement is Theorem 1.3 in [4], and also follows from Theorem 8.1.

The second part follows from consideration of the Galois group G of the minimal polynomial $m(t)$ over $\mathbb{Q}$. If $(a_n)_{n \geq 1}$ is s-nondegenerate then

$m^{(s)}(t)$ has distinct roots, and therefore $m(t)$ must have distinct roots also. The action of G on the roots of $m(t)$ and $m^{(s)}(t)$ must be the same. For, clearly $\sigma(\alpha) = \beta$ implies $\sigma(\alpha^s) = \beta^s$. On the other hand suppose two roots of $m^{(s)}(t)$, say α^s and β^s, lie in the same orbit, but α and β do not lie in the same orbit. Then there exists $\sigma \in G$ with $\sigma(\alpha^s) = \beta^s$ and $\sigma(\alpha) \neq \beta$. But then $\sigma(\alpha)^s = \beta^s$, which contradicts the assumption that $(a_n)_{n\geq 1}$ is s-nondegenerate. It follows that $m^{(s)}(t)$ has the same degree as $m(t)$. $\square$

Example 8.5 We continue Example 7.2. We will see that the sequence is s-nondegenerate for $s = 1, 2, 3$ but not for $s = 4$.

The characteristic polynomial $\chi(t) = t^2 + 2t + 2$ with the two initial values $a_1 = -2, a_2 = 0$ defines the sequence of C_1:

$$-2, 0, 4, -8, 8, 0, -16, 32, -32, 0, 64, -128, 128, 0, -256, 512...$$

The polynomial whose roots are the squares of the roots of $\chi(t)$ is $t^2 + 4$. With initial values $0, -8$, this characteristic polynomial gives the sequence $0, -8, 0, 32, 0, -128, ...$ which is the subsequence $(a_{2n})_{n\geq 1}$. With initial values $-2, 4$ this gives a sequence $-2, 4, 8, -16, -32, ...$ which is the subsequence $(a_{2n+1})_{n\geq 0}$.

The polynomial whose roots are the cubes of the roots of $\chi(t)$ is $t^2 - 4t + 8$. With initial values $-2, -8$ this sequence is $-2, -8, -16, 0, 128, 512, ...$ With initial values $0, 8$ this sequence is $0, 8, 32, 64, 0, ...$ With initial values $4, 0$ this sequence is $4, 0, -32, -128, -256, ...$ These are the subsequences with subscripts in the three arithmetic progressions $3n, 3n + 1$, and $3n + 2$.

The polynomial whose roots are the fourth powers of the roots of $\chi(t)$ is $(t + 4)^2$. In other words, -4 is the fourth power of both roots of $\chi(t)$. The subsequence $(a_{4n})_{n\geq 1}$ is $-8, 32, -128, 512,...$ which has minimal polynomial $t + 4$.

$$-2, 0, 4, -\mathbf{8}, 8, 0, -16, \mathbf{32}, -32, 0, 64, -\mathbf{128}, 128, 0, -256, \mathbf{512}...$$

The sequence of C_1 is s-nondegenerate for $s = 1, 2, 3$ but not for $s = 4$.

Note: $t + 4$ is the minimal polynomial of all the subsequences $(a_{4n+j})_{n\geq 1}$, $j = 0, 1, 2, 3$. Only one initial value is needed.

Example 8.6 Consider the curve of genus 2 defined by

$$y^2 + y = x + \frac{1}{x} + \frac{1}{x+1} \qquad (8.1)$$

over $\mathbb{F}_2$. This curve is an ordinary curve of genus 2 with characteristic polynomial $t^4 - t^2 + 4$. Since the characteristic polynomial is a polynomial in t^2 the roots come in $\pm$ pairs, and the squares of the roots are not distinct. The polynomial whose roots are the squares of the roots of $\chi(t)$ is $(t^2 - t + 4)^2$. Therefore the minimal polynomial of the subsequences $(a_{2n+1})_{n \geq 0}$ and $(a_{2n})_{n \geq 1}$ is $t^2 - t + 4$. The sequence of the curve is

$$0, -2, 0, 14, 0, 22, 0, -34, 0, -122, 0, 14, 0, 502, 0, \ldots$$

We may generalize Example 8.6 by considering the class of curves whose L-polynomials are polynomials in t^s, for some $s > 1$. The characteristic polynomials of the sequences of such curves will not be s-nondegenerate, because if α is a root and ζ_s is an s-th root of unity, then $\alpha \zeta_s$ is also a root. The splitting field of such a polynomial will contain the s-th roots of unity.

9   The Skolem–Mahler–Lech Theorem

In this section we consider what happens when two curves have the same number of points over infinitely many extensions. We will use the following theorem.

Theorem 9.1 (Skolem–Mahler–Lech) *Let $\mathcal{A}$ be a linear recurring sequence of complex numbers. If $\mathcal{A}$ contains infinitely many zeros, then the set of indices n for which $a_n = 0$ is the union of a finite set and a finite number of arithmetic progressions.*

See [4] for further discussion and a proof.

We will use the Skolem–Mahler–Lech Theorem to prove a theorem (Theorem 9.4) about divisibility of the L-polynomials of two curves that have the same number of points infinitely many times. To put this result in some context, the other techniques that imply divisibility of L-polynomials are algebraic in nature. One well-known approach is to use a theorem of Kleiman [8], also sometimes attributed to Serre, which implies divisibility whenever there is a covering map $C \longrightarrow C'$.

Theorem 9.2 (Kleiman–Serre) *If there is a morphism of curves $C \longrightarrow C'$ that is defined over $\mathbb{F}_q$ then $L_{C'}(t)$ divides $L_C(t)$.*

Another approach is to use the Kani–Rosen decomposition [7], which applies when a curve has an automorphism group of a suitable type. This shows, roughly speaking, that the L-polynomial of the curve decomposes into the L-polynomials of quotient curves.

A paper by Bombieri and Katz [1] uses the Skolem–Mahler–Lech theorem in a discussion of whether $|a_n| \to \infty$ over the nonzero a_n where $a_n = \#E(\mathbb{F}_{p^n}) - (p^n + 1)$ is the sequence corresponding to the elliptic curve E defined over the finite field $\mathbb{F}_p$ with p elements.

9.1 Basic application of Skolem–Mahler–Lech

The theorem of Skolem–Mahler–Lech can be applied to subsequences of linear recurring sequences as follows. Here $f^{(s)}(t)$ denotes the polynomial whose roots are the s-th powers of the roots of $f(t)$.

Theorem 9.3 *Let $\mathcal{A} = (a_n)_{n \geq 1}$ and $\mathcal{B} = (b_n)_{n \geq 1}$ be two linear recurring sequences, with given characteristic polynomials $\chi_{\mathcal{A}}(t)$ and $\chi_{\mathcal{B}}(t)$. Assume that*

   1. $\chi_{\mathcal{A}}(t)$ is the minimal polynomial of $(a_n)_{n \geq 1}$,

   2. $\mathcal{A}$ is nondegenerate,

   3. $a_n = b_n$ for infinitely many n.

Then there exists a positive integer s such that $\chi_{\mathcal{B}}^{(s)}(t)$ is divisible by $\chi_{\mathcal{A}}^{(s)}(t)$.

Proof The first two hypotheses imply that all subsequences $(a_{ns+j})_{n \geq 1}$ of $\mathcal{A}$ have a minimal polynomial whose degree is the same as the degree of $\chi_{\mathcal{A}}(t)$, by Corollary 8.4.

The third hypothesis says that the sequence $(a_n - b_n)$ contains 0 infinitely many times. The Skolem–Mahler–Lech theorem implies that the n for which $a_n = b_n$ form a union of arithmetic progressions (apart from possibly a finite number of exceptions). Let $a_{ns+j} = b_{ns+j}$ be one of these arithmetic progressions, so $a_{ns+j} = b_{ns+j}$ for all $n \geq 1$. This means that the minimal polynomial of $(b_{ns+j})_{n \geq 1}$ must equal the minimal polynomial of $(a_{ns+j})_{n \geq 1}$, which is $\chi_{\mathcal{A}}^{(s)}(t)$. By the general fact that the minimal polynomial divides the characteristic polynomial, which in this case is $\chi_{\mathcal{B}}^{(s)}(t)$, the proof is complete.   □

In the theorem above, if the subsequence $(b_{ns+j})_{n \geq 1}$ has a minimal polynomial of degree $< \deg(\chi_{\mathcal{B}})$, then the s-th powers of the roots of $\chi_{\mathcal{B}}(t)$ must not be distinct, and hence $\mathcal{B}$ will be degenerate.

Now let $(a_n)_{n \geq 1}$ be the sequence for the curve C_1 defined over $\mathbb{F}_q$ and let $(b_n)_{n \geq 1}$ be the sequence for the curve C_2 defined over $\mathbb{F}_q$. Then applying

Theorem 9.3 we get the following result.

Theorem 9.4 *Let C_1 and C_2 be two curves defined over $\mathbb{F}_q$ such that*

1. *the minimal polynomial of the sequence of C_1 is the reverse of the L-polynomial of $C_1(\mathbb{F}_q)$,*

2. *the L-polynomial of $C_1(\mathbb{F}_{q^r})$ has no repeated roots, for each $r \geq 1$,*

3. *$C_1(\mathbb{F}_q)$ and $C_2(\mathbb{F}_q)$ have the same number of points over infinitely many extensions $\mathbb{F}_{q^r}$ of $\mathbb{F}_q$.*

Then there exists a positive integer s such that the L-polynomial of $C_2(\mathbb{F}_{q^s})$ is divisible by the L-polynomial of $C_1(\mathbb{F}_{q^s})$.

We also remark that we do not require the full strength of the second hypothesis. We only require that the L-polynomial of C_1 has no repeated roots over *one* extension of $\mathbb{F}_q$, namely the extension of degree s where s is the integer arising in the proof. However there is no way to predict what s will be, so to be on the safe side we use this hypothesis.

9.2 Value of s

The value of the integer s in the statement of Theorem 9.4 is clearly of importance. Sometimes $s = 1$. However, s cannot always be 1 as the following example shows. Let C_1 be an elliptic curve defined over $\mathbb{F}_q$, with L-polynomial $qt^2 + at + 1$, and let C_2 be its quadratic twist, with L-polynomial $qt^2 - at + 1$. Then C_1 and C_2 have the same number of rational points over all even degree extensions of $\mathbb{F}_q$, but clearly the L-polynomials over $\mathbb{F}_q$ do not divide each other if $a \neq 0$. In this case, $s = 2$.

The next corollary gives some conditions under which $s = 2$.

Corollary 9.5 *Let C_1 and C_2 be two curves defined over $\mathbb{F}_q$ such that*

1. *the minimal polynomial of the sequence of C_1 is the reverse of the L-polynomial of $C_1(\mathbb{F}_q)$,*

2. *the L-polynomial of $C_1(\mathbb{F}_{q^r})$ has no repeated roots, for each $r \geq 1$,*

3. *$C_1(\mathbb{F}_q)$ and $C_2(\mathbb{F}_q)$ have the same number of points over infinitely many extensions $\mathbb{F}_{q^r}$ of $\mathbb{F}_q$.*

Let M be the splitting field of the characteristic polynomial $\chi_{C_2}(t)$. Suppose that there exist two primes p_1, p_2, distinct from each other and p which is the characteristic of $\mathbb{F}_q$, such that $\gcd(p_1 - 1, p_2 - 1) = 2$ and the ideals

$(p_1), (p_2)$ *both split completely in* M. *Then the L-polynomial of* $C_2(\mathbb{F}_{q^2})$ *is divisible by the L-polynomial of* $C_1(\mathbb{F}_{q^2})$.

Proof Let $L_i^{(r)}(t)$ be the L-polynomial of C_i over $\mathbb{F}_{q^r}$, for $i = 1, 2, r \geq 1$. Let g_i be the genus of C_i, for $i = 1, 2$. We may write

$$\#C_1(\mathbb{F}_{q^n}) = q^n + 1 - \sum_{j=1}^{2g_1} \tau_j^n \tag{9.1}$$

for some complex numbers $\tau_1, \ldots, \tau_{2g_1}$. Similarly for C_2 we may write

$$\#C_2(\mathbb{F}_{q^n}) = q^n + 1 - \sum_{k=1}^{2g_2} \sigma_k^n \tag{9.2}$$

for some complex numbers $\sigma_1, \ldots, \sigma_{2g_2}$. We want to prove that each τ_j^2 is one of the σ_k^2. From the proof of the Skolem–Mahler–Lech theorem in [4] we know that for each τ_i, there is σ_j such that the ratio τ_i/σ_j is an $(\ell - 1)$-th root of unity, where ℓ is any prime such that the τ_i and σ_j are units in the ℓ-adic numbers $\mathbb{Q}_\ell$, and (ℓ) splits completely in M.

Let ℓ be a prime different from p. It is well known (see [3] for example) that the τ_i and σ_j are units in $\mathbb{Q}_\ell$.

Therefore, using the hypotheses, we have that τ_i/σ_j are $(p_1 - 1)$-th roots of unity, and also $(p_2 - 1)$-th roots of unity. Since $\gcd(p_1 - 1, p_2 - 1) = 2$ the ratios are square roots of unity. So for each i, $\tau_i^2 = \sigma_j^2$ for some j, and so the roots of the L-polynomial of $C_1(\mathbb{F}_{q^2})$ are a subset of the roots of the L-polynomial of $C_2(\mathbb{F}_{q^2})$. $\qquad\square$

The same argument with $\ell = 2$ when p is odd shows the following case where we have $s = 1$.

Corollary 9.6 *Let* q *be odd. Let* C_1 *and* C_2 *be two curves defined over* $\mathbb{F}_q$ *such that*

1. *the minimal polynomial of the sequence of* C_1 *is the reverse of the L-polynomial of* $C_1(\mathbb{F}_q)$,

2. *the L-polynomial of* $C_1(\mathbb{F}_{q^r})$ *has no repeated roots, for each* $r \geq 1$,

3. $C_1(\mathbb{F}_q)$ *and* $C_2(\mathbb{F}_q)$ *have the same number of points over infinitely many extensions* $\mathbb{F}_{q^r}$ *of* $\mathbb{F}_q$.

Let M *be the splitting field of the characteristic polynomial* $\chi_{C_2}(t)$, *and suppose that the ideal* (2) *splits completely in* M. *Then the L-polynomial of* $C_2(\mathbb{F}_q)$ *is divisible by the L-polynomial of* $C_1(\mathbb{F}_q)$.

In the remark at the start of this subsection, we said that s cannot be 1 for an elliptic curve and its twist. Corollary 9.6 does not contradict this remark, because for an elliptic curve E with $\chi_E(t) = t^2 + at + q$, the ideal (2) never splits completely in the splitting field $M = \mathbb{Q}(\sqrt{a^2 - 4q})$ because $a^2 - 4q$ cannot be 1 modulo 8.

10 Recovery

In the light of Theorem 9.4 one may wonder when we can recover the characteristic polynomial of the original sequence from the characteristic polynomial of a subsequence. In other words, we wonder when it is possible to recover the L-polynomial of $C(\mathbb{F}_q)$ given the L-polynomial of $C(\mathbb{F}_{q^s})$. In those situations, whenever Theorem 9.4 tells us that the L-polynomial of $C_1(\mathbb{F}_{q^s})$ divides the L-polynomial of $C_2(\mathbb{F}_{q^s})$, we may conclude that the L-polynomial of $C_1(\mathbb{F}_q)$ divides the L-polynomial of $C_2(\mathbb{F}_q)$.

Lemma 10.1 *Fix a positive integer $r > 1$. Let $T_r : \mathbb{Q}[x] \longrightarrow \mathbb{Q}[x]$ be defined by the rule that $T_r(f(x))$ is the polynomial whose roots are the r-th powers of the roots of $f(x)$, and $T_r(f(x))$ has the same leading coefficient as $f(x)$. Let R_r be the set of all monic irreducible polynomials in $\mathbb{Q}[x]$ whose splitting field does not contain any nontrivial r-th roots of unity. Then*

(i) the restriction of T_r to R is injective, and therefore bijective onto its image, and

(ii) for $f(x)$ and $g(x)$ in R, if $T_r(f(x))$ divides $T_r(g(x))$, we may conclude that $f(x)$ divides $g(x)$.

Proof Assume $T = T_r$ is restricted to $R = R_r$. Every element of R is irreducible and therefore separable, and so has distinct roots. If $f(x) \in R$ then the definition of R implies that $T_r(f(x))$ has distinct roots, because if α and β are roots of $f(x)$ with $\alpha^r = \beta^r$ then α/β is an r-th root of unity in the splitting field, and is therefore equal to 1. The definition of R also implies that r is odd (if r is even then the splitting field contains -1).

(i) We first prove that $f(x)$ and $T(f(x))$ have the same splitting field. Let $M = \mathbb{Q}(\alpha_1, \ldots, \alpha_n)$ be the splitting field of $f(x)$, and let $N = \mathbb{Q}(\alpha_1^r, \ldots, \alpha_n^r)$ be the splitting field of $T(f(x))$. If $M \neq N$ choose a nontrivial automorphism $\sigma \in Gal(M/N)$. The existence of a nontrivial automorphism follows from the fact that M is a splitting field and N is a proper subfield of it. Suppose $\sigma(\alpha_i) = \alpha_j$ where $i \neq j$. Then

$$\alpha_j^r = \sigma(\alpha_i)^r = \sigma(\alpha_i^r) = \alpha_i^r$$

so α_i/α_j is an r-th root of unity in M. Since M contains no nontrivial r-th roots of unity we get $\alpha_i = \alpha_j$, a contradiction.

Each root of $T(f(x))$ has exactly one r-th root in the splitting field M, because if there were two r-th roots their quotient would be an r-th root of unity in M. So we can recover $f(x)$ from $T(f(x))$ (the roots are the unique r-th roots).

(ii) From (i) we know that T is bijective as a function from R to its image $T(R)$.

We next observe that $T(fg) = T(f)T(g)$, which is clear when we write $f(x) = (x - \alpha_1)\cdots(x - \alpha_n)$ and $g(x) = (x - \beta_1)\cdots(x - \beta_m)$.

It follows from this property that $f|h$ implies $T(f)|T(h)$. Finally, since T is bijective it follows that T^{-1} also has this property. (Let $a = T^{-1}(x)$, $b = T^{-1}(y)$, $c = T^{-1}(xy)$. Then $T(c) = xy = T(a)T(b) = T(ab)$ which implies $c = ab$.) So we conclude that $f|h$ implies $T^{-1}(f)|T^{-1}(h)$ for $f, h \in T(R)$. $\qquad\square$

Corollary 10.2 *Let C_1 and C_2 be two curves defined over $\mathbb{F}_q$ such that*

1. *the minimal polynomial of the sequence of C_1 is the reverse of the L-polynomial*

2. *the L-polynomial of $C_1(\mathbb{F}_{q^r})$ has no repeated roots, for each $r \geq 1$.*

3. *$C_1(\mathbb{F}_q)$ and $C_2(\mathbb{F}_q)$ have the same number of points over infinitely many extensions $\mathbb{F}_{q^r}$ of $\mathbb{F}_q$.*

Let s be the positive integer arising in the proof of Theorem 9.4. and assume that the splitting fields of the L-polynomials of $C_1(\mathbb{F}_q)$ and $C_2(\mathbb{F}_q)$ contain no s-th roots of unity. Then the L-polynomial of $C_2(\mathbb{F}_q)$ is divisible by the L-polynomial of $C_1(\mathbb{F}_q)$.

The proof follows from Theorem 9.4 and Lemma 10.1.

Here is an algorithm when r is a prime for computing f given $T_p(f)$, which uses a polynomial factorization algorithm and relies on the assumption that $Spl(f) \cap \mathbb{Q}(\zeta_p) = \mathbb{Q}$ (i.e., there are no p-th roots of unity in the splitting field).

Algorithm

Input: An odd prime p and $g(x) \in \mathbb{Z}[x]$, where $g(x) = T_p(f(x))$.

Output: $f(x)$

    1. Compute $G(x) := g(x^p)$

2. Factor $G(x)$ into irreducible factors over the cyclotomic field $\mathbb{Q}(\zeta_p)$ where ζ_p is a primitive complex p-th root of unity

3. Choose the unique irreducible factor that is defined over $\mathbb{Q}$ and output this factor.

Proof that algorithm works:

This algorithm works because $G(x)$ is equal to $\prod_{i=1}^{p} f(\zeta_p^i x)$, and has one irreducible factor defined over $\mathbb{Q}$ and $p-1$ conjugate factors defined over $\mathbb{Q}(\zeta_p)$. Since $Spl(f) \cap \mathbb{Q}(\zeta_p) = \mathbb{Q}$ and f is irreducible over $\mathbb{Q}$, the factors do not factor further over $\mathbb{Q}(\zeta_p)$.

Example 10.3 Suppose we start with the curve C of genus 2

$$y^2 = x^5 + x^2 - x$$

defined over $\mathbb{F}_3$. The L-polynomial of C is $9t^4 + 6t^3 + 4t^2 + 2t + 1$. The sequence of the curve is

$$2, 4, 2, 4, -58, 52, 86, -68, 2, -716, \dots$$

with characteristic polynomial $\chi(t) = t^4 + 2t^3 + 4t^2 + 6t + 9$ (which is irreducible over $\mathbb{Q}$).

Let ζ_7 be a primitive 7-th root of unity, then ζ_7 is not in the splitting field of $\chi(t)$ because the splitting field is a degree 8 extension of $\mathbb{Q}$. The polynomial whose roots are the seventh powers of the roots of $\chi(t)$ is $g(t) = t^4 + 86t^3 + 3700t^2 + 188082t + 4782969$.

On the other hand, suppose we were to start with $g(t)$. Then we could recover $\chi(t)$ using the algorithm. We compute

$$g(t^7) = t^{28} + 86t^{21} + 3700t^{14} + 188082t^7 + 4782969$$

and then we factor this polynomial over $\mathbb{Q}(\zeta_7)$ into irreducible factors. There is one factor defined over $\mathbb{Q}$, which is $f(x)$, and six other Galois-conjugate factors defined over $\mathbb{Q}(\zeta_7)$:

$$g(t^7) = (t^4 + 2t^3 + 4t^2 + 6t + 9)(t^4 + 2\zeta_7 t^3 + 4\zeta_7^2 t^2 + 6\zeta_7^3 t + \zeta_7^4)(\dots).$$

Remark Assuming we know the characteristic polynomial, can we get the initial values back too? Yes. If we know every s-th value of the sequence, we need to find any s consecutive values of the original sequence and then we know the whole sequence. Knowing every s-th value will give us $(s-1)^2$ linear equations for $a_{s+1}, \dots, a_{s^2-s+1}$ in $(s-1)^2$ unknowns. Solving those gives s consecutive values.

218 O. Ahmadi and G. McGuire

11 Acknowledgements

We thank Rod Gow and Igor Shparlinski for many helpful discussions.

References

[1] E. Bombieri and N. Katz, A note on lower bounds for Frobenius traces, *L'enseignement Mathematique*, Volume 56, Issue 3/4, 2010, pp. 20317227. DOI: 10.4171/LEM/56-3-1

[2] W. Bosma, J. Cannon and C. Playoust, The magma algebra system. i. the user language. *J. Symbolic Comput.*, 24(3-4):235–265, 1997.

[3] A.-S. Elsenhans and J. Jahnel, On the characteristic polynomial of the Frobenius on etale cohomology, ArXiv e-prints 1106.3953, 2011.

[4] G. Everest, A. van der Poorten, I. Shparlinski and T. Ward, Recurring Sequences, Mathematical Surveys and Monographs 104. Providence, RI: American Mathematical Society.

[5] W. Fulton, Algebraic Curves, Benjamin, 1969.

[6] S. W. Golomb and G. Gong, Signal Design for Good Correlation: For Wireless Communication, Cryptography, and Radar, Cambridge University Press, 2005.

[7] E. Kani and M. Rosen, Idempotent relations and factors of Jacobians, *Math. Ann.*, 284(2):307–327, 1989.

[8] S. L. Kleiman, Algebraic cycles and the Weil conjectures, in Dix Exposés sur la Cohomologie des Schémas, Advanced Studies Pure Math. 3, 359–386 (1968).

[9] J. Lahtonen, G. McGuire and H. Ward, Gold and Kasami-Welch Functions, quadratic forms, and bent functions, *Advances in Mathematics of Communications*, 1, (2) (2007).

[10] R. Lidl and H. Niederreiter, Finite Fields, Addison-Wesley, 1983.

[11] G. Musiker, Combinatorial aspects of elliptic curves, *Sem. Lothar. Combin.*, 56 (2006/2007), Art. B56f. 2006.

[12] J. H. Silverman, The Arithmetic of Elliptic Curves, 2nd printing, Springer, 1992.

[13] H. Stichtenoth, Algebraic Function Fields and Codes, 2nd ed, Springer, 2008.

[14] P. T. Young, On lacunary sequences, *The Fibonacci Quarterly*, 41.1 (2003), 41-47.

School of Mathematics
Institute for Research in Fundamental Sciences (IPM)
Tehran, Iran
oahmadid@ipm.ir

School of Mathematical Sciences
University College Dublin
Ireland
gary.mcguire@ucd.ie

New tools and results in graph minor structure theory

Sergey Norin[1]

Abstract

Graph minor theory of Robertson and Seymour is a far reaching generalization of the classical Kuratowski–Wagner theorem, which characterizes planar graphs in terms of forbidden minors. We survey new structural tools and results in the theory, concentrating on the structure of large t-connected graphs, which do not contain the complete graph K_t as a minor.

1 Introduction

Graphs in this paper are finite and simple, unless specified otherwise. A graph H is a *minor* of a graph G if H can be obtained from a subgraph of G by contracting edges. Numerous theorems in structural graph theory describe classes of graphs which do not contain a fixed graph or a collection of graphs as a minor. A classical example of such a description is the Kuratowski–Wagner theorem [92, 93].

Theorem 1.1 *A graph is planar if and only if it does not contain K_5 or $K_{3,3}$ as a minor.*

(We will say that G *contains* H *as a minor*, if H is isomorphic to a minor of G, and we will use the notation $H \leq G$ to denote this. The notation is justified as the minor containment is, indeed, a partial order. We say that G is H-*minor free* if G does not contain H as a minor.)

Clearly a graph is a forest if and only if it does not contain K_3 as a minor. In [16] Dirac proved that a graph does not contain K_4 as a minor if and only if it is series-parallel. In [93] Wagner characterizes graphs which do not contain K_5 as a minor, as follows.

Theorem 1.2 *A graph does not contain K_5 as a minor if and only if it can be obtained by 0-, 1- and 2- and 3-clique sum operations from planar graphs and V_8. (The graph V_8 is shown on Figure 1.)*

One can simplify the statement of the above theorem by restricting attention to 4-connected graphs, where a graph G is k-*connected* if

---

[1]The author was supported by an NSERC Discovery grant.

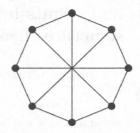

Figure 1: The Wagner graph W_8

$|V(G)| \geq k + 1$, and $G - X$ is connected for every set $X \subseteq V(G)$ with $|X| \leq k$. Theorem 1.2 implies the following.

Corollary 1.3 *A 4-connected graph does not contain K_5 as a minor if and only if it is planar.*

Characterization of K_6-minor free graphs appears to be a daunting task. However, Jørgensen [34] conjectured that Corollary 1.3 extends, as follows. A graph G is *an apex* if $G - v$ is planar for some vertex $v \in V(G)$.

Conjecture 1.4 *A 6-connected graph does not contain K_6 as a minor if and only if it is an apex.*

In [38, 39] Kawarabayashi et al. proved that Conjecture 1.4 holds for large graphs.

Theorem 1.5 *There exists an absolute constant N such that every 6-connected graph on at least N vertices with no K_6 minor is an apex.*

Note that if $K_t \leq G$ for a graph G then $K_{t-1} \leq G - \{v\}$ for every $v \in V(G)$. It follows that one can construct t-connected K_t-minor free graphs for $t \geq 5$ by adding $t - 5$ vertices to a 5-connected planar graph. Unfortunately, for $t \geq 8$ not every t-connected graph K_t-minor free graph can be constructed in this manner. (For $t = 8$, consider for example a graph obtained from K_{10} by deleting the edges of a perfect matching.) Thus Jørgensen's conjecture does not naturally extend to general t. However, an extension of Theorem 1.5 holds. Norin and Thomas [55] proved the following, verifying a conjecture of Thomas.

Theorem 1.6 *For every $t \geq 6$ there exists a constant $N = N(t)$ such that every t-connected graph G on at least N vertices and no K_t minor contains a set $X \subseteq V(G)$ with $|X| = t - 5$ such that $G - X$ is planar.*

The proof of Theorem 1.6 relies on the results from the Graph Minor series of Robertson and Seymour [61–64, 66–84], as well as recent extensions of these results. In this paper we mainly survey these techniques as well as other recent results in the area.

Several excellent surveys on related topics have appeared in recent years. The excluded minor theorems were surveyed by Thomas [88]. A high level overview of the graph minor theory is given by Lovász [46]. A detailed survey of the structural aspects of the graph minor theory by Karawabayshi and Mohar [37] is closest in spirit to our work. We try to avoid unnecessary overlap with [37] by concentrating on the techniques employed in the proof of Theorem 1.6, as well as recent advances in the field. This survey at times delves deep into technical details of this technical subject and is primarily intended for readers already familiar with the main concepts.

In Sections 2, 3 and 4 we discuss the main classical notions in the graph minor theory, namely treewidth, structure theorems and well-quasi ordering, resepctively, highlighting recent results related to these concepts. In Sections 5 and 6 we discuss some of the new tools used in the proof of Theorem 1.6. Finally, in Sections 7 and 8 we consider applications of the techniques described in this survey to separators and extremal problems, respectively.

We use standard graph theoretical notation and terminology, see [11] for an excellent reference.

2   Treewidth

Treewidth of a graph is an important graph parameter, introduced by Halin [30] and rediscovered by Robertson and Seymour [67]. A *tree decomposition* of a graph G is a pair $(T, \mathcal{W})$, where T is a tree and $\mathcal{W}$ is a family $\{W_t \mid t \in V(T)\}$ of vertex sets $W_t \subseteq V(G)$, such that the following three conditions hold:

(W1) $\cup_{t \in V(T)} W_t = V(G)$,

(W2) every edge of G has both ends in some W_t,

(W3) If $t, t', t'' \in V(T)$ are such that t' lies on the path in T between t and t'', then $W_t \cap W_{t''} \subseteq W_{t'}$.

If T is a path then we say that $(T, \mathcal{W})$ is a *path decomposition* of G. The *width* of a tree decomposition $(T, \mathcal{W})$ is defined as $\max_{t \in V(T)} (|W_t| - 1)$, and the *treewidth* of G is defined as the minimum width of a tree decomposition of G. We denote the treewidth by $\mathrm{tw}(G)$.

One of the reason for the prominence of the notion of treewidth in structural and algorithmic graph theory is its relation to several other important graph parameters. We say that two graph parameters t_1 and t_2 are *tied* if there exists a function f such that $t_1(G) \leq f(t_2(G))$ and $t_2(G) \leq f(t_1(G))$ for every graph G. Parameters tied to treewidth were recently surveyed by Harvey and Wood [31]. Let us present some key examples.

2.1 Brambles

A collection of subsets $\mathcal{B}$ of the vertex set of a graph G is called a *bramble* if for all $B, B' \in \mathcal{B}$ the subgraph $G[B \cup B']$ of G induced by $B \cup B'$ is connected. (In particular, $G[B]$ is connected for every $B \in \mathcal{B}$.) The *order* of $\mathcal{B}$ is the minimum size of the set $S \subseteq V(G)$ such that $S \cap B \neq \emptyset$ for every $B \in \mathcal{B}$. The *bramble number* $\mathrm{bn}(G)$ of G is the maximum order of the bramble in G. The following duality characterization of treewidth by Seymour and Thomas [86] provides an important tool in the area.

Theorem 2.1 *For every graph* G

$$\mathrm{bn}(G) = \mathrm{tw}(G) + 1.$$

2.2 Grid minors

A key component of the graph minor theory is the result that either a graph has "small" treewidth or it has a "large" planar minor, in the following sense, described in the Grid Minor Theorem [69]. For a graph G, let $\mathrm{gr\text{-}min}(G)$ denote the maximum k such that G contains a $k \times k$-grid as a minor.

Theorem 2.2 *There exists a function* f *such that*

$$\mathrm{gr\text{-}min}(G) \leq \mathrm{tw}(G) \leq f(\mathrm{gr\text{-}min}(G)).$$

The optimum function f has been the subject of significant research effort over the years. For a long time the best known upper bound was $f(r) = 20^{64r^5}$ by Robertson, Seymour and Thomas [65]. A similar bound was given by Diestel, Gorbunov, Jensen, and Thomassen [12] with a much simpler proof. Recently, however, the bound was successively improved, first, by Kawarabayashi and Kobayashi [36] and, independently, by Leaf and Seymour [43] to $2^{O(r^2 \log r)}$, and then in a breakthrough result by Chekuri and Chuzhoy [7] to a polynomial $O(r^{98})$. The best known lower bound on $f(r)$ is $\Omega(r^2 \log r)$ [65].

2.3 Tangles

Another notion dual to the notion of treewidth is that of a tangle. A *g-separation* of a graph G is a pair of subgraphs (A, B) of G such that $A \cup B = G$ and $E(A) \cap E(B) = \emptyset$. The *order* of a g-separation (A, B) is $|V(A) \cap V(B)|$. A *tangle* $\mathcal{T}$ of order $\theta \geq 1$ in G is a collection of g-separations of G, satisfying the following:

(i) for every g-separation (A, B) of G of order $< \theta$ either $(A, B) \in \mathcal{T}$, or $(B, A) \in \mathcal{T}$,

(ii) if $(A_1, B_1), (A_2, B_2), (A_3, B_3) \in \mathcal{T}$ then $A_1 \cup A_2 \cup A_3 \neq G$, and

(iii) if $(A, B) \in \mathcal{T}$ then $V(A) \neq V(G)$.

The tangle number $\mathrm{tn}(G)$ is the maximum order of a tangle in G. The relation between the tangle number and the treewidth of a graph is captured in the following theorem of Robertson and Seymour [75].

Theorem 2.3 *Let G be a graph with $\mathrm{tn}(G) \geq 2$. Then the treewidth $\mathrm{tw}(G)$ of G satisfies*

$$\mathrm{tn}(G) \leq \mathrm{tw}(G) + 1 \leq \frac{3}{2} \mathrm{tn}(G).$$

2.4 Erdős–Pósa property

In this subsection we consider an application of the concepts and results introduced in the previous two.

We say that a graph G *packs k H-minors* if there exist vertex disjoint subgraphs $H_1, \ldots, H_k$ of G such that $H \leq H_i$ for $1 \leq i \leq k$. We say that *H-minors have the Erdős–Pósa property* if for every positive integer k there exists an integer $\mathrm{ep}(k, H)$ such that in every graph G, which does not pack k H-minors, there exists a set of vertices $X \subseteq V(G)$ such that $|X| \leq \mathrm{ep}(k, H)$ and $G - X$ contains no H minor.

Robertson and Seymour show in [69] that H-minors have the Erdős–Pósa property if and only if H is planar. Let us sketch their argument for connected planar graphs H.

Theorem 2.4 *Let H be a connected planar graph. Then H-minors have the Erdős–Pósa property.*

Proof The proof is by induction on k. The base case $k = 1$ is trivial.

For the induction step, note that it is not hard to deduce from Theorem 2.2 that there exists w such that every graph G with $\mathrm{tw}(G) \geq w$ packs k H-minors. We will show that $\mathrm{ep}(k, H) \leq 2\,\mathrm{ep}(k - 1, H) + 3w$. Suppose for a contradiction that there exists a graph G, which does not pack k H-minors, such that $H \leq G - X$ for every $X \subseteq V(G)$ with $|X| \leq 2\,\mathrm{ep}(k - 1, H) + 3w$. We will define a tangle $\mathcal{T}$ in G of order $w + 1$. It will follow from Theorem 2.3, that $\mathrm{tw}(G) \geq \mathrm{tn}(G) - 1 \geq w$, in contradiction with the choice of G.

It remains to define the tangle $\mathcal{T}$. Consider a g-separation (A, B) of G of order at most w. Let $Z = V(A) \cap V(B)$. Then either $A - Z$, or $B - Z$ has an H-minor, as otherwise, $G - Z$ contains no H minor, and $|Z| \leq w + 1$, as desired. On the other hand, if both $A - Z$ and $B - Z$ contain H as a minor, then neither $A - Z$, nor $B - Z$ packs $k - 1$ H-minors, by the assumption on G. Thus, by the induction hypothesis, there exist $Y_1 \subseteq V(A) - Z$ and $Y_2 \subseteq V(B) - Z$ such that $|Y_1|, |Y_2| \leq \mathrm{ep}(k - 1, H)$ and $A - Y_1 - Z$ and $B - Y_2 - Z$ contain no H-minor. It follows that $G - Y_1 - Y_2 - Z$ contains no H minor, once again contradicting the choice of G. Thus exactly one of $A - Z$ and $B - Z$ contains H as a minor. If $B - Z$ contains an H minor, let $(B, A) \in \mathcal{T}$, and, otherwise, let $(A, B) \in \mathcal{T}$.

It is not hard to verify that $\mathcal{T}$ satisfies the conditions (i), (ii) and (iii) in the definition of a tangle. Let us do this for condition (ii). If $(A_1, B_1), (A_2, B_2), (A_3, B_3) \in \mathcal{T}$ are such that $A_1 \cup A_2 \cup A_3 = G$ then $G' = G - \cup_{i=1}^{3}(V(A_i) \cap V(B_i))$ contains no H minor, as each of components of G' is a subgraph of $A_i - V(B_i)$ for some $i = 1, 2, 3$. As

$$\sum_{i=1}^{3} |V(A_i) \cap V(B_i)| \leq 3w \leq 3\,\mathrm{ep}(k - 1, H) + 3w,$$

we obtain a contradiction to the choice of G, and so (ii) holds. $\square$

For non-planar graphs H, H-minors do not have the Erdős–Pósa property for the following reason. Let Σ be a surface of minimum genus on which H embeds. Then any graph embeddable on Σ does not pack two H minors. On the other hand, for every n there exists a graph G_n embeddable on Σ such that $H \leq G - X$ for every $X \subseteq V(G)$ with $|X| \leq n$. In particular, one can construct G_n with these properties by taking $2n + 1$ different drawings of H in Σ, such that no point of Σ belongs to more than two of these drawings, and adding a vertex to G_n for every point of intersection between the drawings that is not already a vertex of G_n. This construction motivates the following definition and conjecture of Thomas, stated e.g. in [37]. We say that G *packs* k *H-minors half-integrally* if there exist subgraphs $H_1, \ldots, H_k$ of G such that $H \leq H_i$ for $1 \leq i \leq k$ and no vertex of G belongs to more than two of these subgraphs.

Conjecture 2.5 *For every graph H and an integer k there exists an integer* $\mathrm{ep}_{1/2}(k, H)$ *such that in every graph G, which does not pack k H-minors half-integrally, there exists a set of vertices $X \subseteq V(G)$ such that* $|X| \leq \mathrm{ep}_{1/2}(k, H)$ *and $G - X$ contains no H minor.*

Tools used in the proof of Theorem 1.6 can be used to approach this conjecture, and we will return to discussing it in the next section. In [14] it is shown that complete minors satisfy the (integral) Erdős–Pósa property, if one restricts one's attention to sufficiently highly connected graphs.

Theorem 2.6 *For all positive integers p and k there exists a positive integer* $\mathrm{ep_c}(k, p)$ *such that in every $(k(p-3) + 14p + 14)$-connected graph G which does not pack k H-minors there exists a set of vertices $X \subseteq V(G)$ such that $|X| \leq \mathrm{ep}_c(k, p)$ and $G - X$ contains no H minor.*

The connectivity bound in Theorem 2.6 is close to being tight, as in [14] the authors construct $(k(p-3) - (p-3)(p-4)/2 - 6)$ connected graphs $G_{k,n,p}$ for all positive integers $k \geq p \geq 5$ and n, such that $G_{k,n,p}$ does not pack k H-minors, but $H \leq G - X$ for every $X \subseteq V(G)$ with $|X| \leq n$.

2.5 Separation numbers

A *separation* of a graph G is a pair (A, B) of subsets of $V(G)$ such that $A \cup B = V(G)$ and no edge of G has one end in $A - B$ and the other in $B - A$. The *order* of the separation (A, B) is $|A \cap B|$. A separation (A, B) of a graph G on n vertices is *balanced* if $|A \setminus B| \leq 2n/3$ and $|B \setminus A| \leq 2n/3$. The *separation number* $\mathrm{sn}(G)$ of a graph G is a smallest number s such that every subgraph of G has a balanced separation of order at most s. The relation between the separation number and the treewidth has been explored starting with Robertson and Seymour [69], who have shown that $\mathrm{sn}(G) \leq \mathrm{tw}(G) + 1$ for every graph G. On the other hand, Bodlaender et al. [5] have proved that $\mathrm{tw}(G) \leq 1 + \mathrm{sn}(G) \log(|V(G)|)$. Fox [26] stated without proof that $\mathrm{sn}(G)$ and $\mathrm{tw}(G)$ are tied. Finally, in [23] a linear relation between $\mathrm{sn}(G)$ and $\mathrm{tw}(G)$ is established.

Theorem 2.7 $\mathrm{tw}(G) \leq 105\,\mathrm{sn}(G)$ *for every graph G.*

In the next two subsections of this section we mention a couple of surprising recent results relating treewidth to parameters from other areas of mathematics.

2.6 Gonality

In algebraic geometry *gonality* of an algebraic curve is the minimum degree of a non-constant rational map from the curve to the projective line. Thus, in some sense, gonality of a curve is a parameter measuring how much the curve resembles the projective line. In [4] Baker and Norin consider graphs as discrete analogues of algebraic curves.

The gonality of a finite graph can be defined by the means of a simple solitaire chip-firing game, as follows. A *chip configuration* D on a graph G is a function $D : V(G) \to \mathbb{Z}_+$ assigning to every vertex of G a non-negative number of chips. A *move* in the game consists of selecting a set $U \subseteq V(G)$ and for every edge joining a vertex of $u \in U$ to a vertex of $v \in V(G) - U$ transferring one chip from u to v. (A move is only legal if it results in a configuration with a non-negative number of chips on every vertex.) A chip configuration D is *winning* if for every $v \in V(G)$ it is possible to transfer a chip to v by a sequence of legal moves. The *gonality* $\text{gon}(G)$ of a graph G is the minimum total number of chips in a winning configuration.

Within the framework of [4] trees serve as analogues of the projective line. Therefore by analogy with algebro-geometric setting one might expect gonality to be related to treewidth. And indeed, in [91], van Dobben de Bruyn and Gijswijt prove the following.

Theorem 2.8 $\text{tw}(G) \leq \text{gon}(G)$ *for every graph G. Moreover, $\text{tw}(G) = \text{gon}(G)$ if G is a tree, a cycle, a two-dimensional grid, or a complete multipartite graph.*

Several fundamental questions related to graph gonality remain open. To the best of our knowledge it is unknown whether gonality is tied with treewidth. Further, we do not know whether it is true that $\text{gon}(H) \leq \text{gon}(G)$ if H is a minor (or, even, a subgraph) of G. The positive answer to the second question would, of course, imply a positive answer to the first by Theorem 2.8.

2.7 Poset dimension

A *poset* P is a set equipped with an irreflexive, antisymmetric and transitive binary relation $<_P$. A poset P is a *linear order* if every two elements of P are comparable, that is $x <_P y$, or $y <_P x$, or $x = y$ for all

$x, y \in P$. A subposet of P which is a linear order is called a *chain*. The *height* of P is the maximum cardinality of a chain in P.

A linear order L is an *extension* of P if L and P have the same ground set and $x <_P y$ implies $x <_L y$. The *dimension* $\dim(P)$ of P is the minimum integer d such that there exist extensions $L_1, L_2, \ldots, L_d$ of P such that $x <_P y$ if and only if $x <_{L_i} y$ for all $1 \leq i \leq d$. The dimension and the height are fundamental parameters of a poset. It is not hard to see that they are not in general related. However, in some cases one can bound the dimension by a function of height, and this is where the treewidth comes into play.

We say that x *covers* y in P, if $x >_P y$ and there does not exist $z \in P$ such that $x >_P z >_P y$. One can associate a *cover graph* $G(P)$ with a poset P with $V(G(P)) = P$ and $xy \in E(G)$ if and only if either x covers y, or y covers x in P. Joret et al. [33] have recently proved the following theorem relating height and the dimension of a poset when the treewidth of the cover graph is bounded.

Theorem 2.9 *For every pair of positive integers w and h there exists a least positive integer $d = d(w, h)$ such that if P is a poset of height at most h and the treewidth of the cover graph of P is at most w, then $\dim(P) \leq d$.*

It is conjectured in [33] that Theorem 2.9 can be extended as follows.

Conjecture 2.10 *Let H be a graph. Then for every integer h, there is a least positive integer $d = d(H, h)$ so that if P is a poset of height h and the cover graph of P does not contain H as a minor, then $\dim(P) \leq d$.*

3   Global structure

The central result in the graph minor theory is an approximate topological characterization of graphs which do not contain a given graph as a minor. There are several variants of such a characterization. The more refined ones are rather unwieldy to state, but are needed for the more involved applications.

Fix a graph H. Informally, every graph G that does not contain H as a minor can be obtained by gluing graphs, which can be "almost" embedded in some surface in which H cannot be embedded, in a "tree-like fashion". Clarifying the notions of "almost" and "tree-like fashion" is our next goal.

3.1   Clique sums

Let $k \geq 1$ be an integer. Let G_1 and G_2 be two graphs with disjoint vertex sets. For $i = 1, 2$ let $X_i \subseteq V(G_i)$ be a clique of size k in G_i, that is a vertex set of a complete subgraph. Let G be a graph obtained from G_1 and G_2 by identifying X_1 and X_2 and possibly deleting some of the edges with both ends among the identified vertices. Then we say that G is *a clique-sum of order* k, or a k-*sum*, of G_1 and G_2.

Clique-sums have already appeared in the statement of Theorem 1.2 in the introduction. If H is a $(k+1)$-connected graph and G_1 and G_2 are graphs not containing H as minor, then a $\leq k$-sum of G_1 and G_2 does not contain H as a minor. This observation explains why structural descriptions of graphs not containing a given graph as a minor frequently use the notion of clique-sum (e.g. [48, 49]).

3.2 Vortices

A proper minor-closed class of graphs can be obtained by considering graphs embedded on a fixed surface, except that in a bounded number of faces additional edges can be drawn, crossing each other in a controlled manner. In fact, we need to consider a more complex construction, which we now describe.

A *vortex* is a pair (G, Ω), where G is a graph and Ω is a cyclic permutation of some set of vertices G, which we denote by $V(\Omega)$. For $x, y \in V(\Omega)$ we denote by $\Omega[x, y]$ and $\Omega[y, x]$ the two intervals in Ω with ends x and y.

A *vortical decomposition* $\mathcal{V}$ of (G, Ω) is a notion closely related to the path decomposition, defined as follows. The set $\mathcal{V}$ is a family of vertex sets $\{V_x \mid x \in V(\Omega)\}$ such that the following four conditions hold:

(V1) $\cup_{x \in V(\Omega)} V_x = V(G)$,

(V2) $x \in V_x$,

(V3) every edge of G has both ends in some V_x,

(V4) For $x, y \in V(\Omega)$ every vertex of $V_x \cap V_y$ either lies in $\cap_{z \in \Omega[x,y]} V_z$ or $\cap_{z \in \Omega[x,y]} V_y$.

The *depth* of $\mathcal{V}$ is $\max_{x \in \Omega} |V_x|$, and, naturally, the *depth* of a vortex is the minimum width of its vortical decomposition. One can similarly define the *adhesion* of $\mathcal{V}$, as $\max_{x,y \in \Omega, x \neq y} |V_x \cap V_y|$, and the *adhesion* of a vortex as the minimum adhesion of its vortical decomposition.

Vortices of small depth and of small adhesion are considered in different versions of the graph minor structure theorem. While vortices of small depth do not seem to allow a dual characterization along the lines of Theorem 2.2, vortices of small adhesion do. Let us present this characterization, which partially motivates the definition of the vortical decomposition above. A *bump* in a vortex (G, Ω) is a path P with ends in $V(\Omega)$

and otherwise disjoint from $V(\Omega)$. A *transaction of order k* in (G, Ω) is a sequence of pairwise vertex disjoint bumps $P_1, \ldots, P_k$ such that there exist $x, y \in V(\Omega)$ so that for every $1 \leq i \leq k$ the path P_i has one end in $\Omega[x, y]$ and the other in $\Omega[y, x] - \{x, y\}$. The following is shown in [73].

Theorem 3.1 *Let (G, Ω) be a vortex containing no transaction of order k. Then (G, Ω) has adhesion at most k. Moreover, if (G, Ω) has adhesion at most k then it contains no transaction of order $2k + 1$.*

In graph minor theory one frequently considers minors of labeled graphs, where the definition of a minor depends on the choice of labeling. Here we introduce such a definition for vortices. For a vortex (G, Ω) and a subgraph of G' of G, we say that $(G', \Omega|_{V(G')})$ is a *subvortex* of (G, Ω). We say that (G'', Ω'') is *a vortex minor* of (G, Ω) if it can be obtained from a subvortex (G', Ω') of (G, Ω) by repeatedly contracting edges with at least one end outside of Ω'. As an application of the concept of vortex minor let us present a theorem of Robertson and Seymour, characterizing 4-connected vortices with no cross. A vortex (G, Ω) is *rural* if G can be drawn in a disk with $V(\Omega)$ drawn on the boundary of the disk in the order given by Ω. We say that a vortex (G, Ω) is *k-connected* if there is no separation (A, B) of G of order at most k with $V(\Omega) \subseteq A$ and $B - A \neq \emptyset$. A *cross* is a vortex (C, Ω_C), where $V(C) = \{v_1, v_2, u_1, u_2\}$, $E(C) = \{v_1 u_1, v_2 u_2\}$ and $\Omega = (v_1, v_2, u_1, u_2)$. The following is proved in [73].

Theorem 3.2 *Let (G, Ω) be a 4-connected vortex, then either (G, Ω) is rural, or (G, Ω) contains a cross as a vortex minor.*

A *multivortex* is a tuple $(G, \Omega_1, \ldots, \Omega_r)$ such that (G, Ω_i) is a vortex for every $i = 1, \ldots, r$ and $V(\Omega_i) \cap V(\Omega_j) = \emptyset$ for $i \neq j$. An *embedding of a multivortex* $(G, \Omega_1, \ldots, \Omega_r)$ in a surface Σ with cuffs $\Delta_1, \ldots, \Delta_r$ is an embedding $\sigma : G \hookrightarrow \Sigma - \cup_{i=1}^{r} \Delta_i$, where $\Delta_1, \ldots, \Delta_r$ are pairwise disjoint interiors of disks in Σ and for every $1 \leq i \leq r$ we have

- $\partial \Delta_i \cap \sigma(G) = \sigma(V(\Omega_i))$, and

- the clockwise cyclic order of the vertices of $\sigma(V(\Omega_i))$ on Δ_i corresponds to Ω_i if Σ is orientable, and is Ω_i or its reverse, if Σ is not orientable.

Multivortex minors can be defined analogously to the vortex minors above. Vortices and multivortices are central to the definition of graphs almost embeddable on a surface, as seen in the next subsection.

3.3 The clique sum structure

We are now ready to define graphs almost embeddable on a surface. A *segregation of a graph* G is a tuple $(G_0, V_1, V_2, \ldots, V_r)$, such that

(S1) $V_i = (G_i, \Omega_i)$ is a vortex for $i = 1, \ldots, r$,

(S2) $G_0, G_1, \ldots, G_r$ are subgraphs of G,

(S3) $G = G_0 \cup G_1 \ldots \cup G_r$,

(S4) $V(\Omega_i) = V(G_i) \cap V(G_0)$, and

(S5) $G_1, \ldots, G_r$ are pairwise vertex disjoint.

One can consider a segregation as a partition of a graph into a "central part" G_0 and disjoint "attachments" $G_1, \ldots, G_r$, where a cyclic order is prescribed on the set of vertices each attachment shares with the central part.

Let Σ be a surface and k a positive integer. A *near embedding of G in* Σ is a tuple $(G_0, X, \mathcal{V}, \sigma)$, such that

(E1) $\mathcal{V} = (V_1, \ldots, V_r)$ for some positive integer r, where $V_i = (G_i, \Omega_i)$ is a vortex for every $1 \leq i \leq r$,

(E2) $(G_0, V_1, \ldots, V_r)$ is a segregation of $G - X$, and

(E3) σ is an embedding of the multivortex $(G_0, \Omega_1, \ldots, \Omega_r)$ in Σ.

Essentially, a near-embedding describes an embedding of the central part of a segregation of a graph, after first deleting a specified set of vertices. We say that a near embedding *has depth* $\leq k$, if $r \leq k$, $|X| \leq k$, and V_i has depth at most k for every $1 \leq i \leq k$.

The graph minor structure theorem proved in [82] can now be stated as follows.

Theorem 3.3 *For every graph H there exists an integer k such that every graph not containing H as a minor can be obtained by $\leq k$-sums from graphs which allow a near embedding of depth $\leq k$ in some surface, in which H cannot be embedded.*

Theorem 3.3 is used, in particular, to prove that for every graph H there exists an integer w such that every graph G not containing H as a minor allows a two coloring with both color classes inducing a subgraph of treewidth at most w. We discuss this result in more detail in Section 7. In fact, a proof of a certain refinement of this result discussed there requires a strengthening of Theorem 3.3, which only very recently was explicitly stated in [21]. We say that a near embedding $(G_0, X, \mathcal{V}, \sigma)$ of G *has a essential apices* if

- there exist $Z \subseteq X$ with $|Z| \leq a$ such that every vertex in $X - Z$ is adjacent only to vertices in $X \cup V(G_1) \cup \ldots \cup V(G_r)$.

Theorem 3.4 *For every graph H there exists an integer k such that every graph G not containing H as a minor can be obtained by $\leq k$-sums from graphs $G_1, \ldots, G_s$, and for each $1 \leq i \leq s$ there exists an integer a_i and a surface Σ_i such that the graph G_i allows a near embedding in Σ_i of depth $\leq k$ with a_i essential apices, while $H - Z$ cannot be embedded on Σ_i for every $Z \subseteq V(H)$ with $|Z| \leq a_i$.*

Note that Theorem 3.4 implies that if H is an apex then every H-minor free graph can be obtained by clique-sums from graphs almost embedded on surfaces on which H cannot be embedded with no essential apices.

3.4   Structure with respect to a tangle

The dichotomy between graphs of bounded treewidth and graphs containing a large tangle is at the center of many arguments in graph minor theory, but is not captured by Theorem 3.3.

Consider, for example, the argument in the proof of Theorem 2.4. It is shown there that for all integers m, k and w and all connected graphs H, if the graph G does not pack k H-minors, while $H \leq G - X$ for every $X \subseteq V(G)$ with $|X| \leq 2m + 3w$, then

- either G contains a tangle $\mathcal{T}$ of order $\geq w + 1$, such that for every g-separation (A, B) of G of order at most w, we have $(A, B) \in \mathcal{T}$ if and only if $H \leq B - V(A \cap B)$, or

- there exists a g-separation (A, B) of G of order at most w such that neither $A - V(A \cap B)$, nor $B - V(A \cap B)$, pack $k - 1$ H-minors.

If we are investigating the structure of graphs which do not pack k H-minors, then in the second case we can apply an inductive argument, while in the first case, we are interested in the structure of G relative to the tangle $\mathcal{T}$. To describe this structure we need to refine the definition of the k-almost embedding as follows.

A vortex (G, Ω) is called *small* if $|V(\Omega)| \leq 3$ and *large*, otherwise. A near embedding $(G_0, X, (V_1, \ldots, V_r), \sigma)$ of G in Σ is a *k-near embedding* if

(E4) $|X| \leq k$,

(E5) there exists $k' \leq k$ such that V_i is small for $i > k'$, and

(E6) V_i has adhesion $\leq k$ for $i \leq k'$.

Given a tangle $\mathcal{T}$ in G of order θ and $X \subseteq V(G)$ with $|X| < \theta$, let $\mathcal{T} - X$ denote the set of all g-separations (A, B) of $G - X$ of order less than $\theta - |X|$ such that there exists a g-separation $(A', B') \in \mathcal{T}$ with $X \subseteq V(A) \cap V(B)$, $A' - X = A$ and $B' - X = B$. It is shown in [75] that $\mathcal{T} - X$ is a tangle in $G - X$ of order $\theta - |X|$. We say that a near embedding $(G_0, X, \mathcal{V}, \sigma)$ *captures* $\mathcal{T} - X$ if for every $(A, B) \in \mathcal{T} - X$ and every $W \in \mathcal{V}$ we have $B \not\subseteq W$. (That is the "large" side of a separation in $\mathcal{T} - Z$ is never contained in a vortex.) We now state the structure theorem relative to a tangle from [82].

Theorem 3.5 *For every graph H there exist integers $\theta, k \geq 0$ such that the following holds. Let G be an H-minor free graph, and let $\mathcal{T}$ be a tangle in G of order at least θ. Then G has a k-near embedding $(G_0, X, \mathcal{V}, \sigma)$ in some surface in which H cannot be embedded, which captures $\mathcal{T} - X$.*

Further refinements of this theorem are required in most applications. To introduce these refinements we require a number of additional technical definitions.

Given an embedding $\sigma : G \hookrightarrow \Sigma$ of a graph G in a surface Σ, we say that a curve C in Σ is σ-*normal* if it intersects the drawing of G in vertices only. Given σ, *the distance between points* $x, y \in \Sigma$ is the minimum value of $|\sigma(G) \cap C|$ over all σ-normal curves C in Σ joining x and y.

Given a near-embedding $(G_0, X, \mathcal{V}, \sigma)$ of a graph G, *the distance between vortices* $(G_i, \Omega_i), (G_j, \Omega_j) \in \mathcal{V}$ is the minimum distance with respect to σ between points in $\sigma(V(\Omega_i))$ and $\sigma(V(\Omega_j))$. The *representativity of the embedding* $\sigma : G \hookrightarrow \Sigma$ in a surface $\Sigma \neq S^2$ is the minimum value of $|\sigma(G) \cap C|$ over all σ-normal genus reducing closed curves C in Σ.

We say that a vortex (G_i, Ω_i) is d-*free* in a tangle $\mathcal{T}$ if for every separation $(A, B) \in \mathcal{T}$ such that $G_i \subseteq A$ we have $|V(A) \cap V(B)| \geq \min\{d, |V(\Omega)|\}$.

Finally, we need one last technical definition, which will be important for the applications in the next subsection. We say that a multivortex (H, Ω_1, Ω_2) is an *annulus* if it can be embedded in the plane. We say that an annulus is d-*fat* if the distance between $V(\Omega_1)$ and $V(\Omega_2)$ in this embedding is at least d. We say that a vortex (G', Ω) is d-*fat in a tangle* $\mathcal{T}$ if there exist subgraphs G_1, G_2 of G' and a circular order Ω' with $V(\Omega') = V(G_1) \cap V(G_2)$ such that $V(\Omega) \subseteq V(G_1)$, (G_1, Ω, Ω') is a d-fat annulus, and (G_2, Ω) is d-free in $\mathcal{T}$. Essentially, a d-fat vortex can be considered as a d-free vortex surrounded by a thick planar fringe.

We say that a near embedding $(G_0, X, \mathcal{V}, \sigma)$ of a graph G in a surface Σ is $(\mathcal{T}, d)$-*rich* if

- $(G_0, X, \mathcal{V}, \sigma)$ captures $\mathcal{T} - X$,

- either $\Sigma = S^2$ or the embedding $\sigma : G_0 \hookrightarrow \Sigma$ has representativity at least d,

- every vortex in $\mathcal{V}$ is d-free in $\mathcal{T} - X$,

- the distance between any two large vortices is at least d in σ, and

- every large vortex is d-fat in $\mathcal{T} - X$.

The following refinement of Theorem 3.5 is a consequence of the main result of [13].

Theorem 3.6 *For every graph H and an integer $d \geq 0$, there exist integers $k, \theta \geq 0$ such that the following holds. Let G be an H-minor free graph, and let $\mathcal{T}$ be a tangle in G of order at least θ. Then G has a $(\mathcal{T}, d)$-rich k-near embedding $(G_0, X, \mathcal{V}, \sigma)$ in some surface in which H cannot be embedded.*

3.5 Building a minor

Let us now demonstrate an application of Theorem 3.6. Given multivortices $\mathcal{H}_1 = (H_1, \Omega_1, \Omega_2, \ldots, \Omega_r)$ and $\mathcal{H}_2 = (H_2, \Omega_1, \Omega_2, \ldots, \Omega_r)$, such that $V(H_1) \cap V(H_2) = \cup_{i=1}^r V(\Omega_i)$, let $\mathcal{H}_1 \times \mathcal{H}_2$ be defined as the graph $H_1 \cup H_2$.

Given a near embedding $(G_0, X, \mathcal{V}, \sigma)$ of a graph G such that $\{(G_i, \Omega_i)\}_{i=1}^r$ is the set of all large vortices in $\mathcal{V}$, define *the residual multivortex* of this near-embedding as $(G_1 \cup G_2 \cup \ldots \cup G_r, \Omega_1, \ldots, \Omega_r)$. Thus the residual multivortex captures the structure of the set of the large vortices. The following is a special case of the main result of [53], which is a crucial ingredient of the proof of Theorem 1.6.

Theorem 3.7 *For every positive integer N there exists a positive integer d such that the following holds. Let H, G be graphs $|V(H)| + |E(H)| \leq N$. Let $\mathcal{T}$ be a tangle in G of order at least d. Let $(G_0, \emptyset, \mathcal{V}, \sigma)$ be an $(\mathcal{T}, d)$-rich near embedding of the graph G. Let $\mathcal{H}_1, \mathcal{H}_2$ be multivortices such that*

(a) *H is isomorphic to $\mathcal{H}_1 \times \mathcal{H}_2$,*

(b) *$\mathcal{H}_1$ is a minor of the residual multivortex of $(G_0, \emptyset, \mathcal{V}, \sigma)$,*

(c) *$\mathcal{H}_2$ can be embedded in Σ.*

Then H is a minor of G.

Essentially, Theorem 3.7 shows that, given a sufficiently robust near embedding of a graph G, one can construct a given small graph as a minor of G from a part of it embeddable in the surface, and parts which can be found as minors of large vortices of the near embedding. The result of [53] in full generality is more technical: It allows for the incorporation of the apex vertices (the set X in the definition of near embedding), for rooted minors, and obtains H as a subdivision, rather than a minor under certain additional technical conditions.

Theorem 3.7 can be used, in particular, to establish Conjecture 2.5 for graphs H which can be drawn in the plane with a single crossing. Let us present a very informal sketch of the argument. Let kH denote the union of k vertex disjoint copies of H. Following the argument in the beginning of Section 3.4, for any given θ, we can restrict our attention to graphs which contain a tangle $\mathcal{T}$ of order $\geq \theta + 1$, such that for every g-separation (A, B) of G of order at most θ, we have $(A, B) \in \mathcal{T}$ if and only if $H \leq B - V(A \cap B)$. Applying Theorem 3.6 we obtain a near-embedding $(G_0, X, \mathcal{V}, \sigma)$ of G in the plane satisfying the conclusion of that theorem for some integer d, which we can choose to satisfy Theorem 3.7. If none of the large vortices of $(G_0, X, \mathcal{V}, \sigma)$ contains a union of k disjoint crosses as a vortex minor, then one can show that G contains a bounded set of vertices Z, such that none of the vortices contains a cross disjoint from Z. Applying a variant of Theorem 3.2 we can deduce that $G - Z$ has a k-near embedding $(G_0', X', \mathcal{V}', \sigma')$ in the plane capturing $\mathcal{T} - Z - X'$ with no large vortices (for bounded k). It now follows that $G - Z - X'$ contains no H minor, as H cannot be drawn in the plane or use the small vortices. If on the other hand one of the vortices of $(G_0, X, \mathcal{V}, \sigma)$ contains a union of k disjoint crosses as a vortex minor, then one can find a half-integral packing of k H-minors in G using Theorem 3.7.

Let us emphasize once again that the argument presented here omits some crucial technical details, but hopefully it outlines a setting in which the combination of Theorems 3.6 and 3.7 is useful.

4   Well-quasi-ordering

4.1   Well-quasi-ordering for minors

The following result from [62] is the pinnacle of the Robertson–Seymour graph minor theory.

Theorem 4.1 *Let $G_1, G_2, \ldots, G_n, \ldots$ be an infinite sequence of graphs. Then there exist indices $i < j$ such that $G_i \leq G_j$.*

Equivalently, Theorem 4.1 states that graphs are well-quasi-ordered under the minor relation. We say that a class of graphs $\mathcal{F}$ is *minor-closed* if it is closed under isomorphism and taking minors. A class of graphs is *proper* if it does not include all graphs. One particularly useful corollary of Theorem 4.1 is that any minor-closed class of graphs $\mathcal{F}$ is characterized by finitely many *excluded minors*.

Corollary 4.2 *Let $\mathcal{F}$ be a minor closed class of graphs. Then there exists a finite collection of graphs $H_1, H_2, \ldots, H_k$ such that a graph G does not belong to $\mathcal{F}$ if and only if G contains at least one of $H_1, H_2, \ldots, H_k$ as a minor.*

For several applications a refinement of Theorem 4.1 to rooted graphs is needed, one of which, Corollary 5.1, is presented later in this survey. Let us present this refinement.

For an integer $k \geq 0$, a k-rooted graph is a pair (G, ρ), where G is a graph and $\rho : \{1, 2, \ldots, k\} \to V(G)$ is an injective function. We say that a k-rooted graph (H, ρ') is *a minor* of a k-rooted graph (G, ρ) if there exists a map α mapping vertices of H to disjoint subsets of $V(G)$ satisfying the following properties:

1. $G[\alpha(v)]$ is connected for every $v \in V(H)$;

2. if $v, w \in V(H)$ are joined by an edge then some edge of G joins a vertex of $\alpha(v)$ to a vertex in $\alpha(w)$;

3. $\rho(i) \in \alpha(\rho'(i))$ for every $i \in k$.

Note that for 0-labeled graphs H and G, H is a minor of G using the above definition, if and only if $H \leq G$.

In [63] Robertson and Seymour prove that k-rooted graphs are well-quasi-ordered under the minor relation.

Theorem 4.3 *Let $k \geq 0$ be an integer $(G_1, \rho_1), (G_2, \rho_2), \ldots, (G_n, \rho_n), \ldots$ be an infinite sequence of k-rooted graphs. Then there exist indices $i < j$ such that (G_i, ρ_i) is a minor of (G_j, ρ_j).*

In fact, one of the main results of [63] is a generalization of Theorem 4.3, allowing labels from any well-order, rather than a finite set.

4.2   Well-quasi-ordering for topological minors

An intriguing recent result of Liu and Thomas characterizes graph classes well-quasi-ordered under the topological minor relation. Let us

present the necessary definitions. A graph is a *subdivision* of another if
the first can be obtained from the second by replacing each edge by a non-
zero length path with the same ends, where the paths are disjoint, except
possibly for shared ends. We say that a graph H is a *topological minor* of
a graph G, and write $H \leq_t G$, if G contains a subgraph isomorphic to a
subdivision of G. Clearly, $\leq_t$ is a partial order on graphs.

It is however not a well-quasi-order. Let us describe an example. For
simplicity of presentation, let us allow parallel edges in graphs in this
subsection. For every positive integer k, a *Robertson chain* R_k *of length*
k is obtained from a path of length k by doubling the edges. *The ends* of
R_k are the end vertices of the original path. Let a graph G_k be obtained
from R_k by adding four additional vertices of degree one, two adjacent to
each of the ends of the chain. It is easy to see that $G_i \not\leq_t G_j$ for all $i \neq j$.
However, Robertson chains represent the only essential obstruction to the
well-quasi-ordering. Liu and Thomas [45] prove the following, confirming
a conjecture of Robertson.

Theorem 4.4 *Let $k \geq 1$ be an integer. $G_1, G_2, \ldots, G_n, \ldots$ be an infinite
sequence of graphs, such that $R_k \not\leq_t G_i$ for every i. Then there exist
indices $i < j$ such that $G_i \leq_t G_j$.*

The proof of Theorem 4.4 uses, in particular, a recent theorem of
Dvořák [19], which gives a rough structural characterization of graphs not
containing a given graph as a topological minor, which refines an earlier
characterization due to Grohe and Marx [28].

5 Patches

Many arguments in structural graph minor theory proceed by considering the cases when a graph has bounded treewidth separately. As we have
seen in Section 2 graphs of bounded treewidth enjoy a variety of structural
properties. In the proof of Theorem 1.6 however we use a different type of
decomposition, which we describe in this section.

5.1 Definitions

Denote $\{1, 2, \ldots, q\}$ by $[q]$. A *q-patch* or simply a *patch* H is a triple
consisting of a graph, which we will denote by the same letter H, and injective maps $a_H, b_H : [q] \to V(H)$. Two patches H and H' are *isomorphic*
if there exists an isomorphism $\psi : V(H) \to V(H')$ between the corresponding graphs, such that $\psi a_H = a_{H'}$ and $\psi b_H = b_{H'}$. We generally do
not distinguish between isomorphic patches. We denote $\text{Im}(a_H) \cup \text{Im}(b_H)$

New tools in graph minor structure theory

by ∂H. We think of $\text{Im}(a_H)$ as the left boundary of H, and $\text{Im}(b_H)$ as its right boundary.

Let H_1 and H_2 be patches. We define $H_1 \times H_2$ to be a patch obtained from the disjoint union of H_1 and H_2 by identifying the vertices $b_{H_1}(i)$ and $a_{H_2}(i)$ for all $i \in [q]$, and setting $a_{H_1 \times H_2} := a_{H_1}$, $b_{H_1 \times H_2} = b_{H_2}$. Intuitively, the product $H_1 \times H_2$ is obtained by gluing the left boundary of H_2 on to the right boundary of H_1. This product operation is associative, but not commutative. Let id denote the patch with $|V(\text{id})| = q$, $E(\text{id}) = \emptyset$ and $a_{\text{id}}(i) = b_{\text{id}}(i)$ for every $i \in [q]$. The patch id acts as an identity with respect to the product operation defined above.

A patch H is *a minor* of a patch G if there exists a map α mapping vertices of H to disjoint subsets of $V(G)$ satisfying the following properties:

1. $G[\alpha(v)]$ is connected for every $v \in V(H)$;

2. if $v, w \in V(H)$ are joined by an edge then some edge of G joins a vertex of $\alpha(v)$ to a vertex in $\alpha(w)$;

3. $a_G(i) \in \alpha(a_H(i))$ and $b_G(i) \in \alpha(b_H(i))$ for every $i \in [q]$.

Note that the map α satisfying the properties (1) and (2) above exists if and only if the graph G contains H as a minor, and that this definition is essentially identical to the definition of a minor of a rooted graph in Section 4.1. We will write $H \leq G$ if the patch H is a minor of a patch G. It is easy to verify that the minor relation is transitive, and that if H_1, H_2, G_1, G_2 are patches, $H_1 \leq G_1$ and $H_2 \leq G_2$ then $H_1 \times H_2 \leq G_1 \times G_2$. Thus the set of isomorphism classes of patches forms an ordered monoid with respect to multiplication operation and minor quasiorder defined in this section. Note that Theorem 4.3 immediately implies that the minor relation is a well-quasi-order and, in particular, the following corollary holds.

Corollary 5.1 *Let $\mathcal{F}$ be a class of q-patches, closed under taking minors. Then there exists a finite collection of graphs q-patches $H_1, H_2, \ldots, H_k$ such that a q-patch G does not belong to $\mathcal{F}$ if and only if G contains at least one of $H_1, H_2, \ldots, H_k$ as a minor.*

A patch H is *linked* if it contains vertex disjoint paths $P_1, P_2, \ldots, P_q$, such that P_i has ends $a_H(i)$ and $b_H(i)$ for $i \in [q]$. Equivalently, H is linked if and only if id $\leq H$. It follows that if H is linked then $H' \leq H \times H'$ and $H' \leq H' \times H$ for every patch H'.

Let $G, H_1, H_2, \ldots, H_l$ be q-patches, such that $G = H_1 \times H_2 \times \ldots \times H_l$ then we say that $\mathcal{H} = (H_1, H_2, \ldots, H_l)$ is a *q-patch decomposition of G length l*. We say that $\mathcal{H}$ is *non-degenerate* if $|V(H_i)| > q$ for $i = 1, \ldots, l$.

For a patch decomposition $\mathcal{H} = (H_1, H_2, \ldots, H_l)$, let $W_i = V(H_i)$, for $i = 1, 2, \ldots, l$, let $W_0 = \mathrm{Im}(a_{H_1})$ and let $W_{l+1} = \mathrm{Im}(b_{H_l})$. Let $\mathcal{W} = (W_0, \ldots, W_{l+1})$. If $\mathcal{H}$ is a patch decomposition of a patch G, then $\mathcal{W}$ is a *linear decomposition* of the graph G, as defined, for example, in [38]. The following result is proved in [58] using a different terminology.

Theorem 5.2 *For all integers l, w there exists an integer n such that for every graph G with $|V(G)| \geq n$ and $\mathrm{tw}(G) \leq w$ then for some $q \leq w$ there exists a non-degenerate linked q-patch decomposition of G of length l.*

For a patch G let *the toolbox* $T(G)$ of G be the set of all equivalence classes of minors of G on at most $2q$ vertices. We say that a linked patch G is *universal* if for all patches H, H_1, H_2 such that $H_1, H_2 \in T(G)$, $H \leq H_1 \times H_2$ and $|V(H)| \leq 2q$, we have $H \in T(G)$. Let $\mathcal{H} = (H_1, H_2, \ldots, H_l)$ be a non-degenerate patch decomposition of G. For $1 \leq i \leq j \leq l-1$ define $H_{ij} = H_i \times H_{i+1} \times \ldots \times H_j$. We say that $\mathcal{H}$ is *universal* if the toolbox $T(H_{ij})$ is the same for all $i, j \in \{1, 2, \ldots, l\}$ and, moreover, for all such pairs i, j the patch H_{ij} is universal. The advantage of considering universal patch decompositions is the following: If a patch K with $|V(K)| \leq 2q$ is a minor of a patch G, and G admits a universal patch decomposition of length l then K^l obtained by multiplying the patch K by itself l times is a minor of G. A Ramsey theoretic argument is used in [54] to refine Theorem 5.2 to produce universal patch decompositions as follows.

Lemma 5.3 *For all integers $q, l \geq 0$ there exists an integer L with the following properties. Let $\mathcal{H}$ be a non-degenerate linked q-patch decomposition of a q-patch G of length L, then there exist linked patches H, H', H'' such that $G = H' \times H \times H''$ and H admits a universal patch decomposition of length l.*

5.2 Patch decompositions of t-connected graphs

We say that a patch H is *internally t-connected* if for every separation (A, B) of H with $|A \cap B| < t$, $B - A \neq \emptyset$, $A - B \neq \emptyset$, we have $\partial H \cap (B - A) \neq \emptyset$ and $\partial H \cap (A - B) \neq \emptyset$. We now describe a family of patches, such that each internally t-connected patch H, admitting a sufficiently long patch decomposition will necessarily contain a member of this family as a minor. We say that a q-patch K is *t-quasicomplete* if there exist $S_1, S_2, \ldots, S_t \subseteq V(K)$, such that

- there exists an edge of K with one end in S_i and another in S_j for all $1 \leq i < j \leq t$;

- for every $i \in [t]$ we have $|S_i| \le 2$ and, if $|S_i| = 2$, then $S_i = \{a_H(j), b_H(j)\}$ for some $j \in [q]$;

- If $\{i \in [t] \mid |S_i| = 2\} \ne \emptyset$ then

$$q - |\{j \in [q] \mid a_K(j) = b_K(j) \notin \cup_{i \in t} S_i\}| + |\{i \in [t] \mid |S_i| = 2\}| \le t.$$

A patch P is a t-*pinwheel* if there exist distinct $i_1, i_2, j_1, j_2, \ldots, j_{t-5} \in [q]$ and $x \in V(P)$ such that

- $a_P(i_s)b_P(i_s), a_P(i_s)x, a_P(j_r)x \in E(P)$ for every $s \in \{1,2\}, r \in [t-5]$;

- $a_P(k) = b_P(k)$ for $k \in [q] - \{i_1, i_2\}$.

Finally, a patch F is a t-*fan* if there exist distinct $i_1, i_2, j_1, j_2, \ldots, j_{t-4} \in [q]$ such that

- $a_F(i_1)b_F(i_1), a_F(i_2)b_F(i_2), a_F(i_1)a_F(i_2) \in E(F)$;

- $a_F(i_1)a_F(j_r) \in E(F)$ for every $r \in [t-4]$;

- $a_F(k) = b_F(k)$ for $k \in [q] - \{i_1, i_2\}$.

We are now ready to state the main technical result of [54]. It is used in the proof of Theorem 1.6 for graphs of bounded treewidth.

Theorem 5.4 *For all positive integers t, q, d there exists a positive integer L, satisfying the following. Let G be an internally t-connected q-patch. Suppose that there exists a non-degenerate patch decomposition of G of length L. Then G contains a t-quasicomplete patch, a t-pinwheel pinwheel or a t-fan patch as a minor.*

To obtain a K_t minor from the patches produced as outcomes of Theorem 5.4, an additional tool is necessary. It requires the t-connectivity of the graph, as opposed to the internal t-connectivity of a patch. The following lemma is a reformulation of [38, Lemma 3.7].

Lemma 5.5 *For all positive integers k, q, t there exists an integer $L \ge 3$ such that the following holds. Let G be a t-connected graph. Suppose that G admits a non-degenerate patch decomposition $\mathcal{H} = (H_1, H_2, \ldots, H_L)$ such that the decomposition $(H_2, \ldots, H_{L-1})$ is universal. Let R be a linked minor of H_2 such that for some integers $s_1 \le s_2$ we have*

- $s_1 + s_2 \le t$

- $a_R(i) \ne b_R(i)$ for $1 \le i \le s_1$,

- $a_R(i) = b_R(i)$ *for* $s_2 < i \leq q$.

Then G contains as a minor a graph obtained from R^k (the patch obtained by multiplying R by itself $k-1$ times) by deleting $\{a_{R'}(i)\}_{i>s_2}$, and joining vertices $\{a_{R'}(i)\}_{i=1}^{s_1}$ to vertices $\{b_{R'}(i)\}_{i=1}^{s_1}$ by a matching of size s_1.

While Lemma 5.5 does not immediately apply to the case, when R is a quasicomplete minor, different techniques allow us to find a K_t minor contained in G in this case. If G contains a t-pinwheel or a t-fan patch as a minor, however, then applying Lemma 5.5 with $s_1 = 2$, $s_2 = t - 2$, in combination with the preceding arguments allows one to deduce the following useful theorem, which is preceded by a few definitions.

Let C_1 and C_2 be two vertex-disjoint cycles of length $k \geq 3$ with vertex-sets $\{x_1, x_2, \ldots, x_k\}$ and $\{y_1, y_2, \ldots, y_k\}$ (in order), respectively. For $t \geq 6$, *the t-pinwheel with k vanes and spokes $\{p_1, p_2, \ldots, p_{t-5}\}$* is the graph obtained from the union of C_1 and C_2 by adding a vertex z_i adjacent to x_i and y_i for each $i = 1, \ldots, k$, and, finally, adding vertices $\{p_1, p_2, \ldots, p_{t-5}\}$ adjacent to each of the vertices $\{z_1, z_2, \ldots, z_k\}$. Similarly, we obtain a *t-fan with k blades* from the union of C_1 and C_2 by joining x_i to y_i for each $i = 1, \ldots, k$, and adding vertices $\{f_1, f_2, \ldots, f_{t-4}\}$ adjacent to each of the vertices $\{x_1, x_2, \ldots, x_k\}$. The results described in this section imply the following.

Theorem 5.6 *For all positive integers $t \geq 6, k \geq 3$ and w there exists an integer $N = N(t, k, w)$ such that every t-connected graph G with $|V(G)| \geq N$ and $\mathrm{tw}(G) \leq w$ contains either K_t, t-pinwheel with k vanes, or t-fan with k blades, as a minor.*

Additional tools can be used to omit the assumption that the treewidth of G is bounded, but their description is exceedingly technical even for this survey. To derive Theorem 1.6 from Theorem 5.6 one needs to consider extensions of fans and pinwheels, which cannot be made planar by removing $t - 5$ vertices. We discuss the techniques for analyzing these extensions in the next section.

6 Breaking planarity

6.1 Non-planar extensions of subdivisions of planar graphs

Recall that a graph is a *subdivision* of another if the first can be obtained from the second by replacing each edge by a non-zero length path with the same ends, where the paths are disjoint, except possibly for shared ends. The replacement paths are called *segments*, and their ends are called *branch-vertices.* Let G, S, H be graphs such that S is a subgraph of H and is isomorphic to a subdivision of G. In that case we say that S is a G-*subdivision* in H. If G has no vertices of degree two (which will be the case in our applications), then the segments and branch-vertices of S are uniquely determined by S. An S-*path* is a path of length at least one with both ends in S and otherwise disjoint from S. A graph G is *internally 4-connected* if it is 3-connected and for every separation (A, B) of order three one of $G[A]$ and $G[B]$ has at most three edges.

Let a non-planar graph H have a subgraph S isomorphic to a subdivision of a planar graph G. For various problems in structural graph theory it is useful to know the minimal subgraphs of H that have a subgraph isomorphic to a subdivision of G and are non-planar. Robertson, Seymour and Thomas (see [56]) have shown that under some mild connectivity assumptions these "minimal non-planar extensions" of G are quite nice:

Theorem 6.1 *Let G be an internally 4-connected planar graph on at least seven vertices, let H be an internally 4-connected non-planar graph, and let there exist a G-subdivision in H. Then there exists a G-subdivision S in H such that one of the following conditions holds:*

(i) *there exists an S-path in H joining two vertices of S not incident with the same face, or*

(ii) *there exist two disjoint S-paths with ends s_1, t_1 and s_2, t_2, respectively, such that the vertices s_1, s_2, t_1, t_2 belong to some face boundary of S in the order listed. Moreover, for $i = 1, 2$ the vertices s_i and t_i do not belong to the same segment of S, and if two segments of S include all of s_1, t_1, s_2, t_2, then those segments are vertex-disjoint.*

The connectivity assumptions guarantee that the face boundaries in a planar embedding of S are uniquely determined, and hence it makes sense to speak about incidence with faces.

If G is a graph and $u, v \in V(G)$ are not adjacent, then by $G + uv$ we denote the graph obtained from G by adding an edge with ends u and v. The following is a consequence of Theorem 6.1.

Theorem 6.2 *Let G be an internally 4-connected triangle-free planar graph, and let H be an internally 4-connected non-planar graph such that H has a subgraph isomorphic to a subdivision of G. Then there exists a graph G' such that G' is isomorphic to a minor of H, and either*

(i) *$G' = G + uv$ for some vertices $u, v \in V(G)$ such that no facial cycle of G contains both u and v, or*

(ii) $G' = G + u_1v_1 + u_2v_2$ *for some distinct vertices* $u_1, u_2, v_1, v_2 \in V(G)$ *such that* u_1, u_2, v_1, v_2 *appear on some facial cycle of* G *in the order listed.*

While the statement of Theorem 6.2 is nicer, it has the drawback that we assume that H has a subgraph isomorphic to a *subdivision* of G, and deduce that it has only a *minor* isomorphic to G'.

A close relative of Theorem 6.1 is used in [15] to show that for every positive integer k, there is an integer N such that every 4-connected non-planar graph with at least N vertices has a minor isomorphic to the complete bipartite graph $K_{4,k}$, or the graph obtained from a cycle of length $2k + 1$ by adding an edge joining every pair of vertices at distance exactly k, or the graph obtained from a cycle of length k by adding two vertices adjacent to each other and to every vertex on the cycle. Using this Bokal, Oporowski, Richter and Salazar [6] proved that, except for one well-defined infinite family, there are only finitely many graphs of crossing number at least two that are minimal in a specified sense.

6.2 Non-apex extensions of apex graphs

Let G be a graph. By a *mold* for G we mean a collection $Z = (Z_e : e \in F)$ of (not necessarily disjoint) sets, where $F \subseteq E(G)$ and each Z_e is disjoint from $V(G)$. Given a mold Z, for G we define a new graph L, as follows. We add the elements of $\bigcup_{e \in F} Z_e$ as new vertices. We subdivide each edge $e \in F$ exactly once, denoting the new vertex by $\hat{e}$. Finally, for every $e \in F$ and every $z \in Z_e$ we add an edge between z and $\hat{e}$. We say that L is the graph *determined by G and Z*.

Suppose that there exists a subdivision of L in H, assume that G is planar, but that the graph obtained from H by deleting the vertices that correspond to $\bigcup_{e \in F} Z_e$ is not. In the proof of Theorem 1.6 we need to extend the results of Section 6.1 to this scenario. It is done in [56] as follows.

Let $\eta : G \hookrightarrow S \subseteq H$ denote that S is a G-subdivision in H, and η maps vertices of G to vertices of S and edges of G to the corresponding paths of S.

Theorem 6.3 *Let G be an internally 4-connected triangle-free planar graph not isomorphic to the cube, let $F \subseteq E(G)$ be such that no two elements of F belong to the same facial cycle of G, let $Z = (Z_e : e \in F)$ be a mold for G, and let L be the graph determined by G and Z. Let H be a graph, and let $\eta : L \hookrightarrow S \subseteq H$. Let $H' := H - \bigcup_{e \in F} \eta(Z_e)$. If H' is internally 4-connected and non-planar, then there exists a set $F' \subseteq F$ with $|F - F'| \leq 1$*

such that the graph L' determined by G and $(Z_e : e \in F')$ satisfies one of the following conditions:

- *there exist vertices $u, v \in V(L) - V(Z)$ that do not belong to the same facial cycle of $L \backslash V(Z)$ such that $L + uv$ is isomorphic to a minor of H,*

- *there exists a facial cycle C of $L - V(Z)$ and distinct vertices $u_1, u_2, v_1, v_2 \in V(C)$ appearing on C in the order listed such that $L + u_1v_1 + u_2v_2$ is isomorphic to a minor of H, and $u_iv_i \notin E(G)$ for $i = 1, 2$.*

Theorem 6.3 closely parallels Theorem 6.2, except for the technical condition that the edges of G, that the elements of Z attach to, are required to be non-cofacial, and breaking planarity requires us to sacrifice attachments of Z to one of the edges.

We are finally ready to demonstrate an application of the theory outlined in this section to the proof of Theorem 1.6.

For $k \geq 3$, let G be *a planar ladder of length $2k$*, that is G obtained from vertex-disjoint cycles C_1 and C_2 of length $2k$ with vertex-sets $\{x_1, x_2, \ldots, x_{2k}\}$ and $\{y_1, y_2, \ldots, y_{2k}\}$ (in order), respectively, by joining x_i to y_i for $1 \leq i \leq 2k$. Let $F = \{x_1y_1, x_3y_3, \ldots, x_{2k-1}y_{2k-1}\}$, and let $Z_e = \{z_1, z_2, \ldots, z_{t-5}\}$ for some $t \geq 6$ and for every $e \in F$. Then the graph L determined by G and $Z = (Z_e : e \in F)$ contains a subdivision of a t-pinwheel with k vains as a subgraph. Theorem 6.3 is applicable to G and allows one to deduce the following, which is an important step in the proof of Theorem 1.6.

Corollary 6.4 *For every $t \geq 6$ there exists k such that the following holds. Let H be a graph, and let $\eta : P \hookrightarrow S \subseteq H$ be an embedding of a t-pinwheel P with k vanes and spokes $\{p_1, \ldots, p_{t-5}\}$ in H. If $H' = H - \eta(\{p_1, \ldots, p_{t-5}\})$ is internally 4-connected, then either H' is planar, or H contains K_t as a minor.*

7   Separators and expansion

An important property of proper minor-closed classes of graphs is the existence of sublinear separators. Recall that a separation (A, B) of a graph G on n vertices was defined to be balanced in Section 2.5 if $|A \backslash B| \leq 2n/3$ and $|B \backslash A| \leq 2n/3$.

We say that a set $X \subseteq V(G)$ is *a balanced separator of a graph G* if there exists a balanced separation (A, B) of G with $X = A \cap B$. The first

major result on existence of separators in minor-closed classes is a theorem
of Lipton and Tarjan [44] for planar graphs.

Theorem 7.1 *Let G be a planar graph then G contains a balanced separator of size at most $2\sqrt{2|V(G)|}$.*

This result was later generalized to arbitrary minor-closed classes by
Alon, Seymour and Thomas in [1] as follows.

Theorem 7.2 *Let G be a graph with no K_h minor then G contains a balanced separator of size at most $h^{3/2}\sqrt{|V(G)|}$.*

Theorem 7.2 has numerous algorithmic applications, but we will present
an application of a slightly different kind, to enumeration of graphs in a
minor-closed class. A class of graphs $\mathcal{F}$ is called *small* if there exists
a constant c such that the number of graphs in $\mathcal{F}$ with the vertex set
$\{1, 2, \ldots, n\}$ is at most $c^n n!$ for all integers $n \geq 1$. In [57] the authors
show that every proper minor-closed class of graphs is small, answering a
question of Welsh. The proof in [57] does not use separators, but rather a
bound on the number of cliques in a graph in a minor closed class. In [22],
Dvořák and Norin, however, prove the following stronger result.

Theorem 7.3 *Let $\mathcal{F}$ be a class of graphs closed under taking induced subgraphs such, such that every graph $G \in \mathcal{F}$ with $n \geq 3$ vertices has a balanced separator of size at most $\frac{cn}{(\log n \log \log n)^2}$. Then $\mathcal{F}$ is small.*

Small graph classes are of interest from the point of view of probabilistic
graph theory. McDiarmid, Steger and Welsh [50] describe a number of
properties of a random labelled n-vertex graph in a small *addable* graph
class. By the result of [57] this includes those minor-closed graph classes
for which every excluded minor is 2-connected, in particular the class of
all graphs which do not contain K_t as a minor, for $t \geq 3$.

Let us now briefly describe one potential approach to a proof of Theorem 7.2. The original proof does not explicitly rely on the Graph Minor
Structure theorem, although it uses many ideas which are used in its proof.
The following result, however, essentially strengthens Theorem 7.2, and is
proved in [10] using Theorem 3.3.

Theorem 7.4 *Let $\mathcal{F}$ be a proper minor-closed class of graphs. Then there
exist constants a and b such that for every integer $l > 0$ every graph $G \in \mathcal{F}$
has a vertex l-coloring so that the subgraph of G induced by the union of
any $j < l$ colors has treewidth at most $aj + b$.*

In [51] a variant of Theorem 7.4 is extended to a more general context as follows. In [51] a class of graph $\mathcal{F}$ is defined to have *a low treewidth coloring* if for every $l \geq 1$ there exists an integer $N(l)$ such that every graph $G \in \mathcal{F}$ has a vertex $N(l)$-coloring in which subgraphs induced by the union of any $1 \leq j < l$ colors have treewidth at most $j-1$. It is proved in [10] that proper minor-closed classes have a low treewidth coloring.

A *depth r minor of a graph* G is obtained from G by contracting vertex disjoint subgraphs of radius at most r. The $\nabla_r(G)$ is defined as

$$\max\left\{ \frac{|E(H)|}{|V(H)|} \ : \ G \text{ is a depth } r \text{ minor of } G \right\}.$$

We say that a class of graphs $\mathcal{F}$ *has bounded expansion* if for every integer $r \geq 0$ there exists an integer $f(r)$ such that $\nabla_r(G) \leq f(r)$ for every $G \in \mathcal{F}$. As we will discuss in Section 8, if $\mathcal{F}$ is a proper minor-closed class then there exists an integer c such that $\nabla_r(G) \leq c$ for every $G \in \mathcal{F}$. Thus proper minor-closed graph classes have bounded expansion, but so do many other classes, e.g. graphs which can be drawn in the plane with crossings in such a way that every edge crosses at most one other edge. (See [52] for further examples.)

We are now ready to state a result from [51].

Theorem 7.5 *A class of graphs has a low treewidth coloring if and only if it has bounded expansion.*

In [51] several other characterizations of classes of bounded expansion are given.

Let us return to our earlier discussion and show that Theorem 7.4 indeed implies a version of Theorem 7.2. Let $\mathcal{F}$ be a proper minor-closed class of graphs, and let a and b be as in Theorem 7.4. For $G \in \mathcal{F}$ with $|V(G)| = n$, let $l = \lceil \sqrt{n} \rceil$, and let $V_1, \ldots, V_l$ be the color classes in the vertex coloring of G satisfying the conditions of Theorem 7.4. Without loss of generality $|V_1| \leq \sqrt{n}$. As mentioned in Section 2.5, we have

$$\mathrm{sn}(G \setminus V_1) \leq \mathrm{tw}(G \setminus V_1) + 1 \leq a\sqrt{n} + a + b + 1.$$

Thus G has a balanced separator of size at most $(a+1)\sqrt{n} + a + b + 1$.

Theorem 7.4 in turn allows for an elegant proof via the notion of *tree-breadth*, recently introduced by Dujmovic, Morin and Wood [18], which is defined as follows. A *layering* of a graph G is a partition $(L_0, L_1, \ldots, L_k)$ of $V(G)$ such that for every edge vw of G with $v \in L_i$, $w \in L_j$, we have $|i-j| \leq 1$. The set V_i are called *layers*. The *tree-breadth* $\mathrm{tb}(G)$ of a graph G is a minimum integer l such that there exists a tree-decomposition $(T, \mathcal{W})$, and a layering $(L_0, L_1, \ldots, L_k)$ of G, such that $|W_t \cap L_i| \leq l$ for every $t \in V(T)$ and every layer L_i. Essentially, tree-breadth of a graph G is small if it has a layering and a tree-decomposition which are close to being "orthogonal", in a sense that the intersection of every layer with every bag of the tree decomposition is small. Clearly, $\mathrm{tb}(G) \leq \mathrm{tw}(G) + 1$, as one can consider a layering of G with one layer. One can readily see, however, that $\mathrm{tb}(G)$ and $\mathrm{tw}(G)$ are not tied, as grid graphs have arbitrarily large treewidth and tree-breadth at most 2.

Recall that a graph H is an apex if $H - v$ is planar for some $v \in V(H)$. Minor-closed classes of bounded tree-breadth are characterized in [18] as follows.

Theorem 7.6 *Let $\mathcal{F}$ be a minor-closed class of graphs. Then the following are equivalent*

(a) *there exists an integer k such that $\mathrm{tb}(G) \leq k$ for every $G \in \mathcal{F}$, and*

(b) *there exists an apex H such that $H \notin \mathcal{F}$.*

The proof of Theorem 7.6 uses the tree-cotree decomposition of graphs embedded on surfaces, inspired by the ideas of Eppstein [24], and the apex-minor free case of Theorem 3.4.

To see a connection between the tree-breadth and Theorem 7.4 note that if G is a graph and $l > 0$ an integer then G admits a vertex l-coloring so that the subgraph of G induced by the union of any $j < l$ colors has treewidth at most $\mathrm{tb}(G)j + 1$. Indeed, one can consider the layering $(L_0, L_1, \ldots, L_k)$ of G, which achieves the tree breadth bound and color the vertices of L_i in the color $i \pmod l$. Thus Theorem 7.6 implies Theorem 7.4 for minor-closed classes which exclude some apex. However, as shown in [10] using Theorem 3.3, it is not hard to reduce the proof of Theorem 7.4 in general to such classes.

Finally, let us note that Dvořák [20] considered the connection between existence of sublinear separators and decompositions similar to the one obtained in Theorem 7.4 for classes of bounded expansion. As a consequence he proved the following result.

Theorem 7.7 *Let $\mathcal{F}$ be a class of graphs with maximum degree bounded by some constant Δ. Suppose that $\mathcal{F}$ has strongly sublinear separators, that is there exist constants $1 > \delta \geq 0$ and C such that every graph $G \in \mathcal{F}$ has a balanced separator of size $\leq C|V(G)|^\delta$. Then there exists $\gamma > 0$ such that*

$$\nabla_r(G) \leq \gamma e^{r^{3/4}},$$

for every $G \in \mathcal{F}$.

8   Densities of minor-closed graph families

We denote the average degree of a graph G by $d(G) := 2|E(G)|/|V(G)|$. The maximum average degree of graphs in minor closed classes has been the subject of a substantial amount of research in Graph Minor Theory. Let us introduce the necessary notation. For a family of graphs $\mathcal{F}$ we define *the limiting density* as

$$\text{limd}(\mathcal{F}) := \limsup_{n \to \infty} \max_{G \in \mathcal{F}, |V(G)|=n} d(G),$$

that is $\text{limd}(\mathcal{F})$ is the minimum constant c such that every graph G in $\mathcal{F}$ satisfies $|E(G)| \leq (c + o_{|V(G)|}(1))|V(G)|/2$. Let $Ex_m(H)$ denote the family of all H-minor free graphs. For brevity, we denote $\text{limd}(Ex_m(H))$ by $f(H)$ in this section.

Investigation of the function $f(H)$ originated with works of Wagner [94] and Mader [47]. In [47] it is shown that $f(H)$ is finite, meaning that the average density of graphs in minor closed classes is upper bounded by a constant. Kostochka [40,41] and Thomason [89] independently determined the approximate order of magnitude of $f(H)$ for complete graphs. They showed that there exist constants $c_1 > c_2 > 0$ such that every graph G with $d(G) \geq c_1 t \sqrt{\log t}$ contains a K_t minor, while there exists a K_t-minor free graph G_t with $d(G_t) \geq c_2 t \sqrt{\log t}$ for every $t \geq 1$. Thomason [90] further improved this result by precisely determining the asymptotics of $f(K_t)$:

$$f(K_t) = (\alpha + o_t(1))t\sqrt{\log t},$$

for an explicit constant $\alpha = 0.638....$

The quantity $\text{limd}(\mathcal{F})$ have been determined exactly for several other minor-closed graph classes. Let us start by stating several classical results using the terminology of this section. We have $\text{limd}(\mathcal{T}) = 2$, $\text{limd}(\mathcal{S}) = 4$ and $\text{limd}(\mathcal{P}) = 6$, where $\mathcal{T}, \mathcal{S}$ and $\mathcal{P}$ are classes of forests, series-parallel graphs and planar graphs, respectively. Dirac [17] has shown that $f(t) = 2t - 4$ for $t \leq 5$, and Mader [47] extended this result to $t \leq 7$, while Jorgensen [34], and Song and Thomas [87] further extended it to $t = 8$ and 9, respectively. In [8] Chudnovsky, Reed and Seymour have shown that $f(K_{2,t}) = t + 1$, for every $t \geq 2$. Kostochka and Prince [42] have shown that $f(K_{3,t}) = t + 3$, for every $t \geq 6300$.

An intriguing extension of this line of research has recently been instigated by Eppstein [25]. Rather than studying the value of $f(H)$ for a particular graph H, Eppstein proposed an investigation of the set of possible values of $\text{limd}(\mathcal{F})$ over proper minor-closed graph classes, which we will denote by $\mathcal{D}$. As a tool for this investigation, Eppstein considers density-minimal graphs, where a graph G is *density-minimal* if $d(H) \leq d(G)$ for every minor H of G. The following theorem from [25] demonstrates the relevance of density-minimal graphs.

Theorem 8.1 *The set $\mathcal{D}$ is the closure of the set of densities of density-minimal graphs.*

We include the proof by Eppstein of the folllowing result on the structure of $\mathcal{D}$, as an illustration of an elegant application of Corollary 4.2.

Theorem 8.2 *The set $\mathcal{D}$ is well-ordered.*

Proof It suffices to show that for every $\Delta \geq 0$ there exists $\delta > 0$ such that $\mathcal{D}$ does not intersect the open interval $(\Delta, \Delta + \delta)$. By Theorem 8.1 it suffices to show that this interval contains no densities of density-minimal graphs.

Let $\mathcal{F}$ be the family of all graphs F such that $\mathrm{d}(F') \leq \Delta$ for every minor F' of F. Clearly, $\mathcal{F}$ is minor-closed, and thus by Corollary 4.2 there exists a finite collection of graphs $H_1, H_2, \ldots, H_k$ such that every graph not belonging to $\mathcal{F}$ contains one of them as a minor. In particular $\mathrm{d}(H_i) > \Delta$ for every $1 \leq i \leq k$. Let $\delta := \min\{\mathrm{d}(H_i) - \Delta : \ |1 \leq i \leq k\}$. Then every density-minimal graph either belongs to $\mathcal{F}$ and has density at most Δ, or contains one of $H_1, H_2, \ldots, H_k$ as a minor and has density at least $\Delta + \delta$, as desired. $\qquad\Box$

Very recently, answering a question of Eppstein from [25], Kapadia and Norin [35] proved the following.

Theorem 8.3 *The density $\mathrm{limd}(\mathcal{F})$ is rational for every proper minor-closed class of graphs.*

Let us give a glimpse into the proof of Theorem 8.3, which uses the terminology introduced in Section 5. In [35] the following stronger result is proved confirming a conjecture of Norin, stated in [27].

Theorem 8.4 *Let $\mathcal{F}$ be a proper minor-closed class of graphs. Then there exists $q \geq 0$ and a q-patch H with $|V(H)| > q$ such that $H^n \in \mathcal{F}$ for every positive integer n and*

$$\mathrm{limd}(\mathcal{F}) = \lim_{n \to \infty} \mathrm{d}(H^n).$$

Note that it is easy to see that Theorem 8.4 indeed implies Theorem 8.3, as one can assume that vertices $\mathrm{Im}(a_H)$ form an independent set in H and in such a case

$$\lim_{n \to \infty} \mathrm{d}(H^n) = \frac{2|E(H)|}{|V(H)| - q}.$$

Theorem 8.4 can be informally restated as implying that the density of every proper minor-closed graph class is achieved by gluing copies of a fixed graph to itself in a linear fashion.

The proof of Theorem 8.4 just as the proof of Theorem 8.2 relies on well-quasi-ordering of graphs under minor relation, but rather than Corollary 4.2, it uses Corollary 5.1.

Several fundamental questions concerning densities of minor-closed families remain open. First, one can refine the notion of density of a minor-closed class defining

$$d(\mathcal{F}, n) = \max_{G \in \mathcal{F}, |V(G)| = n} d(G).$$

Geelen, Gerards and Whittle conjecture a generalization of the following in [27].

Conjecture 8.5 *Let $\mathcal{F}$ be a proper minor-closed class of graphs. Then there exist rational $a, b_0, b_1, \ldots, b_{t-1}$ such that for all sufficiently large n we have*

$$d(\mathcal{F}, n) = a + \frac{b_i}{n},$$

where $i \equiv n \pmod{t}$.

Conjecture 8.5 is implied by the positive answer to the following question, which would also strengthen Theorem 8.4.

Question 8.6 *Let $\mathcal{F}$ be a proper minor-closed class of graphs. Does there necessarily exist a positive integer s, such that for every positive integer n there exist positive integers q and k and q-patches H_l, H and H_r with $|V(H)|, |V(H_l)|, |V(H_r)| \leq s$, $|V(H_l \times H^k \times H_r)| = n$ and*

$$d(\mathcal{F}, n) = \frac{2|E(H_l \times H^k \times H_r)|}{n}?$$

Essentially, in the question above we are asking whether it is true that the graph in $\mathcal{F}$ on n vertices with the maximum number of edges can be obtained as in Theorem 8.4: by gluing some fixed graph H to itself in a linear fashion.

One of the more intriguing questions in [25], which remains open concerns the structure of cluster points of $\mathcal{D}$. (A point in $\Delta \in \mathcal{D}$ is *a cluster point* if it lies in the closure of $\mathcal{D} - \{\Delta\}$.) Eppstein proves that $\Delta + 1$ is a cluster point of $\mathcal{D}$ for every $\Delta \in \mathcal{D}$ and asks whether the converse holds.

Question 8.7 *Is it true that if $\Delta + 1$ is a cluster point of $\mathcal{D}$ then $\Delta \in \mathcal{D}$?*

As noted in [25], it is not hard to derive from Theorem 8.1 that the positive answer to Question 8.7 implies Theorem 8.3.

It appears quite possible that the structure of the set $\mathcal{D}$ is rather involved. In particular we do not know the answers to the following decidability questions.

Question 8.8   *1. Is the problem of the membership in $\mathcal{D}$ decidable?*

2. Does there exist an algorithm which, given graphs $H_1, H_2, \ldots, H_k$ determines limd $\mathcal{F}$, *where $\mathcal{F}$ is the class of graphs containing none of $H_1, H_2, \ldots, H_k$ as a minor?*

Finally, very little is known about limiting densities of graph classes closed under topological minors.

Question 8.9 *Is* limd$(\mathcal{F})$ *rational for every proper class of graphs closed under isomorphisms and taking topological minors?*

9   Further open problems

The following conjecture of Hadwiger [29] is considered by many as one of the most important open problems in graph theory.

Conjecture 9.1 *For every $k \geq 1$, every graph with chromatic number at least k contains K_k as a minor.*

Hadwiger's conjecture was proved for $k = 4$ by Hadwiger [29] and Dirac [16]. Wagner [93] proved that for $k = 5$ the conjecture is equivalent to the Four Color Theorem [2,3,60]. Robertson, Seymour and Thomas [85] have proved Hadwiger's conjecture for $k = 6$, also by reducing it to the Four Color Theorem. Dirac [16] and Mader [47] proved that any minimal counterexample to Hadwiger's conjecture is k-connected for $k \leq 5$, and for $k \leq 7$, respectively. Thus, in particular, Conjecture 1.4 would imply Hadwiger's conjecture for $k = 6$, and the following conjecture from [37] would imply the first open case of Hadwiger's conjecture.

Conjecture 9.2 *Let G be a 7-connected graph. Then $G - \{u, v\}$ is planar for some $u, v \in V(G)$.*

Several weakenings of Hadwiger's conjecture are wide open, in particular, the following conjecture stated in [37] seems to be out of reach.

Conjecture 9.3 *There exists a constant C such that for every integer $k \geq 0$ every graph with chromatic number at least Ck contains K_k as a minor.*

Besides the questions listed at the end of Section 8, several conjectures concern extremal properties of minor-closed classes. In particular, the following conjecture of Seymour and Thomas is closely related to Theorem 1.6.

Conjecture 9.4 *For every $t \geq 1$ there exists constant $N = N(t)$ such that every $(t-2)$-connected graph G on at least N vertices with*

$$|E(G)| \geq (t-2)|V(G)| - \frac{t(t-1)}{2} + 1$$

contains K_t as a minor.

(Note that Theorem 1.6 implies that Conjecture 9.4 holds for t-connected graphs.)

If H is a union of k vertex disjoint triangles then it follows from a result of Corradi and Hajnal [9] that $\text{limd}(Ex_m(H)) = 4k - 2$. Reed and Wood [59] conjectured that this result generalizes, as follows.

Conjecture 9.5 *Let H be a 2-regular graph. Then every graph with average degree at least $\frac{4}{3}|V(H)| - 2$ contains H as a minor. In particular,*

$$\text{limd}(Ex_m(H)) \leq \frac{4}{3}|V(H)| - 2.$$

A recent paper [32] contains several refinements of the above conjecture.

References

[1] N. Alon, P. Seymour, and R. Thomas, *A separator theorem for nonplanar graphs*, J. Amer. Math. Soc. **3** (1990), no. 4, 801–808.

[2] K. Appel and W. Haken, *Every planar map is four colorable. I. Discharging*, Illinois J. Math. **21** (1977), no. 3, 429–490.

[3] K. Appel, W. Haken, and J. Koch, *Every planar map is four colorable. II. Reducibility*, Illinois J. Math. **21** (1977), no. 3, 491–567.

[4] M. Baker and S. Norine, *Riemann–Roch and Abel–Jacobi theory on a finite graph*, Advances in Mathematics **215** (2007), no. 2, 766–788.

[5] H. L. Bodlaender, J. R. Gilbert, H. Hafsteinsson, and T. Kloks, *Approximating treewidth, pathwidth, frontsize, and shortest elimination tree*, J. Algorithms **18** (1995), no. 2, 238–255.

[6] D. Bokal, B. Oporowski, B. Richter, and G. Salazar, *Characterizing 2-crossing-critical graphs*, 2013. arXiv:1312.3712.

[7] C. Chekuri and J. Chuzhoy, *Polynomial bounds for the grid-minor theorem*, 2013. arXiv:1305.6577.

[8] M. Chudnovsky, B. Reed, and P. Seymour, *The edge-density for $K_{2,t}$ minors*, J. Combin. Theory Ser. B **101** (2011), no. 1, 18–46.

[9] K. Corradi and A. Hajnal, *On the maximal number of independent circuits in a graph*, Acta Math. Acad. Sci. Hungar. **14** (1963), 423–439. MR0200185 (34 #84)

[10] M. DeVos, G. Ding, B. Oporowski, D. P. Sanders, B. Reed, P. Seymour, and D. Vertigan, *Excluding any graph as a minor allows a low tree-width 2-coloring*, J. Combin. Theory Ser. B **91** (2004), no. 1, 25–41.

[11] R. Diestel, *Graph theory*, Fourth Edition, Graduate Texts in Mathematics, vol. 173, Springer, Heidelberg, 2010.

[12] R. Diestel, T. R. Jensen, K. Yu. Gorbunov, and C. Thomassen, *Highly connected sets and the excluded grid theorem*, J. Combin. Theory Ser. B **75** (1999), no. 1, 61–73.

[13] R. Diestel, K. Kawarabayashi, T. Müller, and P. Wollan, *On the excluded minor structure theorem for graphs of large tree-width*, J. Combin. Theory Ser. B **102** (2012), no. 6, 1189–1210.

[14] R. Diestel, K. Kawarabayashi, and P. Wollan, *The Erdős-Pósa property for clique minors in highly connected graphs*, J. Combin. Theory Ser. B **102** (2012), no. 2, 454–469.

[15] G. Ding, B. Oporowski, R. Thomas, and D. Vertigan, *Large nonplanar graphs and an application to crossing-critical graphs*, J. Combin. Theory Ser. B **101** (2011), no. 2, 111–121.

[16] G. A. Dirac, *A property of 4-chromatic graphs and some remarks on critical graphs*, J. London Math. Soc. **27** (1952), 85–92.

[17] G. A. Dirac, *Homomorphism theorems for graphs*, Math. Ann. **153** (1964), 69–80.

[18] V. Dujmović, P. Morin, and D. R. Wood, *Layered separators in minor-closed families with applications*, 2013. arXiv:1306.1595.

[19] Z. Dvořák, *A stronger structure theorem for excluded topological minors*, 2012. arXiv:1209.0129.

[20] Z. Dvořák, *Sublinear separators, fragility and subexponential expansion*, 2014. arXiv:1404.7219.

[21] Z. Dvořák and R. Thomas, *List-coloring apex-minor-free graphs*, 2014. arXiv:1401.1399.

[22] Z. Dvořák and S. Norine, *Small graph classes and bounded expansion*, J. Combin. Theory Ser. B **100** (2010), no. 2, 171–175.

[23] Z. Dvořák and S. Norin, *Treewidth of graphs with balanced separators*. manuscript.

[24] D. Eppstein, *Diameter and treewidth in minor-closed graph families*, Algorithmica **27** (2000), no. 3-4, 275–291. Treewidth.

[25] D. Eppstein, *Densities of minor-closed graph families*, Electron. J. Combin. **17** (2010), no. 1, Research Paper 136, 21.

[26] J. Fox, *Constructing dense graphs with sublinear Hadwiger number*. J. Combin. Theory Ser. B, to appear.

[27] J. Geelen, B. Gerards, and G. Whittle, *The highly connected matroids in minor-closed classes*, 2013. arXiv:1312.5012.

[28] M. Grohe and D. Marx, *Structure theorem and isomorphism test for graphs with excluded topological subgraphs*, Proceedings of the forty-fourth annual acm symposium on theory of computing, 2012, pp. 173–192.

[29] H. Hadwiger, *Über eine Klassifikation der Streckenkomplexe*, Vierteljschr. Naturforsch. Ges. Zürich **88** (1943), 133–142.

[30] R. Halin, *S-functions for graphs*, J. Geometry **8** (1976), no. 1-2, 171–186.

[31] D. Harvey and D. Wood, *Parameters tied to treewidth*, 2013. arXiv:1312.3401.

[32] D. J. Harvey and D. R. Wood, *Cycles of given size in a dense graph*, 2015. arXiv:1502.03549.

[33] G. Joret, P. Micek, K. G Milans, W. T Trotter, B. Walczak, and R. Wang, *Tree-width and dimension*, 2013. arXiv:1301.5271.

[34] L. K. Jørgensen, *Contractions to K_8*, J. Graph Theory **18** (1994), no. 5, 431–448.

[35] R. Kapadia and S. Norin, *Densities of minor-closed graph classes are rational.* in preparation.

[36] K. Kawarabayashi and Y. Kobayashi, *Linear min-max relation between the treewidth of h-minor-free graphs and its largest grid*, Lipicsleibniz international proceedings in informatics, 2012.

[37] K. Kawarabayashi and B. Mohar, *Some recent progress and applications in graph minor theory*, Graphs Combin. **23** (2007), no. 1, 1–46.

[38] K. Kawarabayashi, S. Norine, R. Thomas, and P. Wollan, *K_6 minors in 6-connected graphs of bounded tree-width.* submitted.

[39] K. Kawarabayashi, S. Norine, R. Thomas, and P. Wollan, *K_6 minors in large 6-connected graphs.* submitted.

[40] A. V. Kostochka, *The minimum Hadwiger number for graphs with a given mean degree of vertices*, Metody Diskret. Analiz. **38** (1982), 37–58.

[41] A. V. Kostochka, *Lower bound of the Hadwiger number of graphs by their average degree*, Combinatorica **4** (1984), no. 4, 307–316.

[42] A. V. Kostochka and N. Prince, *Dense graphs have $K_{3,t}$ minors*, Discrete Math. **310** (2010), no. 20, 2637–2654.

[43] A. Leaf and P. Seymour, *Treewidth and planar minors* (2012). manuscript.

[44] R. J. Lipton and R. E. Tarjan, *A separator theorem for planar graphs*, SIAM J. Appl. Math. **36** (1979), no. 2, 177–189.

[45] C.-H. Liu and R. Thomas, *Well-quasi-ordering graphs by the topological minor relation: Robertson's conjecture.* in preparation.

[46] L. Lovász, *Graph minor theory*, Bull. Amer. Math. Soc. (N.S.) **43** (2006), no. 1, 75–86 (electronic).

[47] W. Mader, *Homomorphiesätze für Graphen*, Math. Ann. **178** (1968), 154–168.

[48] J. Maharry, *An excluded minor theorem for the octahedron*, J. Graph Theory **31** (1999), no. 2, 95–100.

[49] J. Maharry, *A characterization of graphs with no cube minor*, J. Combin. Theory Ser. B **80** (2000), no. 2, 179–201.

[50] C. McDiarmid, A. Steger, and D. J. A. Welsh, *Random graphs from planar and other addable classes*, Topics in discrete mathematics, 2006, pp. 231–246.

[51] J. Nešetřil and P. Ossona de Mendez, *Grad and classes with bounded expansion. I. Decompositions*, European J. Combin. **29** (2008), no. 3, 760–776.

[52] J. Nešetřil, P. Ossona de Mendez, and D. R. Wood, *Characterisations and examples of graph classes with bounded expansion*, European J. Combin. **33** (2012), no. 3, 350–373.

[53] S. Norin and R. Thomas, *Cultivating a vortex*. in preparation.

[54] S. Norin and R. Thomas, *Linear decompositions of large graphs*. in preparation.

[55] S. Norin and R. Thomas, *Minors and linkages in large graphs*. in preparation.

[56] S. Norin and R. Thomas, *Non-planar extensions of subdivisions of planar graphs*, 2014. arXiv:1402.1999.

[57] S. Norine, P. Seymour, R. Thomas, and P. Wollan, *Proper minor-closed families are small*, J. Combin. Theory Ser. B **96** (2006), no. 5, 754–757.

[58] B. Oporowski, J. Oxley, and R. Thomas, *Typical subgraphs of 3- and 4-connected graphs*, J. Combin. Theory Ser. B **57** (1993), no. 2, 239–257.

[59] B. Reed and D. R. Wood, *Forcing a sparse minor*, 2014. arXiv:1402.0272.

[60] N. Robertson, D. Sanders, P. Seymour, and R. Thomas, *The four-colour theorem*, J. Combin. Theory Ser. B **70** (1997), no. 1, 2–44.

[61] N. Robertson and P. Seymour, *Graph minors. XVIII. Tree-decompositions and well-quasi-ordering*, J. Combin. Theory Ser. B **89** (2003), no. 1, 77–108.

[62] N. Robertson and P. Seymour, *Graph minors. XXI. Graphs with unique linkages*, J. Combin. Theory Ser. B **99** (2009), no. 3, 583–616.

[63] N. Robertson and P. Seymour, *Graph minors XXIII. Nash-Williams' immersion conjecture*, J. Combin. Theory Ser. B **100** (2010), no. 2, 181–205.

[64] N. Robertson and P. Seymour, *Graph minors. XXII. Irrelevant vertices in linkage problems*, J. Combin. Theory Ser. B **102** (2012), no. 2, 530–563.

[65] N. Robertson, P. Seymour, and R. Thomas, *Quickly excluding a planar graph*, J. Combin. Theory Ser. B **62** (1994), no. 2, 323–348.

[66] N. Robertson and P. D. Seymour, *Graph minors. I. Excluding a forest*, J. Combin. Theory Ser. B **35** (1983), no. 1, 39–61.

[67] N. Robertson and P. D. Seymour, *Graph minors. III. Planar tree-width*, J. Combin. Theory Ser. B **36** (1984), no. 1, 49–64.

[68] N. Robertson and P. D. Seymour, *Graph minors. II. Algorithmic aspects of tree-width*, J. Algorithms **7** (1986), no. 3, 309–322.

[69] N. Robertson and P. D. Seymour, *Graph minors. V. Excluding a planar graph*, J. Combin. Theory Ser. B **41** (1986), no. 1, 92–114.

[70] N. Robertson and P. D. Seymour, *Graph minors. VI. Disjoint paths across a disc*, J. Combin. Theory Ser. B **41** (1986), no. 1, 115–138.

[71] N. Robertson and P. D. Seymour, *Graph minors. VII. Disjoint paths on a surface*, J. Combin. Theory Ser. B **45** (1988), no. 2, 212–254.

[72] N. Robertson and P. D. Seymour, *Graph minors. IV. Tree-width and well-quasi-ordering*, J. Combin. Theory Ser. B **48** (1990), no. 2, 227–254.

[73] N. Robertson and P. D. Seymour, *Graph minors. IX. Disjoint crossed paths*, J. Combin. Theory Ser. B **49** (1990), no. 1, 40–77.

[74] N. Robertson and P. D. Seymour, *Graph minors. VIII. A Kuratowski theorem for general surfaces*, J. Combin. Theory Ser. B **48** (1990), no. 2, 255–288.

[75] N. Robertson and P. D. Seymour, *Graph minors. X. Obstructions to tree-decomposition*, J. Combin. Theory Ser. B **52** (1991), no. 2, 153–190.

[76] N. Robertson and P. D. Seymour, *Graph minors. XI. Circuits on a surface*, J. Combin. Theory Ser. B **60** (1994), no. 1, 72–106.

[77] N. Robertson and P. D. Seymour, *Graph minors. XII. Distance on a surface*, J. Combin. Theory Ser. B **64** (1995), no. 2, 240–272.

[78] N. Robertson and P. D. Seymour, *Graph minors. XIII. The disjoint paths problem*, J. Combin. Theory Ser. B **63** (1995), no. 1, 65–110.

[79] N. Robertson and P. D. Seymour, *Graph minors. XIV. Extending an embedding*, J. Combin. Theory Ser. B **65** (1995), no. 1, 23–50.

[80] N. Robertson and P. D. Seymour, *Graph minors. XV. Giant steps*, J. Combin. Theory Ser. B **68** (1996), no. 1, 112–148.

[81] N. Robertson and P. D. Seymour, *Graph minors. XVII. Taming a vortex*, J. Combin. Theory Ser. B **77** (1999), no. 1, 162–210.

[82] N. Robertson and P. D. Seymour, *Graph minors. XVI. Excluding a non-planar graph*, J. Combin. Theory Ser. B **89** (2003), no. 1, 43–76.

[83] N. Robertson and P. D. Seymour, *Graph minors. XIX. Well-quasi-ordering on a surface*, J. Combin. Theory Ser. B **90** (2004), no. 2, 325–385.

[84] N. Robertson and P. D. Seymour, *Graph minors. XX. Wagner's conjecture*, J. Combin. Theory Ser. B **92** (2004), no. 2, 325–357.

[85] N. Robertson, P. Seymour, and R. Thomas, *Hadwiger's conjecture for K_6-free graphs*, Combinatorica **13** (1993), no. 3, 279–361.

[86] P. D. Seymour and R. Thomas, *Graph searching and a min-max theorem for tree-width*, J. Combin. Theory Ser. B **58** (1993), no. 1, 22–33.

[87] Z.-X. Song and R. Thomas, *The extremal function for K_9 minors*, J. Combin. Theory Ser. B **96** (2006), no. 2, 240–252.

[88] R. Thomas, *Recent excluded minor theorems for graphs*, Surveys in combinatorics, 1999 (Canterbury), 1999, pp. 201–222.

[89] A. Thomason, *An extremal function for contractions of graphs*, Math. Proc. Cambridge Philos. Soc. **95** (1984), no. 2, 261–265.

[90] A. Thomason, *The extremal function for complete minors*, J. Combin. Theory Ser. B **81** (2001), no. 2, 318–338.

[91] J. van Dobben de Bruyn and D. Gijswijt, *A lower bound on the gonality of finite graphs*, 2014. arXiv:1407.7055.

[92] K. Wagner, *Sur le probléme des courbes gauches en topologie*, Fundamenta Mathematicae **15** (1930), 271–283.

[93] K. Wagner, *Über eine Eigenschaft der ebenen Komplexe*, Mathematische Annalen **114** (1937), 570–590.

[94] K. Wagner, *Beweis einer Abschwächung der Hadwiger-Vermutung*, Math. Ann. **153** (1964), 139–141.

Department of Mathematics & Statistics
McGill University
Burnside Hall, 805 Sherbrooke West
Montreal H3A 2K6, Canada
snorin@math.mcgill.ca

Well quasi-order in combinatorics: embeddings and homomorphisms

Sophie Huczynska and Nik Ruškuc

Abstract

The notion of well quasi-order (wqo) from the theory of ordered sets often arises naturally in contexts where one deals with infinite collections of structures which can somehow be compared, and it then represents a useful discriminator between 'tame' and 'wild' such classes. In this article we survey such situations within combinatorics, and attempt to identify promising directions for further research. We argue that these are intimately linked with a more systematic and detailed study of homomorphisms in combinatorics.

1 Introduction

In combinatorics, indeed in many areas of mathematics, one is often concerned with classes of structures that are somehow being compared, e.g. in terms of inclusion or homomorphic images. In such situations one is naturally led to consider downward closed collections of such structures under the chosen orderings. The notion of partial well order (pwo), or its mild generalisation well quasi-order (wqo), can then serve to distinguish between the 'tame' and 'wild' such classes. In this article we will survey the guises in which wqo has made an appearance in different branches of combinatorics, and try to indicate routes for further development which in our opinion will be potentially important and fruitful.

The aim of this article is to identify major general directions in which wqo has been deployed within combinatorics, rather than to provide an exhaustive survey of all the specific results and publications within the topics touched upon. In this section we introduce the notion of wqo, and present what is arguably the most important foundational result, Higman's Theorem. In Section 2 we attempt a broad-brush picture of wqo in combinatorics, linking it to the notion of homomorphism and its different specialised types. The central Sections 3–5 present three 'case studies' – words, graphs and permutations – where wqo has been investigated, and draw attention to specific instances of patterns and phenomena already outlined in Section 2. Finally, in Section 6, we reinforce the homomorphism view-point, and explore possible future developments from this angle.

A *quasi-order (qo)* is any binary relation which is reflexive ($x \leq x$ for all x) and transitive ($x \leq y \leq z$ implies $x \leq z$). If our quasi-order is also anti-symmetric ($x \le y \le x$ implies $x = y$), it is called a *partial order (po)*. We will write $x < y$ to denote $x \le y$ and $y \not\le x$. Kruskal [42] gives the rule-of-thumb that it is easier to work with partial order than quasi-order "at a casual level", but that "in advanced work, the reverse is true". In fact, many natural "comparison relations", such as via embeddings or homomorphic images, are genuine quasi-orders, and become orders only when we restrict attention to finite structures. The close connection between the two notions is encapsulated by the fact that any qo gives rise to a po on the equivalence classes ($x \equiv y$ if $x \le y \le x$).

A *well quasi-order (wqo)* is a qo which is

- *well-founded*: every strictly decreasing sequence is finite; and

- *has no infinite antichain*: every set of pairwise incomparable elements is finite.

There are various widely-used equivalent formulations of what it means to be wqo, such as:

Theorem 1.1 *The following are equivalent for a quasi-order $\le$ on a set X:*

 (i) $\le$ *is a well quasi-order;*

 (ii) *if $x_0, x_1, \ldots \in X$ then there are $i < j$ with $x_i \le x_j$;*

 (iii) *if $x_0, x_1 \ldots \in X$ then there is an infinite $A \subseteq \mathbb{N}$ such that $x_i \le x_j$ for all $i < j$ in A (every infinite sequence has an infinite non-decreasing subsequence);*

 (iv) *for any $S \subseteq X$, there is a finite $T \subseteq S$ such that $\forall x \in S$, $\exists y \in T$ such that $y \le x$ (every non-empty subset contains only finitely many non-equivalent minimal elements, and every element lies above at least one such minimal element).*

A basic example of a wqo is ($\mathbb{N}, \le$), the natural numbers under the usual ordering. In contrast, ($\mathbb{Z}, \le$) is not a wqo because it is not well-founded, and ($\mathbb{N}, |$), the natural numbers ordered by divisibility, is not a wqo since the prime numbers form an infinite antichain. Another example of a wqo is ($\mathbb{N}^k, \le$), the set of k-tuples of natural numbers with component-wise ordering; the result which asserts that this is a wqo is known as Dickson's Lemma. More generally:

Theorem 1.2 *The class of well quasi-ordered sets is closed under: (i) taking of subsets; (ii) homomorphic images; (iii) finite unions; (iv) finite cartesian products.*

Given an arbitrary finite alphabet A (with at least two elements), the set of words A^* over A is not a wqo under either the lexicographic order nor the factor order (witnessed in both cases by the sequence $ab^i a$, $i = 1, 2, \ldots$). A fundamental result due to G. Higman asserts that A^* is a wqo under the subword (subsequence) ordering.

In fact, Higman proves a vastly more general result, which he couches in terms of *abstract algebras*, i.e. structures of the form (X, F) where X is a set and F is a set of operations on X. Both sets are arbitrary and may be infinite. The operations in F are assumed to be finitary (i.e. take a finite number of arguments), and their arities are assumed to be bounded. In other words, if F_r denotes the set of all basic operations that take r arguments, then all but finitely many F_r are empty. We say that the algebra is *minimal* if it has no proper subalgebras (subsets of X closed under all operations). Suppose that we have a quasi-order $\leq$ on X. This makes (X, F) into an *ordered algebra* if all operations from F are *compatible* with $\leq$, i.e. if for all $f \in F_r$, and all $x_1, \ldots, x_{i-1}, x_i', x_i'', x_{i+1}, \ldots, x_r \in X$ with $x_i' \leq x_i''$ we have

$$f(x_1, \ldots, x_{i-1}, x_i', x_{i+1}, \ldots, x_r) \leq f(x_1, \ldots, x_{i-1}, x_i'', x_{i+1}, \ldots, x_r).$$

We say that $\leq$ is a *divisibility order* if

$$x_i \leq f(x_1, \ldots, x_i, \ldots, x_r)$$

for all $f \in F_r$ and all $x_1, \ldots, x_r \in X$. Finally, suppose that each set F_r is quasi-ordered in its own right. We say that these quasi-orders are *compatible* with the quasi-order on X if

$$f \leq g \Rightarrow f(x_1, \ldots, x_r) \leq g(x_1, \ldots, x_r)$$

for all $f, g \in F_r$ and all $x_1, \ldots, x_r \in X$.

Theorem 1.3 (Higman [30, Theorem 1.1]) *Any minimal algebra (X, F) with a divisibility order $\leq$, such that all F_r are well quasi-ordered and compatible with $\leq$, is itself well quasi-ordered.*

It is fair to say that almost every non-trivial proof of well quasi-orderedness in mathematics utilises, in one way or another, Higman's Theorem (or its sister, Kruskal's Tree Theorem; see discussion in Subsection 4.2). The theorem is seldom applied in its full generality though, but rather in one of a series of specialisations. Firstly, one can assume that the algebra (X, F) comes equipped with a generating set A. By treating the elements of A as nullary operations (constants) we ensure that the algebra is minimal. If in addition we assume that there are only finitely many operations of other arities (thus dispensing with the wqo requirement on F_r), we obtain:

Corollary 1.4 *Suppose that (X, F) is an algebra, ordered by divisibility, generated by a well quasi-ordered set A, and with the set of operations F finite. Then (X, F) is well quasi-ordered.*

Obviously, every finite set is well quasi-ordered, so we obtain a further specialisation:

Corollary 1.5 *Every finitely generated algebra (X, F), ordered by divisibility and with F finite, is well quasi-ordered.*

By specialising Corollary 1.4 in a different direction to the free monoid A^*, consisting of all the words over alphabet A under the operation of concatenation, which clearly is generated by A itself, we obtain two further corollaries:

Corollary 1.6 *If A is a well quasi-ordered alphabet, then the free monoid A^* is well quasi-ordered by the domination ordering:*

$$a_1 \ldots a_m \leq b_1 \ldots b_n \Leftrightarrow (\exists 1 \leq j_1 < \cdots < j_m \leq n)(\forall i = 1, \ldots, m)(a_i \leq b_{j_i}).$$

Corollary 1.7 *If A is a finite alphabet, then the free monoid A^* is well quasi-ordered under the subword ordering:*

$$a_1 \ldots a_m \leq b_1 \ldots b_n \Leftrightarrow (\exists 1 \leq j_1 < \cdots < j_m \leq n)(\forall i = 1, \ldots, m)(a_i = b_{j_i}).$$

Well quasi-order usually makes an appearance when one is interested in certain downward closed classes of mathematical objects. The appropriate order theoretic notions capturing this are ideals, and their duals, filters. Suppose we have quasi-order $(X, \leq)$. A subset I of X is called an *ideal* or *downward closed set* if $y \leq x \in I$ implies $y \in I$. Dually, a subset F of X is called a   if it is upward closed; if $y \leq x$ and $y \in F$ implies $x \in F$. It is clear that the complement of an ideal is a filter, and vice versa. An equivalent condition for X to be a wqo, is that for every filter F of X, there exists a finite set B such that $F = \{x \in X : (\exists b \in B)(b \leq x)\}$ (part (iv) of Theorem 1.1). Here B is said to generate F; Higman refers to this as the *finite basis property*. We will more often use the alternative formulation:

Lemma 1.8 *A quasi-order $(X, \leq)$ is a wqo precisely if, for every ideal I of X, there is a finite set B of forbidden elements such that*

$$I = \mathrm{Av}(B) = \{x \in X : (\forall b \in B)(b \not\leq x)\}.$$

A minimal such B is often called a *basis*.

2   WQO in combinatorics

Although well quasi-order originates in the area of order theory, it appears with particular frequency in combinatorics, and indeed seems to be the appropriate language in which to describe and explore various combinatorial situations. We may naturally wonder why this is so. We have seen in the introduction, that if we have a quasi-order $\leq$ on a class C of combinatorial objects, then various helpful consequences follow if $\leq$ happens to be a well quasi-order. If we have a family X in our class C which is downward-closed under our quasi-order, then we know X can be characterized by avoiding a finite number of forbidden objects from our class. As an immediate implication we know that there are only countably many such downward-closed sets (provided that C itself is countable, which it invariably is in classical combinatorial settings), and thus one at least in principle can hope to explicitly characterise (list) them all. By way of contrast, if the class C is not wqo by virtue of an infinite antichain A, then by noting that every subset of an antichain is also an antichain, we conclude that there are at least continuum many downward closed classes in this case. Thus well quasi-orderedness can be viewed as a demarcation between 'tame' and 'wild' classes of combinatorial objects.

Some typical specific contexts in which wqo has been investigated in combinatorics are:

- words over an alphabet under the subword ordering;

- graphs under the subgraph ordering;

- graphs under the induced subgraph ordering;

- tournaments under the sub-tournament (which coincides with the induced sub-tournament) ordering;

- permutations under (sub)permutation involvement;

- graphs under the minor ordering;

- trees under homeomorphic embedding.

In fact these diverse contexts can be brought under the same umbrella by considering combinatorial structures as (finite) relational structures, i.e. sets with relations defined on them. Thus, for example, in this language an (undirected) graph is a set with a symmetric binary relation. The model can be further refined by requiring the relation to be irreflexive or reflexive. A permutation can be viewed as a set with two linear orders. Similar descriptions can be given for nearly all common combinatorial structures. We will not list all these descriptions here, but refer the reader to [34] for a fairly comprehensive treatment.

In this set-up, all the above orders on combinatorial structures are expressible in terms of homomorphisms. Suppose we have two relational structures $\mathcal{S} = (S, R_i^{\mathcal{S}}\,(i \in I))$ and $\mathcal{T} = (T, R_i^{\mathcal{T}}\,(i \in I))$ in the same signature (so that the arities of $R_i^{\mathcal{S}}$ and $R_i^{\mathcal{T}}$ are the same for every $i \in I$), and let $\phi : S \to T$ be a mapping. We say that ϕ is:

(i) a *homomorphism* if

$$(s_1, \ldots, s_k) \in R_i^{\mathcal{S}} \Rightarrow (\phi(s_1), \ldots, \phi(s_k)) \in R_i^{\mathcal{T}},$$

i.e. if $\phi(R_i^{\mathcal{S}}) \subseteq R_i^{\mathcal{T}}|_{\phi(S)}$;

(ii) a *strong homomorphism* if ϕ is a homomorphism and satisfies $\phi(R_i^{\mathcal{S}}) = R_i^{\mathcal{T}}|_{\phi(S)}$;

(iii) an *M-strong homomorphism* if ϕ is a homomorphism such that

$$(s_1, \ldots, s_k) \in R_i^{\mathcal{S}} \Leftrightarrow (\phi(s_1), \ldots, \phi(s_k)) \in R_i^{\mathcal{T}},$$

i.e. $\phi(R_i^{\mathcal{S}}) = R_i^{\mathcal{T}}|_{\phi(S)}$ and $\phi(\overline{R_i^{\mathcal{S}}}) = \overline{R_i^{\mathcal{T}}}|_{\phi(S)}$, where bars denote complementation.

The 'M' in the final definition refers to the model theoretic definition of a strong embedding (see, for example, [31, Chapter 1]). A homomorphism is an *embedding* if it is injective, and is an *epimorphism* if it is onto. Note that an embedding $\phi : S \to T$ is M-strong if and only if it is strong.

Now, the subgraph ordering $G \leq H$ on the class of graphs becomes simply the existence of an embedding $G \to H$, while the induced subgraph ordering means the existence of such an embedding that is required to be strong. For tournaments, embeddings and strong embeddings coincide, and so the subtournament and induced subtournament orderings are identical. Slightly less obviously, the subword ordering on words over an alphabet A can be interpreted as the existence of an embedding (equivalently, strong embedding), when words are appropriately represented as relational structures. This can be done by viewing a word w as a set X of size $|w|$ (representing the letters of w), with a linear ordering defined on it (specifying the order of letters in w), and a family of unary relations f_a ($a \in A$) such that for every $x \in X$ precisely one $f_a(x)$ is true (thus specifying to which letter of A an element x corresponds). For more details we refer the reader to [34, Subsection 2.2]. Permutation involvement is again the same as existence of (strong) embeddings.

Even less obviously, the graph minor relation can be interpreted in terms of homomorphisms. To do this, we need to consider graphs in their reflexive representation (otherwise a homomorphism could not contract an edge). A graph G is a minor of a graph H if there exists a graph K, an embedding $K \to H$, and an epimorphism $K \to G$ the kernel classes (or fibres) of which are connected.

In this paper we restrict our attention to *finite* combinatorial structures, with a consequence that all the homomorphism-defined orders are in fact partial orders, because they respect size. The single exception is *the* homomorphism order; see Section 6. However, if one extends one's field of study to infinite structures, then they all become genuine quasi-orders. For instance, it is perfectly possible for two non-isomorphic graphs G and H to mutually embed into each other. Generally, we follow the prevailing usage in literature, and favour the term quasi-order in preference to partial order.

Returning to our survey of wqo in combinatorics, we note that the presence of this property offers various algorithmic advantages. Probably the best example of this is membership testing for downward closed subclasses of a class C that is wqo under taking of substructures or induced substructures. Indeed, each such subclass X is given by its finite basis $\{B_1, \ldots, B_k\}$. To test whether an object A belongs to X we need to check whether it avoids each B_i. This can be done by examining all subsets of A having size $|B_i|$, a process that is polynomial in the size of A. So we see that membership in downward closed subclasses of C is of polynomial complexity.

Perusing the literature, we see that historically many combinatorial structures have been approached via the question of whether some of the objects in a given set are 'involved' in the others, which very naturally leads to the topic of wqo. For example, Vazsonyi's conjecture, made in the 1940s, hypothesises that any infinite collection of finite trees must contain two trees such that one is homeomorphically embeddable in the other (topological minor relation); this can equivalently be viewed as conjecturing that the quasi-order of trees under the topological minor relation is wqo (see [42]).

Some natural membership questions, posed early in the study of graph theory, turned out to have answers expressible in terms of whether a graph avoids certain forbidden graphs, also naturally leading to consideration of wqo. For example, in 1930, Kuratowski ([43]) showed that being planar means containing no subdivision of the complete graph K_5 or the complete bipartite graph $K_{3,3}$, while in 1937, Wagner [63] proved that being planar is equivalent to avoiding these two graphs as minors. We may observe that the property of being planar is closed under the taking of minors. This leads naturally to the more general question of whether, for a property P of graphs closed under minors, there exists a finite set of graphs such that a graph possesses property P if and only if it avoids all the graphs in the set as minors? (A result of Robertson and Seymour tells us that the class of graphs under minor order form a wqo, and hence every ideal does indeed have a finite set of forbidden elements.) Another example is that the property of being a cograph (a graph that can be generated from the single-vertex graph by the taking of complements and disjoint union) turns out to be equivalent to being P_4 induced-subgraph-free ([15]).

A possible way to view wqo investigations in combinatorics is as follows. One is originally and naturally interested in combinatorial structures and their substructures (induced or not). Unfortunately, with the exception of words over a finite alphabet, the resulting ordering will not be wqo for any other *full* class of combinatorial objects (e.g. *all* graphs, or *all* permutations). Still, wqo is a desirable property, and one wants to move in its direction. One way of doing it is by 'freeing up' the ordering, such as in the graph minor or homeomorphic embedding contexts. Another way is to try and identify the subclasses of C which are wqo, even if C itself is not.

Another aspect of wqo which can offer new insights and useful methods for combinatorial problems, but as yet has been surprisingly little-studied, is its connection with regular languages and rational generating functions. A natural illustration of this occurs in the setting of words over a finite alphabet. We may naturally ask, whether something similar can be fruitfully carried out in classes of other combinatorial objects.

In the next three sections we will present three case studies, in the form of brief surveys of wqo considerations for words, graphs and permutations. We have chosen these three not only because they comprise much of the existing work on wqo but also because they exhibit different combinations of the viewpoints outlined above. The results for words are classical and can serve as an exemplar and a tool for other, more complicated contexts. In graph theory we have the largest body of results demarcating the boundary between wqo and non-wqo classes. Finally in permutations we encounter the most active attempts at interweaving wqo with structural and enumerative considerations. In Section 6 as an attempt to bring these strands and areas of combinatorics closer together, we will propose a closer and more systematic study of homomorphism orderings.

3   Case study 1: words

Let A be a finite alphabet, and let A^* be the set of all words (i.e. sequences of symbols) over A. Recall that the subword ordering on A^* is defined by $u \leq v$ if u is a subword (i.e. subsequence) of v. Let X be an order ideal in this ordering, i.e. a non-empty set such that $u \leq v \in X$

implies $u \in X$. The complement $A^* \setminus X$ is a filter, i.e. it is upward closed. By Higman's Theorem (Corollary 1.7) the set A^* is wqo, and so $A^* \setminus X$ has finitely many minimal elements $B = \{w^{(i)} = a_1^{(i)} \ldots a_{m_i}^{(i)} : i = 1, \ldots, n\}$. Furthermore, the set X is the avoidance set of B, i.e.

$$X = \mathrm{Av}(B) = \{u \in A^* : w^{(i)} \not\leq u \text{ for } i = 1, \ldots, n\}.$$

Now, the set of all words *containing* a subword $w = a_1 \ldots a_m$ can be expressed as $A^* a_1 A^* a_2 \ldots A^* a_m A^*$. Hence

$$X = A^* \setminus \bigcup_{i=1}^{n} A^* a_1^{(i)} A^* a_2^{(i)} \ldots A^* a_{m_i}^{(i)} A^*.$$

This is a (very simple) example of a *regular expression* from formal language theory, proving that X is a *regular language*. By Kleene's Theorem every regular language is accepted by a *finite state automaton*. It is in turn known that the enumeration sequence of such a language has a rational generating function. This is actually easy to see. Recall that a (deterministic) finite state automaton consists of a finite set of vertices Q, a finite input alphabet A, and a transition function $\tau : Q \times A \to Q$. One vertex $q_I \in Q$ is designated as a start state, and there is a set of final states $Q_F \subseteq Q$. A relationship $\tau(q_1, a) = q_2$ is interpreted as a directed edge from q_1 to q_2 labelled by a. A word $w = a_1 \ldots a_m \in A^*$ is accepted by the automaton if starting at q_I and successively following the edges labelled $a_1, \ldots, a_m$ ends in a final state. The set L of all the accepted words is the language accepted by the automaton. Now, for $q \in Q$, let L_q be the set of words w such that reading of w from q_I terminates in q, and let f_q be the generating function for L_q. Note that a word $w = w'a$ is in L_q if and only if $w' \in L_{q'}$ and $\tau(q', a) = q$. Furthermore, note that $f_q(0) = 1$ if and only if $q = q_I$, and $f_q(0) = 0$ otherwise. This yields the equations

$$f_{q_I} = 1 + x \sum_{\tau(q',a)=q_I} f_{q'},$$

$$f_q = x \sum_{\tau(q',a)=q} f_{q'} \quad (\text{when } q \neq q_I).$$

These equations are clearly linear in the f_q, and the coefficients are polynomial (indeed, linear) in x. Thus, solving this system yields each f_q as a rational function in x, and the generating function for the entire language is $f = \sum_{q \in Q_F} f_q$.

All the above material is folklore, and can be found in standard combinatorics textbooks (see for instance Section 8.1 of [22] or Section 6.5 of [61]) expressed in different degrees of technical sophistication. By presenting it briefly here, in an elementary form, we hope offers the reader a blueprint as to how words, wqo, and Higman's Theorem should combine together to yield structural and enumerative results. We also mention in closing that this has a number of computational and computability consequences. For instance, the above account represents a constructive method for obtaining the generating function from the automaton, or indeed from a finite set of forbidden subwords. Also, the membership problem for a regular language is decidable in linear time (passing the word through the automaton), and a whole other host of properties is decidable; see for instance [32, Section 3.3]. An excellent overview of the uses of formal language theory in combinatorics can be found in [9].

4 Case study 2: graphs

In this section, we consider the graph theoretic context. We mainly have in mind 'standard' graphs, but also discuss digraphs, and special classes such as trees and tournaments. The study of graphs and wqo has a long history, arguably originating with the work of Kuratowski [43] and Wagner [63] in the 1930s, with early key results by Kruskal [41] and Nash-Williams [53] in the 1960s. There is now a wealth of results, covering many different varieties of graphs and possible orderings. Very generally speaking, one is hoping to establish wqo, but the original 'big' class (e.g. all graphs) under the original 'natural' order (e.g. subgraphs) is quite far from possessing this property. This then leads to investigating subclasses, or variations, or different possible orders. As a result, the great majority of the results in the literature are asserting wqo or its absence in one of the contexts arising in this way.

We begin by giving some definitions of graph operations and hence graph relations.

Definition 4.1 Let $G = (V, E)$ be a graph. We define the following operations on G:

- *Removing a vertex v*: removing v from the vertex set V, and removing all edges incident with v from the edge set E.

- *Removing an edge e*: removing e from the edge set E.

- *Suppressing a vertex v of degree two*: removing v from the vertex set V and replacing the two edges incident to v by a single edge.

- *Contracting an edge* $e = uv$: removing u and v from V, and introducing a new vertex z and edges such that z is adjacent to all vertices which were adjacent to u or v.

Now, we may define the following orderings. We present them in a way which shows how each definition increasingly 'frees-up' the orderings ([26]), as described above.

Definition 4.2 Let H, G be graphs.

- H is an *induced subgraph* of G if H can be obtained from G by a sequence of vertex removals.

- H is a *subgraph* of G if H can be obtained from G by a sequence of vertex and edge removals.

- H is a *topological minor* of G if H can be obtained from G by a sequence of vertex removals, edge removals, and suppressions of vertices of degree two.

- H is a *minor* of G if H can be obtained from G by a sequence of vertex removals, edge removals, and edge contractions.

Observe that suppression of a vertex of degree two can be viewed as the contraction of either of the two incident edges, so that this is indeed a hierarchy: H an induced subgraph of G, implies H is a subgraph of G, which in turn implies that H is a topological minor of G, and hence that H is a minor of G. It can be shown that all of these graph relations form quasi-orders on the set of finite graphs. We will see in the ensuing subsections that, initially far from being wqo, they move progressively towards possessing this property. We will also discuss a related ordering by immersions.

4.1 Subgraph order

The class of all graphs is not a wqo under either the subgraph or induced subgraph order; in both cases, an infinite antichain is provided by the set of cycles C_k or the set of double-ended forks F_k; see Figure 1. The latter antichain also witnesses that the class of all trees is not a wqo under subgraph or induced subgraph order.

It is, however, known that the graphs avoiding the path P_k of length k are wqo by the subgraph relation for any value of k ([19]), and that P_4-free graphs are wqo by the induced subgraph relation [17]. Indeed P_4 is the unique maximal graph G such that the class of all G-free graphs is

Figure 1: A typical double-ended fork F_k

wqo under induced subgraph order; in other words, if the class of G-free graphs is wqo then G is necessarily an induced subgraph of P_4 ([17]). Special classes, such as the set of bipartite graphs, have also been investigated under induced subgraph order. For example, it is conjectured in [19] that the P_7-free bipartite graphs are not wqo under the induced subgraph relation; in [36] it is shown that this is indeed the case, but that the P_6-free bipartite graphs are wqo. In [37], wqo classes of graphs defined by more than one induced subgraph obstruction are considered.

In [19], the wqo ideals in the class of graphs under subgraph and induced subgraph order were characterized in terms of forbidden subgraphs.

Theorem 4.3 (Ding [19]) *Let $\mathcal{F}$ be an ideal of graphs with respect to subgraph relation. Then the following are equivalent:*

- *$\mathcal{F}$ is a wqo under the subgraph ordering;*

- *$\mathcal{F}$ is a wqo under the induced subgraph ordering;*

- *$\mathcal{F}$ contains only finitely many graphs C_n (cycles) and F_n (double-ended forks).*

One possible interpretation of this result is that it is algorithmically decidable whether an ideal of graphs defined by finitely many obstructions is wqo or not. Indeed suppose that we have such an ideal $C = \mathrm{Av}(G_1, \ldots, G_k)$. According to Ding's Theorem, C will be wqo if and only if for some n_0 we have $C_n, F_n \notin C$ for all $n \geq n_0$. This in turn is equivalent to each C_n and each F_n $(n \geq n_0)$ containing some G_i. Now, observe that for a graph G with m vertices we have $G \leq C_p$ if and only if $G \leq C_q$ for all $p, q > m$. An analogous assertion holds for the F_p. It follows that C is wqo if and only if $C_{n_1}, F_{n_1} \notin C$, where $n_1 = \max(|G_1|, \ldots, |G_k|) + 5$. This property is clearly algorithmically decidable, by listing all the subgraphs of C_{n_1} and of F_{n_1} and checking they both intersect $\{G_1, \ldots, G_k\}$. It is worth noting that no such algorithm is known for the induced subgraph ordering. Recent related work, which also reflects the position of trees somehow being the key for the wqo/non-wqo distinction for graphs, is the result that graphs of finite tree-depth ordered by the induced subgraph relation form a well quasi-order ([55, Lemma 6.13]).

In the absence of a characterisation of wqo ideals of graphs under the induced subgraph ordering, one can attempt such a generalisation in the presence of additional restrictions. One such approach is to consider labelled graphs (or, equivalently, vertex coloured graphs). A collection of graphs C is said to be n-wqo (by the induced subgraph order) if the set of all n-labellings of members of C is wqo by the induced (labelled) subgraph relation. Clearly, in this terminology, the property of being wqo is equivalent to being 1-wqo. On the other hand, the set of paths $\{P_n\}$ is wqo but not 2-wqo. Rather curiously, every induced subgraph ideal of graphs which is 2-wqo is finitely based [57]. It has been conjectured (by Pouzet [57] in 1972, and Kříž and Thomas [40] in 1990) that for an ideal in the induced subgraph order, being 2-wqo is equivalent to being n-wqo for all n. For more on this, and a possible proof strategy for the conjecture, see [16]. This conjecture is not true in a more general setting of labelled categories, as demonstrated by Kříž and Sgall [39].

Since we do not have a wqo for graphs, we cannot have one for digraphs. We can try restricting our attention to special types of digraphs, for example tournaments, but in fact the subgraph relations do not give rise to a wqo for the class of all tournaments ([29], [45]). The specialization to classes of tournaments avoiding a given tournament as a subgraph were considered by Cherlin and Latka (see [13] for a summary of results in this area):

Theorem 4.4

- Let L be a finite linear tournament (i.e. isomorphic to the structure $(\{1,\ldots,n\},<)$ for some n). Then the L-free tournaments are wqo (indeed of bounded size).

- If L is a non-linear tournament, with at least 7 vertices, then the L-free tournaments are not wqo (two antichains witness this in all cases).

The above theorem is sufficient to conclude that it is algorithmically decidable whether a class of tournaments avoiding a single obstruction is wqo, as it gives a characterisation of wqo for all but finitely many tournaments L (the exceptions being non-linear tournaments of size less than 7). These exceptions have in fact also been analysed; see the discussion in [13, Subsection 2.2]. A question arises whether wqo is decidable for arbitrary ideals

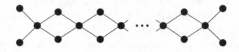

Figure 2: A modified double-ended fork

of tournaments under the subgraph ordering defined by finitely many obstructions. This is at present an open question. We note that work on wqo questions in tournaments was motivated by the analysis of homogeneous directed graphs ([12]). It was shown by Henson in 1972 ([29]) that any antichain of tournaments translates into uncountably many homogeneous digraphs, and in fact this accounts for all but countably many of these digraphs. Various questions about the digraphs can then be mapped back into questions about the structure of the quasi-order of finite tournaments.

The above summary suggests that, if we are hoping to obtain wqo results for the class of all graphs, or classes such as the class of trees or of tournaments, it might be fruitful to consider other orders.

4.2   Topological minor

The *topological minor*, also called the *homeomorphic embedding* relation, has been much-studied historically, and was the setting for some of the earliest results in this area. We defined the topological minor in terms of suppressions of vertices of degree two. An alternative expression is that a graph H is a topological minor of a graph G if a subdivision of H is isomorphic to a subgraph of G, where a subdivision means replacement of edges by paths.

The class of graphs is not wqo by the topological minor ordering. Several antichains witnesses this. One such is F'_k, which can be constructed from the double ended forks F_k by doubling every edge in the central part of the fork (see [48]). This of course is an example in the class of graphs with multiple edges. A suitable modification for the class of simple graphs is shown in Figure 2.

However, this quasi-order is not very far away from being wqo in some sense. For instance, Mader [49] proves that the class of all graphs that do not have k disjoint cycles as topological minors for some $k \in \mathbb{N}$ are wqo.

In a celebrated early result Kruskal [41] proves:

Theorem 4.5 *The class of all trees is wqo under the topological minor ordering.*

This effectively establishes trees as, in a way, being at the boundary between wqo and non-wqo in the world of graphs, a frequently occurring theme in subsequent developments. This could perhaps be elucidated by noticing that words could be interpreted as (labelled) paths, and that topological minor ordering would coincide with the subword ordering in this interpretation. General finite trees could then be viewed as finite sets of overlapping words.

An improved proof of Kruskal's tree theorem was given in 1963 by Nash-Williams [53]. Nash-Williams [54] went on to prove the analogous result for the class of *all* trees (finite or infinite). In fact all these results are proved in greater generality for trees labelled by elements from a wqo set. Nash-Williams also introduces a useful strengthening of wqo which he terms a *better quasi-ordering*. It is in this form that Kruskal's tree theorem is used as a key ingredient in the proof of the graph minor theorem (see below).

It is also worth noting that, as Kruskal himself points out, his tree theorem contains Higman's Theorem 1.4 as a special case. This can be intuitively understood by recalling that the *free* or *term* algebra over a generating set X consists of all formal expressions that can be built from basic operations F and elements of X treated as letters. Then it is easy to see that all elements in such an algebra can be represented as trees, with a vertex representing an operation (and is labelled by that operation), and its children are the arguments. It is an exercise to translate Higman's conditions into the topological minor ordering, and derive Higman's Theorem. The final step is provided by the observation that every algebra of type F is a homomorphic image of the term algebra, and that wqo is preserved under homomorphisms.

Returning to the class of all graphs, it transpires that the antichain F_k' exhibited above is in a sense the only one. Let P_k' be the central doubled path in F_k', or, equivalently, the path P_k with every edge doubled. Clearly, if an ideal I contains infinitely many F_k' (and hence is not wqo) then it also contains *all* P_k'. On the other hand, we have:

Theorem 4.6 *The class of graphs avoiding some P_k' as a topological minor is wqo.*

This was originally conjectured by Robertson in 1980s (unpublished), and only recently proved by Liu [48]. Note that this is very close to a (constructive) characterisation of wqo ideals under the topological minor

276 S. Huczynska and N. Ruškuc

ordering. In fact Ding [20], who proved Robertson's Conjecture in the special case of *minor* (as opposed to topological minor) closed classes, provides such a characterisation in this case: a minor closed class is wqo by the topological minor order if and only if it avoids some graph B'_k, obtained from P'_k by attaching two loops at each of the two ends.

4.3 Minor order

Now we reach the minor order, which is the setting for perhaps the best-known wqo result on graphs. There are various alternative expressions of the definition: for example, a graph G is a minor of a graph H if G can be obtained from a subgraph of H by successively contracting edges, or by collapsing connected subgraphs, see [18, Section 1.7].

A result from the 1930s (often called Kuratowski's Theorem, but in fact due to Wagner in its usually-quoted form) states that a finite graph is planar precisely if it avoids K_5 and $K_{3,3}$ as minors. After a lengthy proof process (in a series of papers spanning 21 years from 1983 to 2004), Robertson and Seymour proved in [59] that:

Theorem 4.7 *The class of graphs is a wqo under minor order.*

The definition of minor can be extended to digraphs in various ways (for digraphs, unlike graphs, contracting edges or connected subgraphs yields different concepts). If we define a digraph minor by saying a digraph H is a minor of a digraph G if H can be obtained from a subdigraph of G by repeatedly contracting a strongly-connected subdigraph to a vertex, we can consider the quasi-order of digraphs under the directed minor relation. This is not a wqo - for example, the directed cycles form an infinite antichain. However, we can consider subclasses; for example, it can be shown ([35]) that

Theorem 4.8 (Kim [35]) *The class of all finite tournaments is a wqo under minor order.*

4.4 Immersion order

Finally, we mention one other minor-type order which has received attention and yielded positive wqo results - the *immersion order* (originally defined by Nash-Williams). Graph H is said to be an *immersion minor* of graph G is there is an injective mapping from $V(H)$ to $V(G)$ such that the images of adjacent elements of H are connected in G by edge-disjoint paths. An alternative definition via graph operations is in terms of lifting: a pair of adjacent edges uv and vw with $u \neq v \neq w$ is *lifted* by deleting the edges uv and vw and adding the edge uw. A graph H is said to be immersed in a graph G precisely if a graph isomorphic to H can be obtained from G by lifting pairs of edges and taking a subgraph [8]. There are also the weak and strong immersion orders (the weak version also allows the operation of vertex-splitting). It was shown in [60] that

Theorem 4.9 *The class of graphs is a wqo under weak immersion order.*

Various consequences of this, in terms of finite bases, are explored in [27]; in particular it is hoped to lead to faster membership algorithms for the general immersion case. It has been conjectured by Seymour that the class of graphs is also a wqo under strong immersion order.

The definition of immersion can be extended to digraphs, by replacing paths with directed paths. It is not the case that the class of all digraphs is a wqo under weak immersion [35], but the class of tournaments is wqo under immersion, and it was shown that in fact the class of tournaments is wqo under strong immersion [14].

5   Case study 3: permutations

Compared to the study of graphs, investigation of wqo in permutations is much younger, with the first results appearing in the early 2000s; see [7]. Furthermore, permutations themselves are fairly restricted combinatorial structures, with no obvious variations available, and with essentially all homomorphism-related orders reducing to the subpermutation ordering; see [34]. Thus the overall body of results is considerably smaller than in the case of graphs. And yet, these restrictions have acted as a catalyst for directing research towards slightly different types of questions. One outcome of this is that interesting links with words and Higman's Theorem have emerged, leading to structural and enumerative consequences seemingly quite remote from the wqo itself. Here we review some developments of this nature.

For two permutations $\alpha = a_1 \ldots a_m$, $\beta = b_1 \ldots b_n$, we say that α is *involved* or *contained* in β (denoted $\alpha \leq \beta$) if β contains a subsequence $b_{k_1} \ldots b_{k_m}$ order isomorphic to α, i.e. satisfying $a_i < a_j$ if and only if $b_{k_i} < b_{k_j}$. If permutations are regarded as relational structures with two linear orders, then this definition coincides with the notion of (induced) substructure. Permutations can also be viewed as (equivalence classes) of sets of n points in a plane (arising from their plots as bijections), in which case involvement becomes simply presence of the corresponding pattern of α in the plot of β; see Figure 3 for illustration. The ideals of permutations under this order are referred to as *pattern (avoidance) classes*.

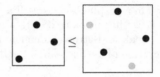

Figure 3: Plot of the permutation 132 and the same permutation as a subpattern of 42513

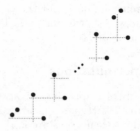

Figure 4: A typical permutation in the antichain $\mathcal{A}$

The poset of all permutations is not wqo, and there are many antichains witnessing this. Perhaps the easiest one is the oscillating antichain $\mathcal{A} = \{A_n : n \in \mathbb{N}\}$, depicted in Figure 4. This antichain, and its left-right reflection, suffice to prove that the only singly-based wqo classes are $\mathrm{Av}(12)$ (increases), $\mathrm{Av}(21)$ (decreases) and $\mathrm{Av}(231)$ and its four symmetries. Murphy's thesis [51] contains an extensive 'Bibliothek' of antichains. A common feature of all the antichains presented there is that the main body of each permutation consists of what was later termed a *pin sequence* [10], with an irregularity at the beginning and the end. This, and some subsequent results on pin sequences [11] give some hope that a (constructive) characterisation of finitely based wqo classes may be possible, but so far this has been elusive.

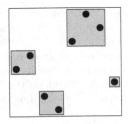

Figure 5: Inflation $3142[12, 21, 132, 1] = 45216873$

5.1 Classes with finitely many simple permutations

The *inflation* of a permutation $\alpha = a_1 \ldots a_m$ by permutations $\delta_1, \ldots, \delta_m$ is the permutation $\alpha[\delta_1, \ldots, \delta_m]$ obtained by replacing each point a_i of α by a set of contiguous points representing δ_i. For an illustration see Figure 5. Observe that the inflation $\alpha[\delta_1, \ldots, \delta_m]$ contains all δ_i, as well as α, as subpermutations. Thus if X is a pattern class and $\alpha[\delta_1, \ldots, \delta_m] \in X$, then necessarily $\alpha, \delta_1, \ldots, \delta_m \in X$. For sets A and B the *inflation* of A by B is

$$A[B] = \{\alpha[\beta_1, \ldots, \beta_m] \, : \, \alpha = a_1 \ldots a_m \in A, \ \beta_1, \ldots, \beta_m \in B\}.$$

When A and B are pattern classes, then so is $A[B]$. The (wreath) closure of a set X is

$$\langle X \rangle = X \cup X[X] \cup X[X[X]] \cup \cdots$$

This is the smallest class containing X closed under inflations.

A permutation is said to be *simple* if it cannot be expressed as an inflation in a non-trivial way. Simple permutations are basic blocks from which all other permutations are built by means of successive inflations. Albert and Atkinson [1] develop a theory of classes with finitely many simple permutations. Let A be such a class and let S be its finite set of simple permutations. Let $W = \langle A \rangle$ be the closure of A and note that $W = \langle S \rangle$. For each $\sigma = a_1 \ldots a_m \in S$ define an m-ary operation f_σ:

$$f_\sigma(\alpha_1, \ldots, \alpha_m) = \sigma[\alpha_1, \ldots, \alpha_m].$$

Clearly, W is closed under all these operations, and is generated by the set S. Now, as S is finite then we have a finitely generated algebra with finitely many basic operations, and it is easy to see that Higman's Theorem (in the form of Corollary 1.5) applies, and W is wqo. But $A \subseteq W$ and so A is wqo as well. Furthermore, because W is defined by a specific construction from a finite set S, it is possible to show that W is finitely based, and then it follows that A is finitely based as well. It is relatively easy to exhibit an algebraic generating function for W. A more detailed analysis of permutations in A, similar in essence to that of words given in Section 3 and likewise dependent on the finiteness of the obstruction set, then yields an algebraic generating function for A itself.

Theorem 5.1 (Albert and Atkinson [1]) *Every pattern class containing only finitely many simple permutations is wqo, finitely based and has an algebraic generating function.*

For instance, it easily follows that all subclasses of Av(231) have algebraic generating functions. In fact, with some further analysis, the authors prove that all proper such subclasses have rational generating functions. All the results in this paper are constructive, allowing, at least in principle, generating functions to be computed from the set of simple permutations and the basis. To complement this, Brignall, Ruškuc and Vatter [11] prove that it is decidable whether the set of simple permutations in a pattern class given by a finite basis is finite.

5.2 Geometric grid classes

The poset of all permutations under involvement is fairly 'wild', and so is the collection of all pattern classes. Considerable effort has therefore gone into trying to identify some 'tame' classes which can be used as building blocks for more general ones. One such attempt, heavily drawing on the geometric intuition of permutations as point plots in the plane, are the so-called grid classes. Let $M = (m_{ij})_{p \times q}$ be a matrix with entries from $\{\pm 1, 0\}$. The *grid class* Grid(M) defined by M is the collection of all permutations whose plot can be partitioned into a $p \times q$ grid such that the content of the cell (i, j) is increasing/decreasing/empty if m_{ij} equals $+1/-1/0$; see Figure 6 for an illustration.

Grid classes are a promising structural tool, but at present the analysis of their structure and properties appears to be difficult. It is conjectured that they are all finitely based and have algebraic generating functions, but no proof is known as yet. However, there is a pleasing result characterising when they are wqo. It is couched in terms of a graph $\Gamma(M)$, the (p, q) bipartite graph with adjacency matrix $(|m_{ij}|)_{p \times q}$.

Theorem 5.2 (Murphy and Vatter [52]) Grid(M) *is wqo if and only if* $\Gamma(M)$ *is a forest.*

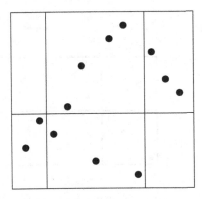

Figure 6: The permutation $(3, 5, 4, 6, 9, 2, 11, 12, 1, 10, 8, 7)$ as a member of
$\mathrm{Grid} \begin{pmatrix} 0 & 1 & -1 \\ 1 & -1 & 0 \end{pmatrix}$

The geometric grid class $\mathrm{Geom}(M)$ is an important subclass of $\mathrm{Grid}(M)$, and is obtained as follows. Fix a $p \times q$ rectangular grid in the plane. In the cell (i, j) draw the SW-NE diagonal if $m_{ij} = 1$, the NE-SW diagonal if $m_{ij} = -1$, and leave it empty if $m_{ij} = 0$. $\mathrm{Geom}(M)$ is the set of all permutations that can be plotted on the resulting set of diagonals. This is illustrated in Figure 7.

The remarkable fact about $\mathrm{Geom}(M)$ is that it admits a natural encoding by words over a finite alphabet, which works as follows. First observe that without loss we can assume that all the diagonals can be oriented, so that those belonging to the same row share the same up-down orientation, while those in the same column have the same left-right orientation. If this is not true for the original matrix M, it can be 'mended' by enlarging M, replacing each each entry 1, -1, 0 by $\begin{pmatrix} 0 & 1 \\ 1 & 0 \end{pmatrix}$, $\begin{pmatrix} 1 & 0 \\ 0 & 1 \end{pmatrix}$, $\begin{pmatrix} 0 & 0 \\ 0 & 0 \end{pmatrix}$, respectively. The alphabet will be $A = \{a_{ij} : m_{ij} = \pm 1\}$. Given a word $w = a_{i_1 j_1} \ldots a_{i_n j_n} \in A^*$, the permutation $\phi(w)$ is obtained as follows. Pick real numbers $0 < d_1 < d_2 < \cdots < d_n < d$, where d is the common length of all the diagonals. Then for each $k = 1, \ldots, n$ place a point on the diagonal in the cell (i_k, j_k) distance d_k from the base point. This process is illustrated in Figure 8.

The assignment $w \mapsto \phi(w)$ defines a surjective, length-preserving, finite-to-one mapping $A^* \to \mathrm{Geom}(M)$. Furthermore, it is order-preserving,

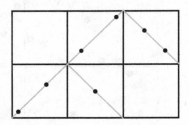

Figure 7: 1352764 as a member of Geom $\begin{pmatrix} 0 & 1 & -1 \\ 1 & -1 & 0 \end{pmatrix}$

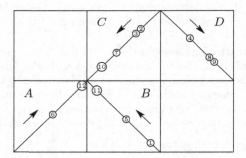

Figure 8: An illustration of the process that yields $\phi(BCCDBACDDCBA) = (3, 5, 4, 6, 9, 2, 11, 12, 1, 10, 8, 7)$. The points are numbered in order in which they are inserted.

in the sense that if $u \leq v$ for $u, v \in A^*$ under the subword ordering then $\phi(u) \leq \phi(v)$ under the permutation containment. It immediately follows from Higman's Theorem (in the form of Corollary 1.7) that Geom(M) is wqo, and a little more work shows it is finitely based as well. By analysing preimages of permutations under ϕ, we obtain a regular language $L \subseteq A^*$ such that ϕ maps L bijectively onto Geom(M). From here it follows that Geom(M) has a rational generating function. Furthermore, one can look at *subclasses* of Geom(M) (termed *geometrically griddable classes*), use wqo for Geom(M) to see they are defined by finitely many obstructions, and conclude that they are finite unions of geometric grid classes and possess the above pleasing properties. To summarise:

Theorem 5.3 (Albert et al. [2]) *Every geometric grid class* Geom(M) *is wqo, finitely based, and has a rational generating function. Subclasses of* Geom(M) *are finite unions of geometric grid classes, and they also are wqo, finitely based and have rational generating functions.*

Remarkably, geometric grid classes seem to arise naturally in structural descriptions of other pattern classes. Some early examples of this are in [6], and some quite advanced ones are in Murphy's thesis [51]. For example, Murphy shows that

$$\mathrm{Av}(132, 4312) = \mathrm{Grid} \begin{pmatrix} 0 & 0 & 1 \\ 1 & 0 & 0 \\ 0 & 1 & 0 \\ 0 & -1 & 0 \\ 0 & 0 & -1 \end{pmatrix}.$$

It should be noted that these early 'applications' of geometric grid classes actually predate their formal introduction, but nonetheless essentially rely on their structural and enumerative properties.

In [5], Albert, Ruškuc and Vatter use geometric grid classes to obtain a general result on *small classes*, defined in terms of growth rates. The *growth rate* for a pattern class C is defined as $\limsup_{n \to \infty} \sqrt[n]{|C_n|}$, which exists by Markus–Tardos Theorem [50]. It is conjectured that the sequence $\sqrt[n]{|C_n|}$ actually always converges, but this is still open. In an earlier work, Vatter [62] showed that there are only countably many pattern classes with growth rate less than κ, and uncountably many of growth rate equal to κ, where $\kappa = 2.20557\ldots$ is the unique positive root of $1 + 2x^2 - x^3$. Incidentally the number κ is related to the antichain $\mathcal{A}$ of Figure 4: it is the growth rate for the smallest class containing $\mathcal{A}$. This reinforces the notion of the antichain $\mathcal{A}$ being in a sense the smallest antichain in the permutation poset. Now, Theorem 5.3 says that geometric grid classes are *strongly rational* in the sense of [4], meaning that all their subclasses have rational generating functions. In [5] it is proved that for a geometric grid class C and a strongly rational class D, the inflation $C[D]$ is again strongly rational. It is also proved that every pattern class of growth rate less than κ is contained in an iterated inflation of the form $C[C[\ldots[C]\ldots]]$ for some geometric grid class. Combining all this together yields:

Theorem 5.4 (Albert, Ruškuc and Vatter [5]) *Each of countably many pattern classes of permutation with growth rate less than κ has a rational generating function.*

Another general application of geometric grid classes exhibited in [5] is to wreath closures: it is shown that the wreath closure of a geometric grid class always has an algebraic generating function. Context free languages and grammars (as opposed to the regular ones for geometric grid classes) make an appearance in this proof, but outlining it is beyond the scope of the present survey.

It is perhaps interesting to point out that, somewhat unexpectedly, the decidability questions one may naturally ask about geometric grid classes seem not to have the obvious positive answers, which is certainly at variance with the situation for words described in Section 3. At present, it is not known how to compute the generating function or the basis from the gridding matrix, and it is also not known how to decide if a finitely based class is in fact a geometric grid class or at least contained in such a class. This is basically due to the non-constructive nature of appeals to Higman's Theorem in the argument. In a rare example of full grid classes behaving better than their geometric subclasses, Huczynska and Vatter [11] give a very nice constructive criterion for a finitely based class to be contained in a grid class. A different example of encoding permutations in a class by a regular language, in order to prove rationality of the generating function, can be found in [3]. It should be noted that the encoding there is not order preserving, and so an appeal to Higman's Theorem is not possible, and indeed the classes obtained are not all wqo.

It is perhaps disappointing that, despite the wealth of wqo results in other areas of combinatorics, there are very few results taking this further into structural, enumeration or computability corollaries. One exception are Petkovšek's *letter graphs* [56]. They are in a way similar to geometric grid classes, in that they rely on a finite specification (the role played by the matrix for permutations) under which every word over A defines a graph. It is proved that the set of all graphs arising in this way from a fixed finite alphabet is wqo and, as a consequence, that it is decidable in polynomial time whether a graph is of this type.

6   Homomorphisms: embeddings and epimorphisms

We have observed in Section 2 that the viewpoint of homomorphisms is a very natural one to take when describing orders on combinatorial structures. Embedding, perhaps the most frequently-encountered order in combinatorics, corresponds to the existence of an injective homomorphism, and the induced substructure ordering corresponds to the existence of a strong injective homomorphism. More complicated orders introduced into the theory, such as graph minor order, still admit natural simple conceptual descriptions via homomorphisms. It is therefore natural to ask wqo type questions for orders defined in terms of the existence of any homomorphism, or of a surjective homomorphism.

It is perhaps rather surprising that there seems to be very little evidence of such investigations in the literature. The homomorphism order itself has been studied, but the kind of questions we have been exploring here have received surprisingly little attention. As for the homomorphic image order, it has been almost entirely unexplored until a recent paper of the current authors ([34]).

The *homomorphism order* is defined by: $G \leq H$ if there exists a homomorphism from G to H (see [28]). Much, though not all, of the literature on orders defined via homomorphism, centres around graphs. This ordering is in fact a genuine quasi-order (i.e. it is not necessarily anti-symmetric) even if one restricts one's attention only to finite structures. For example for any two empty graphs E_m and E_n we have $E_m \leq E_n \leq E_m$. As with any quasi-order, $\leq$ induces a natural equivalence relation: G and H are defined to be *homomorphically equivalent* if $G \leq H \leq G$. Each equivalence class has a unique representative (up to isomorphism) of minimal size, called the *core*. We denote the resulting poset by $\mathcal{G}$; its elements are equivalence classes under the homomorphic equivalence, and the ordering is induced by the homomorphism ordering.

It transpires that this order is radically different from those we have encountered so far. This is probably best encapsulated in the following result:

Theorem 6.1 ([33], [28, Section 3.1]) *The partial order $\mathcal{G}$ is universal, in the sense that it contains every countable partial order as a suborder.*

This has immediate relevant consequences. For example, the poset $\mathcal{G}$ is not even well-founded, in contrast to all the other partial orders we have in this paper. Also, it contains a wealth of antichains. In particular, this poset is very far from being wqo. The significance of these antichains is further reduced by the fact that in the absence of well-foundedness not all ideals are defined as avoidance classes. Perhaps as a consequence, there has been only a limited study of antichains in this poset. It is known (Corollary 3.11, [28]) that there are only two finite maximal antichains, both of size one, in $\mathcal{G}$. The study of infinite antichains in $\mathcal{G}$ has centred around the notion of splitting (see for example [21]). The homomorphism order has been shown to be universal on various other classes of structures: digraphs ([58], oriented paths and trees ([33]), partial orders and lattices ([47]). Locally constrained graph homomorphisms – i.e. homomorphisms with the additional requirement that they restrict to bijections, injections or surjections on the neighbourhood of any vertex – have recently been explored in [24]. Interestingly, these restrictions are sufficient to bring us back from genuine quasi-orders into the world of partial orders.

Since the special case of injective homomorphisms has been so well-studied, it would be natural to expect the embedding order's natural dual, the homomorphic image order, to have been similarly investigated. However, this is not the case. A notable exception is a 1979 paper [44] by Landraitis which considers the homomorphic image order on the set of countable linear order types. The author proves that this is a wqo, which mirrors an earlier result by Laver [46], asserting that the embedding relation is a wqo on this set. In fact, both authors prove that the orders under consideration are *better quasi-orders*.

In [34] the present authors attempt to redress this balance, by focussing on the homomorphic image order in the combinatorial setting, and considering the properties of this order in the realms of digraphs, graphs, words, permutations, posets, and a range of other combinatorial objects. In fact, as outlined in Section 2, we are led to define three homomorphic image orders, corresponding to the three flavours of homomorphism: standard, strong (or induced) and M-strong; let us denote these orderings as $\leq_H$, $\leq_{IH}$, $\leq_{MH}$ respectively. In addition to antichains and wqo property, the paper contains discussions of the joint preimage property and the dual amalgamation property, which are natural analogues of the joint embedding property and amalgamation property in the embedding ordering.

It transpires that, unsurprisingly, under the M-strong homomorphic image order $\leq_{MH}$, none of the classes of structures we consider are wqo. For the other two orderings the situation is more interesting. For digraphs, the collection of linear tournaments is an antichain under $\leq_H$ (and hence $\leq_{IH}$ as well), while the collection of graphs G_n illustrated in Figure 9 is an antichain under $\leq_{IH}$, but not under $\leq_H$. For graphs in the irreflexive representation the collection of all complete graphs K_n forms an obvious antichain under $\leq_H$. In the reflexive representation one can obtain an antichain by taking K_{2n} and removing the edges $\{2i-1, 2i\}$ ($i = 1, \ldots, n$).

Figure 9: An antichain of digraphs under $\leq_{\mathrm{IH}}$ but not under $\leq_{\mathrm{H}}$

Figure 10: An antichain of posets under $\leq_{\mathrm{H}}$

A construction of a collection of reflexive graphs that would be an antichain under $\leq_{\mathrm{IH}}$ but not under $\leq_{\mathrm{H}}$ appears to be harder. For tournaments, the family T_n with vertices $\{1, \ldots, n\}$ and edges

$$
\begin{aligned}
i \to j &\quad \text{if} \quad i \not\equiv j \pmod 2 \\
j \to i &\quad \text{if} \quad i \equiv j \pmod 2
\end{aligned}
$$

for $1 \leq i < j \leq n$, is an antichain under $\leq_{\mathrm{H}}$ in both the reflexive and irreflexive representations. The collection of posets is not wqo, in either the reflexive or non-reflexive representations, as witnessed for example by the family illustrated in Figure 10. All this leaves trees as yet again the boundary class, and here we prove:

Theorem 6.2 (Huczynska and Ruškuc [34]) *The collection of all (reflexive or irreflexive) trees is well quasi-ordered by the homomorphic image ordering (and hence under the strong homomorphic image ordering as well).*

Of the remaining structures, the homomorphic image ordering on permutations is a weakening of the subpermutation involvement, and so is not wqo. Likewise, the ordering on words is a weakening of the subword ordering, to the extent that wqo fails, as witnessed by the collection $(ab)^n$, $n = 1, 2, \ldots$. By way of contrast, the set of all finite equivalence relations is wqo under both $\leq_H$ and $\leq_{IH}$. This is proved by a relatively easy appeal to Higman's Theorem, mirroring the obvious proof that the equivalence relations are wqo under the embedding ordering.

Possible future research directions arising from this work are two-fold. On one hand, one can try to develop an analogue of the existing theory for substructure orderings, asking the equivalent questions in this new setting. Especially worthwhile in this direction would be a study of wqo for classes defined by avoidance conditions, with the aim of obtaining analogues to those previously-discussed for tournaments under subgraph order, or for graphs under topological minor order. On the other hand, motivated by the viewpoint outlined in this paper, it seems to us that a natural new direction is to investigate possible ways of combining the homomorphic image orderings and embedding orderings to produce a richer scheme of orderings for combinatorial structures. This would be guided by the observation that the minor ordering is a *composition* of the subgraph ordering and a special kind of a homomorphic image ordering (with all fibres being connected). In the presence of such a scheme, one would expect at the weak end of the spectrum not to have wqo, at the strong for it to be present, and then one could ask at which point (e.g. after how many compositions) it first occurs. One very concrete question that has occurred to us while writing the article is what happens if one strengthens one term in the above composition for the minor ordering and weakens the other. Specifically: For graphs G and H, let $G \leq H$ if G is a strong homomorphic image of an induced subgraph of H. Is the resulting ordering wqo?

7 Conclusion

We have seen that the property of being well quasi-ordered has been studied in a number of combinatorial settings for over half a century. The greatest body of results is within graph theory, where a variety of classes and orderings have been considered. The literature on wqo in the theory of pattern classes of permutations, while less well developed, contains some promising links with other combinatorial themes such as enumeration and asymptotics. It would appear that the language of homomorphisms, and in particular the interplay between embeddings and epimorphisms, may offer a conceptual link between different areas and methodologies, as well as a welcome extension of the field of investigation.

References

[1] M.H. Albert and M.D. Atkinson, Simple permutations and pattern restricted permutations, *Discrete Math.* 300 (2005), 1–15.

[2] M.H. Albert, M.D. Atkinson, M. Bouvel, N. Ruškuc and V. Vatter, Geometric grid classes of permutations, *Trans. Amer. Math. Soc.* 365 (2013), 5859–5881.

[3] M.H. Albert, M.D. Atkinson and N. Ruškuc, Regular closed sets of permutations, *Theoret. Comput. Sci.* 306 (2003), 85–100.

[4] M.H. Albert, M.D. Atkinson and V. Vatter, Subclasses of the separable permutations, *Bull. London Math. Soc.* 5 (2011), 859–870.

[5] M.H. Albert, N. Ruškuc and V. Vatter, Inflations of geometric grid classes of permutations, *Israel J. Math.* (2014). DOI 10.1007/s11856-014-1098-8.

[6] M.D. Atkinson, Restricted permutations, *Discrete Math.* 195 (1999), 27–38.

[7] M.D. Atkinson, M.M. Murphy and N. Ruškuc, Partially well-ordered closed sets of permutations, *Order* 19 (2002), 101–113.

[8] F.M. Abu-Khzam and M.A. Langston, Graph coloring and the immersion order, in *Computing and Combinatorics*, T. Warnow and B. Zhu (eds.), Lecture Notes in Computer Science 2697, Springer, Berlin, 2003, pp. 394–403.

[9] M. Bousquet-Mélou, Algebraic generating functions in enumerative combinatorics and context-free languages, in *STACS 2005*, V. Diekert and B. Durand (eds.), Lecture Notes in Computer Science 3404, Springer, Berlin, 2005, pp. 18–35.

[10] R. Brignall, S. Huczynska and V. Vatter, Decomposing simple permutations, with enumerative consequences, *Combinatorica* 28 (2008), 385–400.

[11] R. Brignall, N. Ruškuc and V. Vatter, Simple permutations: decidability and unavoidable substructures, *Theoret. Comput. Sci.* 391 (2008), 150–163.

[12] G. Cherlin, *The Classification of Countable Homogeneous Directed Graphs and Countable Homogeneous n-Tournaments*, Mem. Amer. Math. Soc. 131, AMS, Rhode Island, 1998.

290 S. Huczynska and N. Ruškuc

[13] G. Cherlin, Forbidden substructures and combinatorial dichotomies: WQO and universality, *Discrete Math.* 311 (2011), 1534–1584.

[14] M. Chudnovsky and P. Seymour, A well-quasi-order for tournaments, *J. Combin. Theory, Ser. B* 101 (2011), 47–53.

[15] D.G. Corneil, H. Lerchs and L.S. Burlingham, Complement reducible graphs, *Discrete Appl. Math.* 3 (1981), 163–174.

[16] J. Daligault, M. Rao and S. Thomassé, Well-quasi-order of relabel functions, *Order* 27 (2010), 301–315.

[17] P. Damaschke, Induced subgraphs and well quasi-ordering, *J. Graph Theory* 14 (1990), 427–435.

[18] R. Diestel, *Graph Theory*, Graduate Texts in Mathematics 173, Springer, Berlin, 2010

[19] G. Ding, Subgraphs and well quasi-ordering, *J. Graph Theory* 16 (1992), 489–502.

[20] G. Ding, Excluding a long double path minor, *J. Combin. Theory, Ser. B* 66 (1996), 11–23.

[21] D. Duffus, P.L. Erdös, J. Nešetřil and L. Soukup, Antichains in the homomorphism order of graphs, *Comment. Math. Univ. Carolin.* 48 (2007), 571–583.

[22] S. Eilenberg, *Automata, Languages and Machines Vol A*, Academic Press, NY, 1974.

[23] J. Fiala, J. Hubička and Y. Long, Universality of intervals of line graph order, *European J. Combin.* 41 (2014), 221–231.

[24] J. Fiala, D. Paulusma and J. A. Telle, Matrix and graph orders derived from locally constrained graph homomorphisms, in *Mathematical Foundations of Computer Science 2005*, Lecture Notes in Computer Science 3618, Springer Verlag, 2005, pp. 340–351.

[25] R. Fraïssé, *Theory of Relations*, North-Holland, Amsterdam, 1953.

[26] D. Glickenstein, Math 443/543 Graph Theory Notes 11, Department of Mathematics, University of Arizona, 2008.

[27] R. Govindana and S. Ramachandramurthi, A weak immersion relation on graphs and its applications, *Discrete Math.* 230 (2001), 189–206.

[28] P. Hell and J. Nešetřil, *Graphs and Homomorphisms*, Oxford Lecture Series in Mathematics and Its Applications 28, OUP, Oxford, 2004.

[29] C.W. Henson, Countable homogeneous relational systems and N_0-categorical theories, *J. Symbolic Logic* 37 (1972), 494–500.

[30] G. Higman, Ordering by divisibility in abstract algebras, *Proc. London Math. Soc.* 2 (1952), 326–336.

[31] W. Hodges, *Model Theory*, Cambridge University Press, Cambridge, 1993.

[32] J.E. Hopcroft and J.D. Ullman, *Introduction to Automata Theory, Languages, and Computation*, Addison–Wesley, Reading MA, 1979.

[33] J. Hubička and J. Nešetřil, Universal partial order represented by means of oriented trees and other simple graphs, *European J. Combin.* 26 (2005), 765–778.

[34] S. Huczynska and N. Ruškuc, Homomorphic image orders on combinatorial structures, *Order* (2014), to appear.

[35] I. Kim, *On Containment Relations in Directed Graphs*, PhD thesis, Princeton, 2013.

[36] N. Korpelainen and V. Lozin, Bipartite induced subgraphs and well quasi-ordering, *J. Graph Theory* 67 (2011), 235–249.

[37] N. Korpelainen and V. Lozin, Two forbidden induced subgraphs and well quasi-ordering, *Discrete Math.* 311 (2011), 1813–1822.

[38] N. Korpelainen, V. Lozin and I. Razgon, Boundary properties of well quasi-ordered sets of graphs, *Order* 30 (2013), 723–735.

[39] I. Kříž and J. Sgall, Well-quasi-ordering depends on the labels, *Acta Sci. Math.* 55 (1991), 59–65.

[40] I. Kříž and R. Thomas, On well-quasi-ordering finite structures with labels, *Graphs Combin.* 6 (1990), 41–49.

[41] J.B. Kruskal, Well-quasi-ordering, the tree theorem and Vaszsonyi's conjecture, *Trans. Amer. Math. Soc.* 95 (1960), 210–225.

[42] J.B. Kruskal, The theory of well quasi-ordering: a frequently discovered concept, *J. Combin. Theory, Ser. A* 13 (1972), 297–305.

[43] K. Kuratowski, Sur le probléme des courbes gauches en topologie, *Fund. Math.* 15 (1930), 271–283.

[44] C. Landraitis, A combinatorial property of the homomorphism relation between countable order types, *J. Symbolic Logic* 44 (1979), 403–411.

[45] B. Latka, Finitely constrained classes of homogeneous directed graphs, *J. Symbolic Logic* 59 (1994), 124–139.

[46] R. Laver, On Fraïssé's order type conjecture, *Ann. Math. (2)* 93 (1971), 89–111.

[47] E. Lehtonen, Labeled posets are universal, *European J. Combin.* 29 (2008), 493-506.

[48] C-H. Liu, *Graph Structures and Well Quasi-Ordering*, PhD Thesis, Georgia Institute of Technology, 2014.

[49] W. Mader, Wohlquasigeordnete Klassen endlicher Graphen, *J. Combin. Theory, Ser. B* 12 (1972), 105-122.

[50] A. Marcus and G. Tardos, Excluded permutation matrices and the Stanley-Wilf conjecture, *J. Combin. Theory, Ser. A* 107 (2004), 153-160.

[51] M.M. Murphy, *Restricted Permutations, Antichains, Atomic Classes and Stack Sorting*, Ph.D. Thesis, University of St Andrews, 2003.

[52] M.M. Murphy and V. Vatter, Profile classes and partial well-order for permutations, *Electron. J. Combin.* 9 (2003), R17.

[53] C.St.J.A. Nash-Williams, On well-quasi-ordering finite trees. *Math. Proc. Cambridge Philos. Soc.* 59 (1963), 833–835.

[54] C.St.J.A. Nash-Williams, On well-quasi-ordering infinite trees, *Math. Proc. Cambridge Philos. Soc.* 61 (1965), 697–720.

[55] J. Nešetřil and P. Ossona de Mendez, *Sparsity. Graphs, Structures, and Algorithms*, Algorithms and Combinatorics 28, Springer, Heidelberg, 2012.

[56] M. Petkovšek, Letter graphs and well-quasi-order by induced subgraphs, *Discrete Math.* 244 (2002), 375-388.

[57] M. Pouzet, Un bel ordre dábritement et ses rapports avec les bornes d'une multirelation, *C. R. Acad. Sci. Paris Ser AB* 274 (1972), 1677-1680.

[58] A. Pultr and V. Trnková, *Combinatorial, Algebraic, and Topological Representations of Groups, Semigroups, and Categories*, North-Holland Mathematical Library 22, North-Holland, Amsterdam, 1980.

[59] N. Robertson and P. Seymour, Graph minors. XX. Wagners conjecture, *J. Combin. Theory, Ser.B* 92 (2004), 325–357.

[60] N. Robertson and P. Seymour, Graph minors. XXIII. Nash-Williams' immersion conjecture, *J. Combin. Theory, Ser. B* 100 (2010), 181–205.

[61] R.P. Stanley, *Enumerative Combinatorics Vol 2*, Cambridge Studies in Advanced Mathematics 62, CUP, Cambridge, 1999.

[62] V. Vatter, Small permutation classes, *Proc. London Math. Soc.* 103 (2011), 879–921.

[63] K. Wagner, Über eine Eigenschaft der ebenen Komplexe, *Math. Ann.* 114 (1937), 570–590.

School of Mathematics and Statistics
University of St Andrews
St Andrews, U.K.
{sophie.huczynska, nik.ruskuc}@st-andrews.ac.uk

Constructions of block codes from algebraic curves over finite fields[1]

Liming Ma and Chaoping Xing

Abstract

Since the discovery of algebraic geometry codes by Goppa in 1978–1982, mathematicians have been looking for other constructions of block codes from algebraic curves or varieties of higher dimension. In this survey article, we present various constructions of block codes from algebraic curves. As there is a one-to-one correspondence between algebraic curves and function fields, we adopt the function field language throughout this paper; this is advantageous for most of the constructions in this paper.

1 Introduction

The discovery of algebraic geometry codes by Goppa has greatly stimulated research in both coding theory and number theory [1, 4, 5, 6, 7, 8, 9, 12, 16, 21, 24, 25, 33, 34, 44]. As there have been many other constructions of block codes via algebraic geometry, henceforth Goppa's algebraic geometry codes are called the Goppa geometric codes.

The major breakthrough of the Goppa geometric codes is that they improved the long-standing benchmark bound, the Gilbert–Varshamov (GV, for short) bound. Before the Goppa geometric codes, the GV bound had remained for more than 30 years and many people even conjectured that the GV bound is optimal and could not be improved.

The Goppa geometric codes are a natural generalization of the well-known Reed–Solomon codes from projective curves to algebraic curves of higher genus. Due to rich structures of algebraic function fields, there is a great potential to construct block codes via algebraic curves through other methods. Indeed, in the last few decades, various constructions of codes via algebraic curves have been found [2, 3, 8, 9, 17, 24, 26, 29, 37, 39, 40, 41, 42, 43, 44]. In this survey article, we present quite a few constructions of codes through algebraic curves. Many of these constructions have been discovered by the authors and their collaborators.

The paper is organised as follows. In Section 2, we present some preliminaries on codes and algebraic curves. Section 3 is devoted to various constructions of codes.

---

[1]The research is supported by Singapore MoE Tier 1 grant RG20/13.

2 Preliminary

We present some preliminaries on codes and algebraic curves in this section. For more background, the reader can refer to the following books [14, 18, 20, 22, 28, 30]. But we prefer to use the function field language in this survey article (see [22, 30]).

2.1 Codes

In this subsection, we introduce some basic concepts of coding theory which can be found in any standard textbook on error-correcting codes or algebraic geometry codes, for instance see [10, 15, 18, 19, 31].

Definition 2.1 Let A be a finite set of q elements with $q > 1$ and let A^n denote the affine space $\{(a_1, a_2, \ldots, a_n) \mid a_i \in A\}$. Then a *code* of length n over A is a subset $C \subseteq A^n$.

For two vectors $\mathbf{u} = (u_1, u_2, \ldots, u_n)$, $\mathbf{v} = (v_1, v_2, \ldots, v_n) \in A^n$, define the *Hamming distance* as

$$d_H(\mathbf{u}, \mathbf{v}) = |\{1 \leq i \leq n \mid u_i \neq v_i\}|.$$

The Hamming distance is a metric on A^n. The *Hamming weight* is defined by the number of the non-zero coordinates:

$$\mathrm{wt}(\mathbf{u}) = |\{1 \leq i \leq n \mid u_i \neq 0\}|.$$

Moreover, $d(\mathbf{u}, \mathbf{v}) = \mathrm{wt}(\mathbf{u} - \mathbf{v})$. The *minimum distance* of C is defined by

$$d(C) = \min\{d(\mathbf{u}, \mathbf{v}) \mid \mathbf{u} \neq \mathbf{v} \in C\}.$$

We say that C is a *q-ary (n, M, d)-code* if $C \subseteq A^n$, $|C| = M$, and $d(C) = d$. For a code C with minimum distance d, let $t = \lfloor (d-1)/2 \rfloor$. Then C is said to be *t-error correcting*.

From now on, A is taken to be the finite field $\mathbb{F}_q$. Then a subspace C of $\mathbb{F}_q^n$ is called a *linear code*. A q-ary linear code of dimension k in $\mathbb{F}_q^n$ is called an $[n, k]$ *code* over $\mathbb{F}_q$. An $[n, k]$ code with minimum distance d will be referred to as an $[n, k, d]$ *code*. If $C \subseteq \mathbb{F}_q^n$ is a code, then

$$C^\perp = \{u \in \mathbb{F}_q^n \mid \langle u, c \rangle = 0 \text{ for all } c \in C\},$$

where $\langle \cdot, \cdot \rangle$ is the canonical inner product on $\mathbb{F}_q^n$, is called the *dual* of C. Let C be an $[n, k]$ code over $\mathbb{F}_q$. A simple way to describe a specific linear code C is to give a basis of C. A generator matrix of C is a $k \times n$ matrix whose rows are a basis of C. A parity-check matrix of C is a generator matrix H of $C^{\perp}$. A parity-check matrix H of an $[n,k]$ code C is an $(n-k) \times n$ matrix of rank $n-k$; also

$$\mathbf{x} \in C \iff H\mathbf{x}^T = 0.$$

One of the main problems of coding theory is to construct linear codes whose dimension and minimum distance are large in comparison with their length. But there is a trade-off between the dimension and minimum distance of a code. The following proposition provides the simplest bound.

Proposition 2.2 (Singleton bound) *For an $[n,k,d]$ code C, we have*

$$k + d \le n + 1.$$

By the above bound, a code with $k+d = n+1$ is optimal, and such a code is called a *maximum distance separable code*, or MDS code for short. The following proposition shows how the minimum distance d of a linear code can be determined from the generator matrix and the parity matrix of the code.

Proposition 2.3 *Let C be an $[n,k]$ linear code over $\mathbb{F}_q$, and let G be the generator matrix and H the parity-check matrix of C.*

(1) *C has minimum distance d if and only if every $d-1$ columns of H are linearly independent and there exist d columns of H that are linearly dependent.*

(2) *C has minimum distance d if and only if every $n-d+1$ columns of G have rank k and there exist $n-d$ columns of G of rank $k-1$.*

This proposition is a very powerful tool to calculate or estimate the minimum distance of a linear code.

2.2 Algebraic function fields over finite fields

In this survey article, we prefer to use the language of function fields. An extension field F of $\mathbb{F}_q$ is called an *algebraic function field of one variable* over $\mathbb{F}_q$ if there exists an element z of F which is transcendental over $\mathbb{F}_q$ and such that F is a finite extension of the rational function field $\mathbb{F}_q(z)$. If $\mathbb{F}_q$ is algebraically closed in F, then $\mathbb{F}_q$ is called the *full constant field* of F. For brevity, we denote by $F/\mathbb{F}_q$ an algebraic function field with the full constant field $\mathbb{F}_q$. Such an algebraic function field $F/\mathbb{F}_q$ is a global function field.

A *place* P is the maximal ideal of some valuation ring of F, and we denote by $\mathcal{O}_P$ the valuation ring corresponding to the place P. A *normalized discrete valuation* of an algebraic function field $F/\mathbb{F}_q$ is a surjective map $\nu : F \to \mathbb{Z} \cup \{\infty\}$ which satisfies the following properties:

(i) $\nu(x) = \infty$ if and only if $x = 0$;

(ii) $\nu(xy) = \nu(x) + \nu(y)$ for any $x, y \in F$;

(iii) $\nu(x + y) \geq \min\{\nu(x), \nu(y)\}$ for any $x, y \in F$;

(iv) $\nu(a) = 0$ for any $a \in \mathbb{F}_q^*$.

The property (iii) is called the *Triangle Inequality*, and a stronger version of this inequality is given in the following useful lemma.

Lemma 2.4 (Strict Triangle Inequality) *Let ν be a normalized discrete valuation of F/K and $\nu(x) \neq \nu(y)$ for $x, y \in F$. Then*

$$\nu(x + y) = \min\{\nu(x), \nu(y)\}.$$

Usually, we use the notation ν_P for the normalized discrete valuation of F corresponding to P.

Theorem 2.5 (Approximation Theorem) *Let $S \subsetneq \mathbb{P}_F$ be a proper subset of $\mathbb{P}_F$ and let $P_1, P_2, \ldots, P_n \in S$. Suppose there are given n elements $x_1, x_2, \ldots, x_n \in F$ and $r_1, r_2, \ldots, r_n \in \mathbb{Z}$. Then there exists an element $x \in F$ such that $\nu_{P_i}(x - x_i) = r_i$ for $i = 1, \ldots, n$ and $\nu_P(x) \geq 0$ for all $P \in S \setminus \{P_1, \ldots, P_n\}$.*

We denote by $\mathbb{P}_F$ the set of places of F. For a place $P \in \mathbb{P}_F$, the corresponding valuation ring

$$\mathcal{O}_P = \{x \in F : \nu_P(x) \geq 0\}$$

is a local ring and its unique maximal ideal is

$$M_P = \{x \in \mathcal{O}_P : \nu_P(x) > 0\}.$$

Then the residue class field $\mathcal{O}_P/M_P$, denoted by $\widetilde{F}_P$, can be identified with a finite extension of $\mathbb{F}_q$. The degree of this extension is called the *degree* of the place P; that is, $\deg(P) = [\widetilde{F}_P : \mathbb{F}_q]$. A place of degree 1 is called *rational*.

Theorem 2.6 (Hasse–Weil Bound) *Let $F/\mathbb{F}_q$ be a global function field of genus g. Then the number of rational places of $F/\mathbb{F}_q$ satisfies*

$$|N(F) - (q+1)| \leq 2g\sqrt{q}.$$

If $N(F)$ attains the upper Hasse–Weil bound $q + 1 + 2g\sqrt{q}$, then F is called a *maximal function field*. It follows that a maximal function field $F/\mathbb{F}_q$ can exist only if either $g(F) = 0$ or q is a square. For a fixed prime power q and an integer $g \in \mathbb{N}$, let

$$N_q(g) = \max\{N(F) \mid F/\mathbb{F}_q \text{ is a global function field of genus } g\}$$

be the maximum number of rational places that a global function field $F/\mathbb{F}_q$ of genus g has. With an increase in applications of algebraic curves over finite fields in the construction of algebraic geometric codes, global function fields over finite fields with many rational places have been intensively studied; see [1, 4, 5, 6, 7, 16, 21, 44].

2.3 Riemann–Roch Theorem

A *divisor* D of an algebraic function field $F/\mathbb{F}_q$ is a formal sum

$$D = \sum_{P \in \mathbb{P}_F} m_P P$$

with integer coefficients $\nu_P(D) := m_P$ and $m_P \neq 0$ for at most finitely many $P \in \mathbb{P}_F$. The *principal divisor* of x is given by

$$\text{div}(x) = \sum_{P \in \mathbb{P}_F} \nu_P(x)P = (x)_0 - (x)_\infty,$$

where $(x)_0$ and $(x)_\infty$ are positive divisors of F. The degree of the principal divisor $\text{div}(x)$ is equal to 0.

For a divisor D of F, we associate the *Riemann–Roch space* to D:

$$\mathcal{L}(D) = \{x \in F^* \mid \text{div}(x) + D \geq 0\} \cup \{0\}.$$

Then $\mathcal{L}(D)$ is a finite dimensional vector space over $\mathbb{F}_q$ and we denote by $l(D)$ its dimension $\dim_{\mathbb{F}_q}(\mathcal{L}(D))$.

Definition 2.7 The *genus* g of $F/\mathbb{F}_q$ is defined by

$$g = g(F) := \max\{\deg(D) - l(D) + 1 \mid D \in \text{Div}(F)\}.$$

By putting $D = 0$ in the above definition, it is clear that the genus of any function field F is non-negative. The genus g is the most important invariant of the function field F.

Theorem 2.8 (Riemann's Theorem) *Let $F/\mathbb{F}_q$ be an algebraic function field of genus g. Then for any divisor D of F,*

$$l(D) \geq \deg(D) + 1 - g.$$

Moreover, there exists a positive integer c such that, if $\deg(D) \geq c$, then $l(D) = \deg(D) + 1 - g$.

For any divisor A of F, the integer defined by

$$i(A) := l(A) - \deg(A) + 1 - g$$

is called the *index of speciality* of A. If $i(A) = 0$, then the divisor A is called *non-special*.

Let $\mathcal{A}_F$ be the adele space of $F/\mathbb{F}_q$. For any divisor A of F, $\mathcal{A}_F(A)$ is defined by

$$\mathcal{A}_F(A) = \{\alpha \in \mathcal{A}_F \mid \nu_P(\alpha) \geq -\nu_P(A) \text{ for all } P \in \mathbb{F}_F\}.$$

A *Weil differential* of $F/\mathbb{F}_q$ is a $\mathbb{F}_q$-linear map $\omega : \mathcal{A}_F \to \mathbb{F}_q$ vanishing on $\mathcal{A}_F(A) + F$ for some A of F. We call

$$\Omega_F = \{\omega \mid \omega \text{ is a Weil differential of } F/\mathbb{F}_q\}$$

the *module of Weil differentials* of $F/\mathbb{F}_q$. For a divisor A, let

$$\Omega_F(A) = \{\omega \in \Omega_F \mid \omega \text{ vanishes on } \mathcal{A}_F(A) + F\}.$$

For $x \in F$ and $\omega \in \Omega_F$, we define $x\omega : \mathcal{A}_F \to \mathbb{F}_q$ by $(x\omega)(\alpha) := \omega(x\alpha)$. Under this definition, Ω_F is a one-dimensional vector space over F.

For each Weil differential $\omega \neq 0$, we can attach a divisor (ω). The divisor (ω) of the Weil differential ω is the uniquely determined divisor of F satisfying the following conditions:

(a) ω vanishes on $\mathcal{A}_F((\omega)) + F$;

(b) if ω vanishes on $\mathcal{A}_F(A) + F$, then $A \leq (\omega)$.

The divisor W with the form $W = (\omega)$ for some $\omega \in \Omega_F$ is called a *canonical divisor* of $F/\mathbb{F}_q$. Now we can provide the Riemann–Roch Theorem, which is by far the most important theorem in the theory of algebraic function fields.

Theorem 2.9 (Riemann–Roch Theorem) *Let W be a canonical divisor of $F/\mathbb{F}_q$. Then, for each divisor A of $F/\mathbb{F}_q$,*

$$l(A) = \deg(A) + 1 - g + l(W - A).$$

For a divisor A with degree at least $2g - 1$, we have $l(A) = \deg(A) + 1 - g$.

Let P be a place of F. An integer $n \geq 0$ is called a *pole number* of P if there exists an element $x \in F^*$ with $(x)_\infty = nP$. Otherwise, n is called a *gap number* of P.

Theorem 2.10 (Weierstrass Gap Theorem) *Let F have genus $g \geq 1$ and let P be a rational place of F. Then there are exactly g gap numbers $i_1, i_2, \ldots, i_g$ of P and they satisfy $1 = i_1 \leq i_2 \leq \cdots \leq i_g \leq 2g - 1$.*

Let P be a place of F. For $x \in F$, let $\iota_P(x) \in \mathcal{A}_F$ be the adele whose P-component is x and all other components are 0. Now the *local component* $\omega_P : F \to \mathbb{F}_q$ of a Weil differential ω of F can be defined by

$$\omega_P(x) := \omega(\iota_P(x)).$$

Proposition 2.11 *Let $\omega \in \Omega_F$ and $\alpha = (\alpha_P) \in \mathcal{A}_F$. Then*

(1) *the differential $\omega_P(\alpha_P) \neq 0$ for at most finitely many places P;*

(2) *$\omega(\alpha) = \sum_{P \in \mathbb{P}_F} \omega_P(\alpha_P)$;*

(3) *in particular, $\sum_{P \in \mathbb{P}_F} \omega_P(1) = 0$.*

There is a close relationship between the Weil differential and the differential. We can identify Ω_F with the differential module Δ_F. Via this identification the local component of ω at the place P can be evaluated by means of the residue of ω at P, namely $\omega_P(u) = \mathrm{res}_P(u\omega)$ for all $u \in F$. The above proposition is in essence the Residue Theorem for the differentials of $F/\mathbb{F}_q$.

2.4 Local expansion

In this subsection, we introduce the local expansion of an element $f \in F$ at a place P of F. For a given place P of F, there exists an element $t \in F$ with $\nu_P(t) = 1$, which is called a local parameter at P. For a place P and a function $f \in F$ with $\nu_P(f) \geq 0$, we denote by $f(P)$ the residue class $f + P$ of f in $\widetilde{F}_P = \mathcal{O}_P/P$, which can be viewed as an element of a finite field extension of $\mathbb{F}_q$.

L. Ma and C. Xing

Now we choose a sequence $\{t_r\}_{r=-\infty}^{\infty}$ of elements in F such that

$$\nu_P(t_r) = r.$$

For a given function $f \in F$, we can find an integer v such that $\nu_P(f) \geq v$. Hence $\nu_P(f/t_v) \geq 0$. Put $a_v = (f/t_v)(P)$. Then a_v is an element of $\widetilde{F}_P$. It follows that $\nu_P(f/t_v - a_v) \geq 1$. Hence we get

$$\nu_P\left(\frac{f - a_v t_v}{t_{v+1}}\right) \geq 0.$$

Put $a_{v+1} = (f - a_v t_v)/t_{v+1} + P$. Then $a_{v+1} \in \widetilde{F}_P$ and

$$\nu_P(f - a_v t_v - a_{v+1} t_{v+1}) \geq v + 2.$$

Suppose that we have obtained a sequence $\{a_r\}_{r=v}^{m}$ of elements of $\widetilde{F}_P$ such that

$$\nu_P\left(f - \sum_{r=v}^{k} a_r t_r\right) \geq k + 1$$

for all $v \leq k \leq m$. Put $a_{m+1} = (f - \sum_{r=v}^{m} a_r t_r)/t_{m+1} + P$. Then $a_{m+1} \in \widetilde{F}_P$ and $\nu_P(f - \sum_{r=v}^{m+1} a_r t_r) \geq m+2$. In this way, we continue the construction of the a_r. Then we obtain an infinite sequence $\{a_r\}_{r=v}^{\infty}$ of elements of $\widetilde{F}_P$ such that

$$\nu_P\left(f - \sum_{r=v}^{m} a_r t_r\right) \geq m + 1$$

for all $m \geq v$. We summarise the above construction in the formal expansion

$$f = \sum_{r=v}^{\infty} a_r t_r,$$

which is called the *local expansion* of f at P. Usually we take $t_r = t^r$. Then we have the following theorem.

Theorem 2.12 *Let P be a place of $F/\mathbb{F}_q$ of degree one, let $t \in F$ be a local parameter of P and let F_P be the P-adic completion of F. Then every element $z \in F_P$ has a unique representation of the form*

$$z = \sum_{i=n}^{\infty} a_i t^i \text{ with } n \in \mathbb{Z} \text{ and } a_i \in \mathbb{F}_q.$$

*This representation is called the P-adic power series expansion of z with
respect to t. Conversely, if $(c_i)_{i \geq n}$ is a sequence in $\mathbb{F}_q$, then the series
$\sum_{i=n}^{\infty} c_i t^i$ converges in F_P and we have*

$$\nu_P \left(\sum_{i=n}^{\infty} c_i t^i \right) = \min\{i | c_i \neq 0\}.$$

2.5 Narrow ray class groups

Let $F/\mathbb{F}_q$ be a global function field with genus g, where $q = p^r$. Assume
that the number of rational places $N(F) \geq 1$ and distinguish one rational
place which is denoted by ∞. Fix a place P of F with degree e and let
$D = mP$ for simplicity, where $m \in \mathbb{N}_+$. Let $S_\infty = \mathbb{P}_F \setminus \{\infty\}$, $A = \mathcal{O}_{S_\infty}$
and S be a subset of S_∞ containing P such that $S_\infty - S$ is finite.

Let F_∞ be the ∞-completion of F at ∞. There is a sign function
sgn : $F_\infty^* \to \mathbb{F}_q^*$ which is a multiplicative group homomorphism on F_∞^*
such that

(1) sgn$(\alpha) = \alpha$ for any $\alpha \in \mathbb{F}_q^*$;

(2) sgn$(U_\infty^{(1)}) = \{1\}$, where $U_\infty^{(1)} = \{z \in F : \nu_\infty(z - 1) \geq 1\}$.

Let $I_D(S_\infty)$ be the group of the fractional ideals of $\mathcal{O}_{S_\infty}$ that are
prime to P, and let Princ$_D^+(S_\infty)$ be the group of principal ideals zA where
$\nu_P(z - 1) \geq m$ and sgn$(z) = 1$. Then the factor group

$$\mathcal{O}_{S_\infty}/\text{Princ}_D^+(S_\infty),$$

which we denote by Cl$_D^+(A)$, is called the narrow ray class group with
modules D with respect to the sign function sgn. The narrow ray class field
with modulus D, denoted by $F^D(\infty)$, is constructed as the finite abelian
field extension of F corresponding to an open subgroup of the ideal class
group of F such that its Galois group Gal$(F^D(\infty)/F) \cong$ Cl$_D^+(A)$.

Let Cl$_D(\mathcal{O}_S)$ be the S-ray class group with module D. The ray class
field with modulus D is constructed as the finite abelian field extension
of F corresponding to a certain open subgroup of the idele class group of
F of finite index in which the Galois group is isomorphic to Cl$_D(\mathcal{O}_S)$; see
[22]. Such a ray class field with modulus D is denoted by F_S^D. The ray
class field F_S^D with modulus D is the largest finite abelian extension F' of
F in which all places that are not in S split completely in F'/F and the
conductor of F'/F is less than or equal to D.

Now we want to provide the main properties of narrow ray class fields;
here, $F^D(\infty)$ is a finite extension of $F_{S_\infty}^D$ with the following properties.

Theorem 2.13 *Let $F^D(\infty)$ be the narrow ray class field with modulus D and $Cl_D^+(A)$ be its narrow ray class group. Then the following hold.*

(1) $\mathrm{Gal}(F^D(\infty)/F) \cong Cl_D^+(A)$.

(2) $[F^D(\infty) : F_{S_\infty}^D] = q - 1$.

(3) *The field $F_{S_\infty}^D$ is both the decomposition field and the fixed field of ∞ in $F^D(\infty)$. In particular,*

$$\mathrm{Gal}(F^D(\infty)/F_{S_\infty}^D) \cong Cl_D(A)/Cl_D^+(A) \cong \mathbb{F}_q^*.$$

(4) *All places of F apart from P and ∞ are unramified in $F^D(\infty)/F$. For such an unramified place Q, its Artin symbol is given by the residue class in $Cl_D^+(A)$ under the correspondence in (1).*

(5) *The conductor of $F^D(\infty)/F$ is $mP + \min\{q-2,1\}\infty$.*

(6) *$\mathbb{F}_q$ is the full constant field of $F^D(\infty)/\mathbb{F}_q$.*

The narrow ray class group can be employed to construct global function fields with many rational places (see [44]).

Theorem 2.14 *Let G_D be the subgroup of $Cl_D^+(A)$ generated by $\mathbb{F}_q^*$ and the places in $S^* = S_\infty - S$. Let L be the subfield of $F^D(\infty)$ fixed by G_D. Then L has at least $[h(F)\Phi(mP)(|S^*|+1)]/|G_D|$ rational places and the genus of L is given by*

$$2g(L) - 2 = h(F)\left[\frac{\Phi(mP)}{|G_D|}(2g(F) - 2 + me) - e\sum_{i=0}^{m-1}\frac{\Phi(iP)}{|G_{iP}|}\right],$$

where $h(F)$ is the class number of F and

$$\Phi(iP) = |(A/P^i)^*| = (q^e - 1)q^{e(i-1)}.$$

In the following section, we provide various kinds of methods to construct block codes from algebraic curves over finite fields.

3   Constructions

The algebraic geometry method was first introduced into coding theory by V.D. Goppa (see [8, 9]). As a motivation for the construction of the algebraic geometric codes, we consider the Reed–Solomon codes which have long been a well-known class of codes in coding theory. In fact, algebraic geometry codes are a natural generalization of the Reed–Solomon codes.

Let q be a prime power. Let $n = q - 1$ and let $\beta \in \mathbb{F}_q$ be a primitive element of the multiplicative group $\mathbb{F}_q^*$. For any integer $1 \leq k \leq n$, we consider the k-dimensional vector space

$$\mathcal{L}_k := \{f \in \mathbb{F}_q[x] \mid \deg(f) \leq k - 1\} \qquad (3.1)$$

and the evaluation map ev $: \mathcal{L}_k \to \mathbb{F}_q^n$ given by

$$\mathrm{ev}(f) := (f(\beta), f(\beta^2), \ldots, f(\beta^n)) \in \mathbb{F}_q^n. \qquad (3.2)$$

It may be checked that the evaluation map is an $\mathbb{F}_q$-linear injective map, since a non-zero polynomial f with $\deg(f) < n$ has less than n zeros. Therefore

$$C_k := \{(f(\beta), f(\beta^2), \ldots, f(\beta^n)) \mid f \in \mathcal{L}_k\} \qquad (3.3)$$

is an $[n, k]$ code over $\mathbb{F}_q$, and it is called a Reed–Solomon code. By the Singleton bound, the Reed–Solomon codes are MDS codes.

Now we introduce the notion of an algebraic geometry code. Let us fix some notations. Let

(1) $F/\mathbb{F}_q$ be an algebraic function field of genus g;

(2) $P_1, P_2, \ldots, P_n$ be pairwise distinct places of $F/\mathbb{F}_q$ of degree 1, and $D = P_1 + P_2 + \cdots + P_n$;

(3) G be a divisor of F such that $\mathrm{supp}(G) \cap \mathrm{supp}(D) = \emptyset$.

3.1 Goppa's construction

Consider the evaluation map $\mathrm{ev}_D : \mathcal{L}(G) \to \mathbb{F}_q^n$ given by

$$\mathrm{ev}_D(f) := (f(P_1), f(P_2), \ldots, f(P_n)) \in \mathbb{F}_q^n. \qquad (3.4)$$

This evaluation map is $\mathbb{F}_q$-linear, and we define

$$C := \{(f(P_1), f(P_2), \ldots, f(P_n)) \mid f \in \mathcal{L}(G) \subseteq \mathbb{F}_q^n\} \qquad (3.5)$$

as the image of $\mathcal{L}(G)$ under the above evaluation map. This code C is the algebraic geometry code associated with the divisors D and G, and we use the notation $C_{\mathcal{L}}(D, G)$. These algebraic geometry codes were introduced by V.D. Goppa. Therefore these codes are sometimes called Goppa geometric codes as well. The Riemann–Roch theorem can be used to calculate or estimate their parameters n, k and d. Hence, we can obtain the following results.

Theorem 3.1 *The Goppa geometric code $C_{\mathcal{L}}(D, G)$ is an $[n, k, d]$ code with parameters $k = l(G) - l(G - D)$ and $d \geq n - \deg(G)$. Furthermore, if it is assumed that $\deg(G) < n$, then the evaluation map $ev_D : \mathcal{L}(G) \to C_{\mathcal{L}}(D, G)$ is injective and*

$$k = l(G) \geq \deg(G) + 1 - g.$$

Hence, $k + d \geq n + 1 - g$. If, in addition, $2g - 2 < \deg(G) < n$, then $k = \deg(G) + 1 - g$.

Proof The evaluation map $ev_D : \mathcal{L}(G) \to C_{\mathcal{L}}(D, G)$ is a surjective map with kernel

$$\mathrm{Ker}(ev_D) = \{x \in \mathcal{L}(G) \mid \nu_{P_i}(x) > 0 \text{ for } i = 1, \ldots, n\} = \mathcal{L}(G - D).$$

Then the dimension

$$k = \dim(C_{\mathcal{L}}(D, G)) = l(G) - l(G - D).$$

If $\deg(G) < n = \deg(D)$, then $l(G - D) = 0$. Hence, ev_D is injective, and $k = l(G) \geq \deg(G) + 1 - g$ by the Riemann–Roch theorem. Further, if $\deg(G) > 2g - 2$, then $k = \deg(G) + 1 - g$.

Assume that $C_{\mathcal{L}}(D, G) \neq \{0\}$, and consider the minimum distance d of the code $C_{\mathcal{L}}(D, G)$. Choose an element $x \in \mathcal{L}(G)$ with $\mathrm{wt}(ev_D(x)) = d$. Then exactly $n - d$ places $P_{i_1}, \ldots, P_{i_{n-d}}$ in the support of D are zeros of x; so $0 \neq x \in \mathcal{L}(G - (P_{i_1} + \cdots + P_{i_{n-d}}))$. Then

$$0 \leq \deg(G - (P_{i_1} + \cdots + P_{i_{n-d}}) = \deg(G) - n + d.$$

Hence, $d \geq n - \deg(G)$. $\square$

Proposition 3.2 *If $\{f_1, \ldots, f_k\}$ is a basis of the Riemann–Roch space $\mathcal{L}(G)$, then the matrix*

$$M = \begin{pmatrix} f_1(P_1) & f_1(P_2) & \cdots & f_1(P_n) \\ \vdots & \vdots & & \vdots \\ f_k(P_1) & f_k(P_2) & \cdots & f_k(P_n) \end{pmatrix}_{k \times n}$$

is a generator matrix of $C_{\mathcal{L}}(D, G)$.

Example 3.3 Let F be the rational function field $\mathbb{F}_q(x)$ and $G = mP_\infty$. Then $\mathcal{L}(G)$ is the space of polynomials of degree at most m. If we take $D = \sum_{\alpha \in \mathbb{F}_q} P_{x-\alpha}$, then the obtained $[q, m+1, q-m]$ code over $\mathbb{F}_q$ is a Reed–Solomon code.

Remark The lower bound for $k + d$ is very similar to the Singleton Bound. By putting them together, we get

$$n + 1 - g \leq k + d \leq n + 1.$$

If the genus $g = 0$, then $k + d = n + 1$; that is, the Goppa geometric codes constructed from the rational function field are MDS codes. In order to construct good algebraic geometry codes, one has to find algebraic function fields with as many rational places as possible over finite fields. Hence the construction of algebraic geometry codes has inspired a lot of work on algebraic curves over finite fields and global function fields, particularly on the question of number both of rational places and of places.

3.2 Ω-construction

Another construction can be associated with the divisors G and D by using local components of Weil differentials. For a divisor $A \in \mathrm{Div}(F)$, $\Omega_F(A)$ is the space of Weil differentials ω with $(\omega) \geq A$. This is a finite-dimensional vector space over $\mathbb{F}_q$ of dimension $i(A)$. For a Weil differential ω and a place $P \in \mathbb{P}_F$, the map $\omega_P : F \to \mathbb{F}_q$ denotes the local component of ω at P.

Let G and D be the divisors defined as before. Then we define the code $C_\Omega(D, G) \subseteq \mathbb{F}_q^n$ by

$$C_\Omega(D, G) := \{\omega_{P_1}(1), \ldots, \omega_{P_n}(1)) \mid \omega \in \Omega_F(G - D)\}. \tag{3.6}$$

The code $C_\Omega(D, G)$ is also called an algebraic geometry code.

Theorem 3.4 $C_\Omega(D, G)$ is an $[n, k', d']$ code with parameters

$$k' = i(G - D) - i(G) \text{ and } d' \geq \deg(G) - (2g - 2).$$

Under additional hypothesis that $\deg(G) > 2g - 2$,

$$k' = i(G - D) \geq n + g - 1 - \deg(G).$$

Further, if $2g - 2 < \deg(G) < n$, then

$$k' = n + g - 1 - \deg(G).$$

The codes $C_{\mathcal{L}}(D, G)$ and $C_\Omega(D, G)$ are closely related.

Proposition 3.5 The code $C_\Omega(D, G)$ and $C_{\mathcal{L}}(D, G)$ are dual to one other; that is, $C_\Omega(D, G) = C_{\mathcal{L}}(D, G)^\perp$.

Proposition 3.6 *There exists a Weil differential η such that*

$$\nu_{P_i}(\eta) = -1 \text{ and } \eta_{P_i}(1) = 1 \text{ for } i = 1, \ldots, n.$$

Then $C_\Omega(D,G) = C_\mathcal{L}(D,G)^\perp = C_\mathcal{L}(D, D - G + (\eta))$.

We can identify Ω_F with the differential module Δ_F. Via this identification the local component of ω at the place P can be evaluated by means of the residue of ω at P; namely, $\omega_P(u) = \text{res}_P(u\omega)$ for all $u \in F$. In particular, we have $\omega_P(1) = \text{res}_P(\omega)$. Hence we have the following alternative description of the code $C_\Omega(D,G)$.

Theorem 3.7 *The code $C_\Omega(D,G)$ has the following residue representation*

$$C_\Omega(D,G) := \{(\text{res}_{P_1}(\omega), \ldots, \text{res}_{P_n}(\omega)) \mid \omega \in \Omega_F(G - D)\}.$$

This is most commonly used in the literature to define the code $C_\Omega(D,G)$.

Proposition 3.8 *Let t be an element of F such that $\nu_{P_i}(t) = 1$ for $i = 1, \ldots, n$. Then the following hold:*

(a) *the differential $\eta := dt/t$ satisfies $\nu_{P_i}(\eta) = -1$ and $\text{res}_{P_i}(\eta) = 1$ for $i = 1, \ldots, n$;*

(b) *$C_\Omega(D,G) = C_\mathcal{L}(D, D - G + (dt) - (t))$.*

3.3 *H*-construction

This *H*-construction and the following *P*-construction can be found in [32]. Let X be a smooth projective variety over $\mathbb{F}_q$, let $\mathcal{L}$ be a line bundle on X defined over $\mathbb{F}_q$, and let $H^0(\mathcal{L})$ be the space of its sections. Let $P_1, P_2, \ldots, P_n$ be the points of $X(\mathbb{F}_q)$ and $D = \sum_{i=1}^n P_i$. It is impossible to map $H^0(\mathcal{L})$ to $\mathbb{F}_q$ by evaluation at points since the value of a section at a point is not well defined. There is a natural map

$$H^0(\mathcal{L}) \to \oplus_{i=1}^n \overline{\mathcal{L}}_{P_i},$$

where $\overline{\mathcal{L}}_{P_i}$ is the fibre of $\mathcal{L}$ at P_i, that is, a one-dimensional vector space over $\mathbb{F}_q$.

Fix an arbitrary trivialization of the fibres $\overline{\mathcal{L}}_{P_i}$, or equivalently an isomorphism $\overline{\mathcal{L}}_{P_i} \cong \mathbb{F}_q$. We obtain a map

$$\text{Germ}_D : H^0(\mathcal{L}) \to \mathbb{F}_q^n.$$

Then the image $\text{Germ}_D(H^0(\mathcal{L}))$ is a linear code, and we denote it by $C_H(D, \mathcal{L})$. The parameters of this code are given in the following theorem.

Theorem 3.9 *Let X be a curve of genus g with D defined as above, and let $0 \leq \deg(\mathcal{L}) < n$. Then the code $C_H(D, \mathcal{L})$ is an $[n, k, d]$ code with*

$$k \geq \deg(\mathcal{L}) - g + 1, \quad d \geq n - \deg(\mathcal{L}).$$

Corollary 3.10 *Let X be a curve of genus g over $\mathbb{F}_q$, and suppose that $N = |X(\mathbb{F}_q)| > g - 1$. Then, for any $n = g, g + 1, \ldots, N$ and also any $k = 1, \ldots, n-g$, there exists a linear $[n, k, d]$ code whose parameters satisfy*

$$k + d = n - g + 1.$$

3.4 *P*-construction

The equivalence class of a non-degenerate linear $[n, k, d]$ code C is uniquely determined by a projective $[n, k, d]$ system $\mathcal{P}$. If a variety X is given with a fixed embedding $\mathcal{P} \subseteq X(\mathbb{F}_q)$, we obtain a projective $[n, k, d]$ system with $n = |\mathcal{P}|$, $k = m + 1$, and $d = n - \max_H\{|H \cap \mathcal{P}|\}$, where the maximum is taken over all hyperplanes $H \subseteq \mathbb{P}^m$ defined over $\mathbb{F}_q$.

If X is a curve with a fixed embedding, then, by definition, its degree $\deg(X)$ is the number of $\overline{\mathbb{F}_q}$-points counted with appropriate multiplicities in the intersection of X with an arbitrary hyperplane. In any case, $\max_H\{|H \cap \mathcal{P}|\} \leq \deg(X)$. Then the following result is obtained.

Proposition 3.11 *Let a curve $X \subseteq \mathbb{P}^m$ be given and let $N = |X(\mathbb{F}_q)| > 0$. Then, for any n in the interval $N \geq n > \max\{m, \deg(X)\}$, there exists a non-degenerate linear $[n, k, d]$ code with $k = m + 1$ and $d \geq n - \deg(X)$.*

For a projective $[n, k, d]$ system, consider any subset $\mathcal{P} \subseteq X(\mathbb{F}_q)$ mapped into $\mathbb{P}(H^0(\mathcal{L}))$; here $n = |\mathcal{P}|$ and $k = h^0(\mathcal{L})$. The inverse images of hyperplane sections of $\mathbb{P}(H^0(\mathcal{L}))$ are effective divisors D belonging to the class $\mathcal{L}$. Hence, $d = n - \max_D\{|D \cap \mathcal{L}|\}$.

Consider now the case of curves. If X is a curve of genus g and also $a = \deg(\mathcal{L})$, then $k = h^0(\mathcal{L}) \geq a - g + 1$. Furthermore, $d \geq n - a$. Thus we have constructed a linear $[n, k, d]$ code, and we denote it by $C_P(\mathcal{P}, \mathcal{L})$.

3.5 Construction via local expansions

We will give two constructions of linear codes via parity-check matrices which come from the local expansions of n functions at a fixed rational place [41].

3.5.1 The first construction Let $F/\mathbb{F}_q$ be a global function field of genus g with at least $n+1$ rational places. Let $P_\infty, P_1, \ldots, P_n$ be $n+1$ distinct rational places of F. We choose a positive divisor G of degree $2g$ with $P_\infty \notin \operatorname{supp}(G)$. Then, by the Riemann–Roch Theorem,

$$l(G) = g + 1.$$

Now we want to choose a basis of $\mathcal{L}(G)$ in a special way. By the Riemann–Roch Theorem, we have $l(G - P_\infty) = g$. Also, $l(G - (2g+1)P_\infty) = 0$. Hence there exists g integers

$$0 = n_0 < n_1 < \cdots < n_g \le 2g$$

such that $l(G - n_l P_\infty) = l(G - (n_l + 1)P_\infty) + 1$ for $0 \le l \le g$. So we can choose a function

$$w_l \in \mathcal{L}(G - n_l P_\infty) \setminus \mathcal{L}(G - (n_l + 1)P_\infty)$$

for each $0 \le l \le g$. By the Strict Triangle Inequality, the set $\{w_0, \ldots, w_g\}$ forms a basis of the Riemann–Roch space $\mathcal{L}(G)$ over $\mathbb{F}_q$.

We can also choose a non-zero function $f_j \in \mathcal{L}(G + P_j) \setminus \mathcal{L}(G)$, since $l(G + P_j) = 2$ for each $1 \le j \le n$. Then $w_0, \ldots, w_g, f_1, \ldots, f_n$ are linearly independent over $\mathbb{F}_q$.

The first construction via local expansions is given as follows. Let t be a local parameter at P_∞ and put

$$t_r = \begin{cases} t^r, & \text{if } r \notin \{n_0, \ldots, n_g\} \\ w_l, & \text{if } r = n_l \text{ for some } l \in \{0, 1, \ldots, g\}. \end{cases}$$

The local expansion of f_j at P_∞ is given by

$$f_j = \sum_{r=0}^\infty a_{r,j} t_r$$

for some coefficients $a_{r,j} \in \mathbb{F}_q$. For a fixed positive integer m and with $g \le m < n$, we define the vectors $\mathbf{c}_j \in \mathbb{F}_q^m$ for $1 \le j \le n$ by

$$\mathbf{c}_j = (\widehat{a_{n_0,j}}, a_{1,j}, \ldots, \widehat{a_{n_1,j}}, a_{n_1+1,j}, \ldots, \widehat{a_{n_g,j}}, \ldots, a_{m+g,j})$$

where the hat indicates that the corresponding entry is deleted. For simplicity, we write

$$\mathbf{c}_j = (c_{1,j}, \ldots, c_{m,j}).$$

Define the $m \times n$ matrix H over $\mathbb{F}_q$ by

$$H = (\mathbf{c}_1^T, \mathbf{c}_2^T, \ldots, \mathbf{c}_n^T),$$

where $\mathbf{c}_i^T$ is the transpose of $\mathbf{c}_i$. Let $C_m(P_\infty, P_1, \ldots, P_n; G)$ be the linear code with the parity-check matrix H; it is a linear code of length n.

Theorem 3.12 *Let $F/\mathbb{F}_q$ be a global function field of genus g and let $P_\infty, P_1, \ldots, P_n$ be $n+1$ distinct rational places of F. Choose a positive divisor G of degree $2g$ with $P_\infty \notin supp(G)$ and a positive integer m satisfying $g \leq m < n$. Then the linear code $C_m(P_\infty, P_1, \ldots, P_n; G)$ is an $[n, k, d]$ code over $\mathbb{F}_q$ with*

$$k \geq n - m, \quad d \geq m - g + 1.$$

It was shown that the code $C_m(P_\infty, P_1, \ldots, P_n; G)$ is equivalent to an algebraic geometry code in the following sense (see [26]).

Definition 3.13 Two linear codes C_1 and C_2 over $\mathbb{F}_q$ of length n are *equivalent* if there exist n non-zero elements $\lambda_1, \ldots, \lambda_n$ of $\mathbb{F}_q$ such that

$$C_2 = \{(\lambda_1 c_1, \ldots, \lambda_n c_n) \in \mathbb{F}_q^n \mid (c_1, \ldots, c_n) \in C_1\}.$$

3.5.2   The second construction   Let $F/\mathbb{F}_q$ be a global function field of genus g. Let $P_\infty, P_1, P_2, \ldots, P_n$ be $n+1$ distinct rational places of F, where $n > g$. Choose a positive non-special divisor D of F with $l(D) = 1$. Then $l(D + P_i) = 2$ for $1 \leq i \leq n$, and we can choose a function

$$f_i \in \mathcal{L}(D + P_i) \setminus \mathcal{L}(D).$$

Let t be a local parameter at P_∞ and consider the local expansions of the f_i at P_∞ which are of the form

$$f_i = t^{-v} \sum_{i=0}^{\infty} b_{r,i} t^r, \tag{3.7}$$

where $v = \nu_{P_\infty}(D) \geq 0$ and all coefficients $b_{r,i} \in \mathbb{F}_q$. Also, for each $i = 1, 2, \ldots, n$ we define

$$c_{r,i} = \begin{cases} b_{r-1,i} & \text{for } 1 \leq r \leq v, \\ b_{r,i} & \text{for } r \geq v + 1. \end{cases}$$

Now we choose a positive integer m with $g \leq m < n$ and define the $m \times n$ matrix H over $\mathbb{F}_q$ by

$$H = \begin{pmatrix} c_{1,1} & c_{1,2} & \cdots & c_{1,n} \\ \vdots & \vdots & & \vdots \\ c_{m,1} & c_{m,2} & \cdots & c_{m,n} \end{pmatrix}_{m \times n}.$$

We denote by $C(P_\infty, P_1, \ldots, P_n; D; m)$ the linear code over $\mathbb{F}_q$ of length n with parity-check matrix H.

Theorem 3.14 *Let $F/\mathbb{F}_q$ be a global function field with genus g and let $P_\infty, P_1, P_2, \ldots, P_n$ be $n+1$ distinct rational places of F. Choose a positive divisor D with $\deg(D) = g$ and $l(D) = 1$, and a positive integer m satisfying $g \leq m < n$. Then the linear code $C(P_\infty, P_1, \ldots, P_n; D; m)$ defined above is an $[n, k, d]$ code over $\mathbb{F}_q$ with*

$$k \geq n - m \ and \ d \geq m - g + 1.$$

In fact, the code $C(P_\infty, P_1, \ldots, P_n; D; m)$ is also equivalent to an algebraic geometry code.

Theorem 3.15 *Let $F/\mathbb{F}_q$ be a global function field with genus g and let $P_\infty, P_1, P_2, \ldots, P_n$ be $n + 1$ distinct rational places of F, where $n > g$. Let D be a positive non-special divisor of F with $l(D) = 1$ and let m be a positive integer with $g \leq m < n$. Then the code $C(P_\infty, P_1, \ldots, P_n; D; m)$ is equivalent to the algebraic geometric code $C_{\mathcal{L}}(D, G)$, where G is a divisor of F that is equivalent to the divisor $D + \sum_{i=1}^n P_i - (m+1)P_\infty$ and satisfies $supp(G) \cap \{P_1, \ldots, P_n\} = \emptyset$.*

3.5.3 Codes from maximal curves

Let s be a prime power, and put $q = s^2$, and let $F/\mathbb{F}_q$ be a maximal function field of genus g, that is, $N(F) = q + 1 + 2gs$. For a maximal function field $F/\mathbb{F}_q$, the group $\mathrm{Cl}(F)$ of divisor classes of degree zero of F has a nice structure [27] as an abelian group, namely,

$$\mathrm{Cl}(F) \cong (\mathbb{Z}/(s+1)\mathbb{Z})^{2g}.$$

Now let $P_\infty, P_1, P_2, \ldots, P_n$ be $n + 1$ distinct rational places of F. Then $(s+1)P_\infty - (s+1)P_i$ is a principal divisor of F for each i, and there exist $h_i \in F$ for $1 \leq i \leq n$ such that

$$\mathrm{div}(h_i) = (s+1)P_\infty - (s+1)P_i.$$

Next we choose a local parameter t at P_∞ and consider the local expansions of the h_i for $1 \leq i \leq n$ at P_∞ which are of the form

$$h_i = t^s \sum_{r=1}^\infty e_{r,i} t^r, \tag{3.8}$$

where all coefficients $e_{r,i} \in \mathbb{F}_q$. Finally, we select an integer m such that $1 \leq m < n$ and define the $m \times n$ matrix H over $\mathbb{F}_q$ by

$$H = \begin{pmatrix} e_{1,1} & e_{1,2} & \cdots & e_{1,n} \\ \vdots & \vdots & & \vdots \\ e_{m,1} & e_{m,2} & \cdots & e_{m,n} \end{pmatrix}_{m \times n}.$$

We denote by $C(P_\infty, P_1, \ldots, P_n; m)$ the linear code over $\mathbb{F}_q$ of length n with parity-check matrix H. Then we obtain the following result.

Theorem 3.16 *Let $F/\mathbb{F}_q$ with $q = s^2$ be a maximal function field of genus g, let $P_\infty, P_1, P_2, \ldots, P_n$ be $n+1$ distinct rational places of F, and let m be an integer with $1 \leq m < n$. Then $C(P_\infty, P_1, \ldots, P_n; m)$ is a linear $[n, k, d]$ code over $\mathbb{F}_q$ with*

$$k \geq n - m \text{ and } d \geq \left\lceil \frac{m}{s+1} \right\rceil + 1.$$

3.6 NXL codes

Goppa's construction of algebraic geometry codes yields excellent linear codes over $\mathbb{F}_q$ in the asymptotic sense, provided that q is large enough. But, for small q, Goppa's construction is practically useless, since it uses only the rational places and there are too few rational places compared to the genus. Therefore places of arbitrary degree are employed to construct the algebraic geometry codes in [24].

Let $F/\mathbb{F}_q$ be a global function field, let G_1 and G_2 be two divisors of F with $G_1 \leq G_2$, and consider the corresponding Riemann–Roch spaces $\mathcal{L}(G_1)$ and $\mathcal{L}(G_2)$. Note that $\mathcal{L}(G_1)$ is a linear subspace of the vector space $\mathcal{L}(G_2)$ over $\mathbb{F}_q$. Thus, if we choose an ordered basis of $\mathcal{L}(G_2)$, the coordinate vectors of the elements of $\mathcal{L}(G_1)$ form a linear code over $\mathbb{F}_q$ of length $n = l(G_2)$ and dimension $k = l(G_1)$, provided that $l(G_1) \geq 1$. We call such a linear code an *NXL code*.

In this subsection, we will describe a family of NXL codes (see [22]) which is more general than that in [24]. First let us fix the following notations. Let $F/\mathbb{F}_q$ be a global function field over $\mathbb{F}_q$ of genus g. Let $P_1, \ldots, P_r$ be pairwise distinct places of F. Let D be a non-special divisor of F. Let E be a positive divisor of F with $\operatorname{supp}(E) \cap \{P_1, \ldots, P_r\} = \emptyset$ and

$$1 \leq \deg(E - D) \leq \sum_{i-1}^{r} \deg(P_i) - g.$$

We denote $s_i = \deg(P_i)$ and $n = \sum_{i-1}^{r} s_i$. Assume that the degrees $s_1, \ldots, s_n$ can be complete arbitrary, except for the condition $n > g$. Then

$$l(D + P_i) = l(D) + s_i \text{ for } 1 \leq i \leq r,$$

since D is non-special. For each i, we choose a basis

$$\{f_{i,j} + \mathcal{L}(D) : 1 \leq j \leq s_i\}$$

of the factor space $\mathcal{L}(D + P_i)/\mathcal{L}(D)$. Then we can show that

$$\{f_{i,j} + \mathcal{L}(D) : 1 \le j \le s_i, 1 \le i \le r\}$$

is the basis of the factor space $\mathcal{L}(D + \sum_{i=1}^{r} P_i)/\mathcal{L}(D)$.

Now the linear code is constructed as follows. Every

$$f \in \mathcal{L}\left(D + \sum_{i=1}^{r} P_i - E\right) \subseteq \mathcal{L}\left(D + \sum_{i=1}^{r} P_i\right)$$

has a unique representation

$$f = \sum_{i=1}^{r} \sum_{j=1}^{s_i} c_{i,j} f_{i,j} + u \tag{3.9}$$

with all coefficients $c_{i,j} \in \mathbb{F}_q$ and $u \in \mathcal{L}(D)$. Define the $\mathbb{F}_q$-linear map

$$\eta : f \in \mathcal{L}\left(D + \sum_{i=1}^{r} P_i - E\right) \mapsto (c_{1,1}, \ldots, c_{1,s_1}, \ldots, c_{r,1}, \ldots, c_{r,s_r}) \in \mathbb{F}_q^n.$$

The image of η is the linear code $C(P_1, \ldots, P_r; D, E)$ over $\mathbb{F}_q$ of length n.

The linear code $C(P_1, \ldots, P_r; D, E)$ belongs to the family of NXL codes since it is obtained from the general construction principle of these codes by putting

$$G_1 = D + \sum_{i=1}^{r} P_i - E, G_2 = D + \sum_{i=1}^{r} P_i,$$

and puncturing so that only the n coordinates corresponding to the $f_{i,j}$ are kept. The following theorem provides bounds for the dimension and the minimum distance of the linear code $C(P_1, \ldots, P_r; D, E)$.

Theorem 3.17 *Let $F/\mathbb{F}_q$ be a global function field over $\mathbb{F}_q$ of genus g, let $P_1, \ldots, P_r$ be pairwise distinct places of F with degrees $s_1, \ldots, s_r$, and let D be a non-special divisor of F. Furthermore, let E be a positive divisor of F with $supp(E) \cap \{P_1, \ldots, P_r\} = \emptyset$ such that $m := deg(E - D)$ satisfies $1 \le m \le \sum_{i=1}^{r} s_i - g$. Then $C(P_1, \ldots, P_r; D, E)$ is a linear $[n, k, d]$ code over $\mathbb{F}_q$ with*

$$n = \sum_{i=1}^{r} s_i, k = l\left(D + \sum_{i=1}^{r} P_i - E\right) \ge n - m - g - 1, d \ge d_0,$$

where d_0 is the least cardinality of a subset R of $\{1, \ldots, r\}$ for which $\sum_{i \in R} s_i \ge m$. Also, $k = n - m - g - 1$ if $n - m \ge 2g - 1$.

Remark It was pointed out in [26] that any algebraic geometry code $C_{\mathcal{L}}(D, G)$ can be represented as an NXL code with suitable divisors.

3.7   XNL codes

In this subsection, another method of constructing linear codes from places of arbitrary degree of global function fields was introduced by Xing, Niederreiter, and Lam [42]. This method provides a considerable generalization of Goppa's construction of the algebraic geometry codes. This construction uses the following notations:

(1) $F/\mathbb{F}_q$ is a global function field over $\mathbb{F}_q$ of genus g;

(2) $P_1, \ldots, P_n$ are distinct places of F;

(3) G is a divisor of F with $\mathrm{supp}(G) \cap \{P_1, \ldots, P_n\} = \emptyset$;

(4) C_i is an $[n_i, k_i, d_i]$ code over $\mathbb{F}_q$ with $k_i \geq \deg(P_i)$ for $1 \leq i \leq n$;

(5) ϕ_i is the fixed $\mathbb{F}_q$-linear monomorphism from the residue class field of P_i to the linear code C_i for $1 \leq i \leq n$.

We put $n = \sum_{i=1}^{r} n_i$, where n_i is the length of the linear code C_i. Now we define the $\mathbb{F}_q$-linear map

$$\beta : f \in \mathcal{L}(G) \mapsto (\phi_1(f(P_1)), \ldots, \phi_r(f(P_r))) \in \mathbb{F}_q^n, \qquad (3.10)$$

where on the right-hand side we use concatenation of vectors. The image of β is the linear code $C(P_1, \ldots, P_n; G; C_1, \ldots, C_r)$ over $\mathbb{F}_q$ of length n. Such a linear code is called an *XNL code*.

The results on the dimension and the minimum distance of XNL codes can be given in the following form (see [26, 42]).

Theorem 3.18 *Let*

(a) $F/\mathbb{F}_q$ *be a global function field over* $\mathbb{F}_q$ *of genus* g,

(b) $P_1, \ldots, P_n$ *be distinct places of* F,

(c) C_i *be linear* $[n_i, k_i, d_i]$ *codes over* $\mathbb{F}_q$ *with* $k_i \geq \deg(P_i)$,

(d) G *be a divisor of* F *with* $\mathrm{supp}(G) \cap \{P_1, \ldots, P_r\} = \emptyset$ *and with* $g \leq \deg(G) < \sum_{i=1}^{r} k_i$.

Then

$$C(P_1, \ldots, P_r; G; C_1, \ldots, C_1, \ldots, C_r)$$

is a linear $[n, k, d]$ *code with*

$$n = \sum_{i=1}^{r} n_i, k = l(G) \geq \deg(G) - g + 1, \ and \ d \geq d_0,$$

where d_0 is the minimum of $\sum_{i \in M'} d_i$ taken over all the subsets M of $\{1, 2, \ldots, r\}$ for which $\sum_{i \in M} k_i \leq \deg(G)$, with M' denoting the complement of M in $\{1, \ldots, r\}$. Further, if $\deg(G) \geq 2g - 1$, then $k = \deg(G) - g + 1$.

Corollary 3.19 In the special case that $k_i \geq d_i$ for $1 \leq i \leq r$, the minimum distance d of the linear code $C(P_1, \ldots, P_r; G; C_1, \ldots, C_r)$ satisfies

$$d \geq \sum_{i=1}^{r} d_i - \deg(G).$$

Remark It was pointed out that any $[n, k]$ linear code over $\mathbb{F}_q$ can be obtained from the construction of XNL codes; see [26]. Several examples of good XNL codes are listed by Xing, Niederreiter, and Lam in [42], and some more systematic searches for the good XNL codes are given in [2, 43].

3.8 Function-field codes

The function-field code was introduced in [22] and studied further in [11]. A *function-field code* in $F/\mathbb{F}_q$ with respect to a finite non-empty set $\mathcal{P} \subseteq \mathbb{P}_F$ is a non-zero finite-dimensional $\mathbb{F}_q$-linear subspace V of F which satisfies the two conditions $V \subseteq \mathcal{O}_\mathcal{P}$ and $V \cap M_\mathcal{P} = \{0\}$, where $\mathcal{O}_\mathcal{P} := \cap_{P \in \mathcal{P}} \mathcal{O}_P$ and $M_\mathcal{P} := \cap_{P \in \mathcal{P}} M_P$.

By this definition, any non-zero $\mathbb{F}_q$-linear subspace of a function-field code in $F/\mathbb{F}_q$ with respect to $\mathcal{P}$ is again a function-field code in $F/\mathbb{F}_q$ with respect to $\mathcal{P}$. An important family of function-field codes can be provided by some special Riemann–Roch spaces.

Proposition 3.20 Let $F/\mathbb{F}_q$ be a global function field, let $\mathcal{P}$ be a finite non-empty set of places of F, and let G be a divisor of F with $l(G) \geq 1$, $\mathrm{supp}(G) \cap \mathcal{P} = \emptyset$, and $\deg(G) < \sum_{P \in \mathcal{P}} \deg(P)$. Then $\mathcal{L}(G)$ is a a function-field code in $F/\mathbb{F}_q$ with respect to $\mathcal{P}$.

The function-field codes can be used as a tool to generalize the construction of XNL codes. Let V be a function-field code in $F/\mathbb{F}_q$ with respect to $\mathcal{P} = \{P_1, \ldots, P_n\}$, where $P_1, \ldots, P_r$ are distinct places of F. For each $i = 1, \ldots, r$, let C_i be a linear $[n_i, k_i, d_i]$ code over $\mathbb{F}_q$ with $k_i \geq \deg(P_i)$ and let ϕ_i be an $\mathbb{F}_q$-linear map from the residue class field of P_i into C_i, which is injective. Put $n = \sum_{i=1}^{r} n_i$ and define the $\mathbb{F}_q$-linear map $\gamma : V \to \mathbb{F}_q^n$ by

$$\gamma(f) = (\phi_1(f(P_1)), \ldots, \phi_r(f(P_r))) \text{ for all } f \in V,$$

where we use concatenation of vectors on the right-hand side. The image of V under γ is the linear code $C_{\mathcal{P}}(V; C_1, \ldots, C_n)$ over $\mathbb{F}_q$.

Theorem 3.21 *For $I \subset \{1, \ldots, r\}$, let $\overline{I}$ denote the complement of I in $\{1, \ldots, r\}$ and $\mathcal{P}(\overline{I}) = \{P_i : i \in \overline{I}\}$. Then the code $C_{\mathcal{P}}(V; C_1, \ldots, C_n)$ constructed as above is a $[n, k, d]$ linear code over $\mathbb{F}_q$ with*

$$n = \sum_{i=1}^{r} n_i, k = \dim(V), d \geq d_0 = \min_I \sum_{i \in I} d_i,$$

where the minimum is over all $I \subseteq \{1, \ldots, r\}$ for which $V \cap M_{\mathcal{P}(\overline{I})} \neq \{0\}$.

The code $C_{\mathcal{P}}(V; C_1, \ldots, C_n)$ derived from function-field codes form a universal family of linear codes, in the sense that any linear code is obtained by this construction. By using the Approximation Theorem 2.5, we can show the following result (see [23]).

Theorem 3.22 *Let C be an arbitrary $[n, k]$ linear code over $\mathbb{F}_q$. Then there exist a global function field $F/\mathbb{F}_q$, a set $\mathcal{P} = \{P_1, \ldots, P_n\}$ of n distinct rational places of F, and a k-dimensional function-field over V in $F/\mathbb{F}_q$ with respect to $\mathcal{P}$ such that C is equal to the code $C_{\mathcal{P}}(V; C_1, \ldots, C_n)$, where C_i is the linear $[1, 1, 1]$ code over $\mathbb{F}_q$ for $i = 1, \ldots, n$.*

3.9 Construction via narrow ray class groups

The construction via narrow ray class groups was given in [37, 44]. Let $F/\mathbb{F}_q$ be a global function field with genus g, where $q = p^r$. Assume that the number of rational places $N(F) \geq 1$ and distinguish one rational place which is denoted by ∞. Fix a place P of F with degree e and let $D = mP$ for some positive integer m. Now let V_D be the $\mathbb{F}_p$-vector space obtained by taking the quotient of $\mathrm{Cl}_D^+(A)$ by its maximal abelian subgroup, that is,

$$V_D = \mathrm{Cl}_D^+(A)/\mathrm{Cl}_D^+(A)^p.$$

The dimension k of V_D is at least as large as that of $(A/D)^*/((A/D)^*)^p$ which is at least $re(m - 1 - \lfloor (m-1)/p \rfloor)$.

Let $S = \{P_1, P_2, \ldots, P_n\}$ be a finite subset of $S_\infty - \{P\}$. Then each element of S can be viewed as a k-vector in V_D; that is, we can write each element of S as a k-tuple over $\mathbb{F}_q$, and we denote the vectors by $P_1, P_2, \ldots, P_n$ as well. Let $G(S, D)$ be the matrix whose columns are the vectors $P_1, P_2, \ldots, P_n$. Then $G(S, D)$ is a $k \times n$ matrix over $\mathbb{F}_q$ of rank k. Define $C(S, D)$ to be the linear code with generator matrix $G(S, D)$ so that $C(S, D)$ is an $[n, k]$ code over $\mathbb{F}_q$.

Next, let $C(S, D)^\perp$ be the dual code of $C(S, D)$. Then $C(S, D)^\perp$ is an $[n, n - k]$ code over $\mathbb{F}_q$ and

$$C(S, D)^\perp = \left\{ (c_1, c_2, \ldots, c_n) \in \mathbb{F}_q^n \mid \sum_{i=1}^n c_i P_i = 0 \in V_D \right\}.$$

The following theorem provides a lower bound for the minimum distance of $C(S, D)^\perp$.

Theorem 3.23 *Let S consist of all the rational places of F different from ∞ and possibly P when $\deg(P) = 1$. Then the code $C(S, mP)^\perp$ has minimum distance d, where*

$$d \geq \begin{cases} me + 1 - g, & \text{if } p \mid m \\ (m-1)e + 1 - 2g, & \text{otherwise.} \end{cases}$$

Example 3.24 (see [37]) Let $P_1, P_2, \ldots, P_n$ be $n + 1$ distinct rational points of $F/\mathbb{F}_q$ and $D = pQ$ for some closed point Q of degree m such that $\{P, P_1, \ldots, P_n\} \cap \text{supp}(D) = \emptyset$. Then for any positive integer m satisfying $(p - 1)m \log_p q < n$, the linear code $C_D(P; P_1, P_2, \ldots, P_n)$ over $\mathbb{F}_q$ defined by

$$\left\{ (c_1, c_2, \ldots, c_n) \in \mathbb{F}_q^n \mid \overline{\sum_{i=1}^n c_i(P_i - P)} = \overline{0} \text{ in } V_D \right\}$$

has dimension at least $n - (p - 1)m \log_p q$ and minimum distance at least $pm + 1 - 2g$.

3.10 Elkies codes

We will provide two constructions of non-linear codes from function fields in the following two subsections. Elkies' construction uses the infinity to act as an extra symbol of the alphabet in [3]. Xing's construction uses the evaluation of the derivative instead of evaluation at places in the following subsection.

Let $F/\mathbb{F}_q$ be a global function field. For a given divisor G with degree $\deg(G) = 0$ and a number $m < n/2$, define the set

$$\mathcal{L}_m(G) = \{0\} \cup \{f \in F^* \mid (f) + G = G_1 - G_2\}$$

for some effective divisors G_1, G_2 with $\deg(G_1) \leq m$ and $\deg(G_1) \leq m$. We define the map $\widetilde{ev_D}$ on the set $\mathcal{L}_m(G)$ as follows: for each place P_i we choose a rational function ϕ_i whose order at P_i equals the order of G at P_i, and set

$$\widetilde{ev_D}(f) = ((\phi_1 f)(P_1), \ldots, (\phi_n f)(P_n)) \in (\mathbb{F}_q \cup \{\infty\})^n.$$

The image of this map is the non-linear Elkies code $E_m(D, G)$.

Proposition 3.25 *Let G be a divisor of degree 0 of a global function field $F/\mathbb{F}_q$. Let f_1 and f_2 be distinct elements of $\mathcal{L}_m(G)$ of degrees m_1 and m_2. Then the codewords corresponding to f_1 and f_2 coincide in at most $m_1 + m_2$ positions. In particular, the minimum distance of $E_m(D, G)$ is at least $N - 2m$.*

It is difficult to determine the size of the Elkies code $E_m(D, G)$. But we can obtain it in the case that X is an asymptotically optimal curve, that is, X runs over a family of curves of genus $g \to \infty$ with

$$N(X)/g \to \sqrt{q} - 1.$$

Let $M(m, X)$ be the average cardinality of $E_m(D, G)$ over all elements of the Jacobian J_X:

$$M(m, X) = \frac{1}{|J_X|} \sum_{D \in J_X} |E_m(D, G)|.$$

Theorem 3.26 *If X is an asymptotically optimal curve, then for each $\rho > 2q/(q^2 - 1)$, the estimate*

$$M(m, X) = \left(\frac{q+1}{q}\right)^{n+o_\rho(n)} q^{2m-g}$$

holds if $2m/n > \rho$.

3.11 Xing codes

Another non-linear algebraic geometry construction is given in [36] by using the derivatives of functions. Let $F/\mathbb{F}_q$ be an algebraic function field of genus g. Let $P_1, P_2, \ldots, P_n$ be pairwise distinct rational places of $F/\mathbb{F}_q$, and $D = P_1 + P_2 + \cdots + P_n$. Let G be a divisor of F such that $\operatorname{supp}(G) \cap \operatorname{supp}(D) = \emptyset$. Consider the evaluation map

$$ev_D : \mathcal{L}(G) \to \mathbb{F}_q^n, f \mapsto (f(P_1), \ldots, f(P_n)).$$

For an integer r with $0 < r < n$, denote by $B_r(\mathbf{u})$ the sphere in $\mathbb{F}_q^n$ with centre $\mathbf{u}$ and radius r; that is,

$$B_r(\mathbf{u}) = \{\mathbf{x} \in \mathbb{F}_q^n \mid d(\mathbf{x}, \mathbf{u}) \leq r\}.$$

Then the image $\mathrm{ev}_D(f)$ lies in exactly $\sum_{i=0}^{r}(q-1)^i\binom{n}{i}$ spheres of radius r. Hence, there exists a vector $\mathbf{c} \in \mathbb{F}_q^n$ such that the set

$$M_r(\mathbf{c}) := \{f \in \mathcal{L}(G) \mid \mathrm{ev}_D(f) \in B_r(\mathbf{c})\}$$

has cardinality at least

$$|M_r(\mathbf{c})| \geq \frac{|\mathcal{L}(G)|}{q^n} \sum_{i=0}^{r}(q-1)^i\binom{n}{i}. \tag{3.11}$$

Next, let t be a local parameter at a rational place P of F, and let $f \in F$ be a function regular at P. Define the first derivative of f at P with respect to t as

$$f_t'(P) = \frac{f - f(P)}{t}(P) \in \mathbb{F}_q.$$

Now, on the set $M_r(\mathbf{c})$, consider the map

$$\mathrm{ev}_D' : M_r(\mathbf{c}) \to \mathbb{F}_q^n, \quad f \mapsto (f_t'(P_1), \ldots, f_t'(P_n)).$$

We denote the image of the map ev_D' by $C_r(D, G)$, which is a non-linear code.

Theorem 3.27 *For $\deg(G) < 2n - 4r$, the map ev_D' on the set $M_r(\mathbf{c})$ is injective. Also, $C_r(D, G)$ is a q-ary (n, M, d)-code with minimum distance $d \geq 2n - \deg(G) - 4r$ and size*

$$M \geq \frac{|\mathcal{L}(G)|}{q^n} \sum_{i=0}^{r}(q-1)^i\binom{n}{i}.$$

References

[1] R. Auer, *Ray class fields of global function fields with many rational places*, Acta Arith. **95**, 97–122, 2000.

[2] C.S. Ding, H. Niederreiter, and C.P. Xing, *Some new codes from algebraic curves*, IEEE Trans. Inform. Theory **46**, No. 7, 2638–2642, 2000.

[3] N.D. Elkies, *Excellent nonlinear codes from modular curves*, in STOC'01: Proceedings of the 33rd Annual ACM Symposium on Theory of Computing, Hersonissos, Crete, 200–208, 2001.

[4] R. Fuhrmann, A. Garcia, and F. Torres, *On maximal curves*, J. Number Theory **67**, 29–51, 1997.

[5] A. Garcia and H. Stichtenoth, *A tower of Artin–Schreier extensions of function fields attaining the Drinfeld–Vlăduţ bound*, Invert. Math. **121**, 211–222, 1995.

[6] A. Garcia, H. Stichtenoth, and C.P. Xing, *On subfields of the Hermitian function field*, Compositio Math. **120**, 137–170, 2000.

[7] G. van der Geer and M. van der Vlugt, *Tables of curves with many points*, Math. Comp. **69**, 797–810, 2000.

[8] V.D. Goppa, *Codes on algebraic curves*, Soviet Math. Dokl. **24**, 170–172, 1981.

[9] V.D. Goppa, *Algebraic-geometric codes*, Izv. Akad. Nauk. SSSR Ser. Mat. **46**, 762–781, 1982.

[10] V.D. Goppa, *Geometry and Codes*, Kluwer, Dordrecht, 1988.

[11] D. Hachenberger, H. Niederreiter, and C.P. Xing, *Function-field codes*, Appl. Algebra Engrg. Comm. Comput. **19**, 201–211, 2008.

[12] J.P. Hansen and H. Stichtenoth, *Group codes on certain algebraic curves with many rational points*, AAECC **1**, 67–77, 1990.

[13] D.R. Hayes, *Explicit class field theory for rational function fields*, Trans. Amer. Math. Soc. **189**, 77–91, 1974.

[14] J.W.P. Hirschfeld, G. Korchmáros, and F. Torres, *Algebraic Curves over a Finite Field*, Princeton Series in Applied Mathematics, Princeton Univ. Press, 2008.

[15] T. Høholdt, J.H. van Lint, and R. Pellikaan, *Algebraic Geometry Codes, Handbook in Coding Theory*, Vol. **1**, Elsevier, Amsterdam, 871–961, 1998.

[16] Y. Ihara, *Some remarks on the number of rational points of algebraic curves over finite fields*, J. Fac. Sci. Tokyo **28**, 721–724, 1981.

[17] K.H. Leung, S. Ling, and C.P. Xing, *New binary linear codes from algebraic curves*, IEEE Trans. Inform. Theory **48**, 285–287, 2002.

[18] S. Ling and C.P. Xing, *Coding Theory: A First Course*, Cambridge University Press, Cambridge, 2004.

[19] F.J. MacWilliams and N.J.A. Sloane, *The Theory of Error-Correcting Codes*, Amsterdam, the Netherlands: North Holland, 1977.

[20] C.J. Moreno, *Algebraic Curves over Finite Fields*, Cambridge Tracts in Math., Vol. **97**, Cambridge University Press, Cambrige, 1991.

[21] H. Niederreiter and C.P. Xing, *Towers of global function fields with asymptotically many rational places and an improvement on the Gilbert–Varshamov bound*, Math. Nach. **195**, 171–186, 1998.

[22] H. Niederreiter and C.P. Xing, *Rational Points on Curves over Finite Fields: Theory and Applications*, LMS **285**, Cambridge, 2001.

[23] H. Niederreiter and C.P. Xing, *Algebraic Geometry in Coding Theory and Cryptography*, Princeton University Press, 2009.

[24] H. Niederreiter, C.P. Xing, and K.Y. Lam, *A new construction of algebraic-geometry codes*, Appl. Algebra Engr. Comm. Comput. **9**, 373–381, 1999.

[25] R. Pellikaan, B.Z. Shen, and G.J.M. Wee, *Which linear codes are algebraic-geometric?* IEEE Trans. Inform. Theory **37**, 583–602, 1991.

[26] F. Özbudak and H. Stichtenoth, *Constructing codes from algebraic curves*, IEEE Trans. Inform. Theory **45**, 2502–2505, 1999.

[27] H.G. Rück and H. Stichtenoth, *A characterization of Hermitian function fields over finite fields*, J. Reine Angew. Math. **457**, 185–188, 1994.

[28] J.-P. Serre, *Rational Points on Curves over Finite Fields*, Lecture Notes, Harvard University, 1985.

[29] H. Stichtenoth and C.P. Xing, *Excellent non-linear codes from algebraic function fields*, IEEE Trans. Inform. Theory **51**, 4044–4046, 2005.

[30] H. Stichtenoth, *Algebraic Function Fields and Codes*, Graduate Texts in Mathematics **254**, Springer Verlag, 2009.

[31] M.A. Tsfasman and S.G. Vlăduţ, *Algebraic-Geometric Codes*, Dordrecht, The Netherlands: Kluwer, 1991.

[32] M.A. Tsfasman, S.G. Vlăduţ, and D. Nogin, *Algebraic Geometric Codes: Basis Notions*, American Mathematical Soc., 1990.

[33] M.A. Tsfasman and S.G. Vlăduţ, *Modular curves, Shimura curves, and Goppa codes, better than Varshamov–Gilbert bound*, Math. Nach. **109**, 21–28, 1982.

[34] C.P. Xing, *Algebraic-geometry codes with aasymptotic parameters better that the Gilbert–Varshamov and the Tsfasman–Vlăduţ–Zink bounds*, IEEE Trans. Inform. Theory **47**, 347–352, 2001.

[35] C.P. Xing, *Constructions of codes from residue rings of polynomials*, IEEE Trans. Inform. Theory **48**, 2995–2997, 2002.

[36] C.P. Xing, *Nonlinear codes from algebraic curves improving the Tsfasman–Vlăduţ–Zink bound*, IEEE Trans. Inform. Theory **49**, 1653–1657, 2003.

[37] C.P. Xing, *Linear codes from narrow ray class groups of algebraic curves*, IEEE Trans. Inform. Theory **50**, No. 3, 541–543, 2004.

[38] C.P. Xing, *Goppa geometric codes achieving the Gilbert–Varshamov bound*, IEEE Trans. Inform. Theory **51**, 259–264, 2005.

[39] C.P. Xing, *Asymptotically good nonlinear codes from algebraic curves*, IEEE Trans. Inform. Theory **57**, No. 9, 5991–5995, 2011.

[40] C.P. Xing and S. Lin, *A class of linear codes with good parameters from algebraic curves*, IEEE Trans. Inform. Theory **46**, No. 4, 1527–1532, 2000

[41] C.P. Xing, H. Niederreiter, and K.Y. Lam, *Constructions of algebraic-geometry codes*, IEEE Trans. Inform. Theory **45**, No. 4, 1186–1193, 1999.

[42] C.P. Xing, H. Niederreiter, and K.Y. Lam, *A construction of algebraic-geometry codes*, IEEE Trans. Inform. Theory **45**, No. 7, 2498–2501, 1999.

[43] C.P. Xing and S.L. Yeo, *New linear codes and algebraic function fields over finite fields*, IEEE Trans. Inform. Theory **53**, No. 12, 4822–4825, 2007.

[44] C.P. Xing and S.L. Yeo, *Construction of global function fields from linear codes and vice versa*, Trans. Amer. Math. Soc. **361**, 1333–1349, 2009.

School of Mathematical Sciences
Yangzhou University
Yangzhou, China
and
Division of Mathematical Sciences
School of Physical & Mathematical Sciences
Nanyang Technological University
21 Nanyang Link, Singapore
lmma@yzu.edu.cn; xingcp@ntu.edu.sg.

Printed in the United States
By Bookmasters